Semiconductor Sensors
in Physico-Chemical Studies

HANDBOOK OF SENSORS AND ACTUATORS

Series Editor: S. Middelhoek, Delft University of Technology,
The Netherlands

HANDBOOK OF SENSORS AND ACTUATORS 4

Semiconductor Sensors in Physico-Chemical Studies

Edited by
L. Yu. Kupriyanov
Department of Chemical Sensors
Karpov Institute of Physical Chemistry
Moscow, Russia

Translated from Russian by
V. Yu. Vetrov

1996
ELSEVIER
Amsterdam - Lausanne - New York - Oxford - Shannon - Tokyo

ELSEVIER SCIENCE B.V.
Sara Burgerhartstraat 25
P.O. Box 211, 1000 AE Amsterdam,
The Netherlands

Library of Congress Cataloging-in-Publication Data

Poluprovodnikovye sensory v fiziko-khimicheskikh issledovaniĩakh.
  English
    Semiconductor sensors in physico-chemical studies / edited by
L.Yu. Kupriyanov.
        p.   cm. -- (Handbook of sensors and actuators ; v. 4)
    Includes bibliographical references.
    ISBN 0-444-82261-5 (acid-free paper)
    1. Chemical detectors.   2. Semiconductors.   I. Kupriyanov, L. Yu.
  II. Series.
  TP159.C46P6513   1996
  543'.0871--dc20
                                                        96-5829
                                                           CIP

ISBN: 0 444 82261 5

This book is printed on acid-free paper.

Printed in The Netherlands

Introduction to the Series

The arrival of integrated circuits with very good performance/price ratios and relatively low cost microprocessors and memories has had a profound influence on many areas of technical endeavour. Also in the measurement and control field, modern electronic circuits were introduced on a large scale leading to very sophisticated systems and novel solutions. However, in these measurement and control systems, quite often sensors and actuators were applied that were conceived many decades ago. Consequently, it became necessary to improve these devices in such a way that their performance/price ratios would approach that of modern electronic circuits.

This demand for new devices initiated worldwide research and development programs in the field of "sensors and actuators". Many generic sensor technologies were examined, from which the thin- and thick-film, glass fiber, metal oxides, polymers, quartz and silicon technologies are the most prominent.

A growing number of publications on this topic started to appear in a wide variety of scientific journals until, in 1981, the scientific journal Sensors and Actuators was initiated. Since then, it has become the main journal in this field.

When the development of a scientific field expands, the need for handbooks arises, wherein the information that appeared earlier in journals and conference proceedings is systematically and selectively presented. The sensor and actuator field is now in this position. For this reason, Elsevier Science took the initiative to develop a series of handbooks with the name "Handbook of Sensors and Actuators" which will contain the most meaningful background material that is important for the sensor and actuator field. Titles like Fundamentals of Transducers, Thick Film Sensors, Magnetic Sensors, Micromachining, Piezoelectric Crystal Sensors, Robot Sensors and Intelligent Sensors will be part of this series.

The series will contain handbooks compiled by only one author, and handbooks written by many authors, where one or more editors bear the responsibility for bringing together topics and authors. Great care was given to the selection of these authors and editors. They are all well known scientists in the field of sensors and actuators and all have impressive international reputations.

Elsevier Science and I, as Editor of the series, hope that these handbooks will receive a positive response from the sensor and actuator community and we expect that the series will be of great use to the many scientists and engineers working in this exciting field.

Simon Middelhoek

PUBLISHER'S NOTE

We are pleased to present to you this book on Semiconductor Sensors in Physico-Chemical Studies, edited by Dr. Leonid Kupriyanov of the Karpov Institute of Physical Chemistry in Moscow. Translated from Russian by a professional translator, the prose still retains some of its local flavour although, we trust, it is clearly understandable throughout.

Had we further polished the language we would not only have lost at least another six months in bringing the book to you but we would also have incurred costs which would have been reflected in a higher price. We trust you will not find the readability of the text detracts unduly from your appreciation of this work, which we feel makes a most worthwhile contribution to the available literature in this field.

Mark Eligh
Publishing Editor

ABSTRACT

The monograph is devoted to scientific basis of semiconductor chemical sensors technique. Its attention is focused at the usage of semiconductor sensors in the precision physico-chemical studies. The monograph expounds physical and chemical basis underlying the semiconductor sensor method, discusses the mechanism of processes occurring under interaction of gas with semiconductor adsorbent surface, leading to changed electrophysical parameters of the latter.

The monograph is intended for the scientists and engineers specialized in physical-chemistry of heterogeneous and heterogeneous-homogeneous processes and designing of semiconductor chemical sensors.

List of Contributors:

Prof. I.A.Myasnikov (Chapters 3,4)
Chief of Department of Chemical Sensors, Karpov Institute of Physical Chemistry, Moscow, Russia.

Dr. V.Ya. Sukharev (Chapters 1,2)
Senior Researcher, Department of Chemical Sensors, Karpov Institute of Physical Chemistry, Moscov, Russia[1] .

Dr. L.Yu. Kupriyanov (Chapter 5)
Senior Researcher, Department of Chemical Sensors, Karpov Institute of Physical Chemistry, Moscov, Russia.

Dr. S.A. Zavyalov (Chapter 6)
Senior Researcher, Department of Chemical Sensors, Karpov Institute of Physical Chemistry, Moscov, Russia.

---

[1] Present address: LSI Logic Corporation, 3115 Alfred Street, MS J-201, Santa Clara, CA 95054, USA.

CONTENTS

INTRODUCTION

This book is the first attempt to provide a detailed description of scientific basis of the method of semiconductor chemical sensors which are presently widely applied in various domains of industry and in everyday life. The major feature of this book (which distinguishes it from the literature published up to date) is that it mainly examines the use of the method of semiconductor sensors in fine physical and chemical studies.

Arbitrary the book can be divided into two complementary parts. The first one describes the physical and chemical basics leading to description of the method of semiconductor sensors. The mechanisms of underlying processes are given. These processes involve interaction of gas with the surface of semiconductor adsorbent which brings about the change of electric and physics characteristics of the latter. Various models of absorption-induced response of electric and physics characteristics of semiconductor adsorbent are considered. Results of numerous physical and chemical experiments carried out by the authors of this book and by other scientists underlying the method of semiconductor sensors are scrupulously discussed. The possibility of qualitative measurements of ultra-small concentrations of molecules, atoms, radicals as well as excited particles in gases, liquids and on surfaces of solids (adsorbents and catalysts) is demonstrated.

The second part of the book deals with the use of above method in physical and chemical studies. In addition to illustration load, this part of the book has a separate scientific value. The matter is that as examples the book provides a detailed description of the studies of such highly interesting processes as adsorption, catalysis, pyrolysis, photolysis, radiolysis, spill-over effect as well as gives an insight to such problems as behavior of free radicals at phase interface, interaction of electron-excited particles with the surface of solid body, effect of restructuring of the surface of adsorbent on development of different heterogeneous processes.

In Chapter 1 we consider the physical and chemical basis of the method of semiconductor chemical sensors. The items dealing with mechanisms of interaction of gaseous phase with the surface of solids are considered in substantial detail. We also consider in this part the various forms of adsorption and adsorption kinetics processes as well as adsorption equilibria existing in real gas-semiconductor oxide adsorbent systems. We analyze the role of electron theory of chemisorption on development of ideas regarding the effect of adsorption on electrophysical properties of semiconductor adsorbents.

We also address the models of adsorption change in electrophysical characteristics of semiconductor adsorbent caused both by chemisorbed charging of the surface due to the charge transition between surface states and volume bands of adsorbent and by local chemical interaction of adsorbate with electrically active defects of semiconductor.

We consider the existing models of adsorption response of electrophysical characteristics of ideal monocrystalline adsorbent, monocrystal with inhomogeneous surface as well as polycrystal adsorbent characterized by an a priori barrier disorder. The role of recharging of biographic surface states in the process of adsorption charging of the surface of semiconductor is analyzed.

A detailed description is given to the role of point defects available in the volume and on the surface of oxide adsorbents on adsorption-induced change of electrophysical characteristics. We try to deduce the impact of the proper nature of adsorbent as well as the nature of adsorption centers.

In Chapter 2 we touch on theoretical models of adsorption response of electrophysical characteristics of real semiconductor adsorbents used as sensitive elements of chemical sensors.

We scrutinize issues dealing with requirements of high sensitivity and response selectivity of electrophysical parameters in reference to the gas monitored or the type of active particles under study as well as other requirements put forward to adsorbents of chemical sensors. We discuss principles underlying the basis of solving these problems. We dwell on the issue of the type of crystal of adsorbents examined, which is directly linked to the character of intracrystallite contacts.

We also pay attention impact of preparation technique of polycrystal adsorbents and applicability domains of obtained semiconductor sensors.

We consider problems related to electrophysical properties of sintered polycrystalline oxides as well as their adsorption changes. We also analyze the difference in adsorption induced changes of electrophysical characteristics of stoichiometric and non-stoichiometric partially reduced oxide adsorbents.

In Chapter 3 we briefly outline the methods of manufacturing of sensitive elements of semiconductor sensors in order to proceed with the studies of several physical and chemical processes in gases, liquids as well as on the surface of solids. Here we show the peculiarity of preparation of these elements depending on objective pursued and operation conditions. We outline the detection methods (kinetic and stationary), their peculiarities and advantages of their application in various physical and chemical systems.

The same Chapter contains results of studies of effects of adsorption of atom particles as well as simplest free radicals on electric conductivity of semiconductor zinc oxide films.

Using the studies on adsorption of H-atoms as examples (as well as atoms of various metals) by volumetric and beam methods we prove a rigorous proportionality between the values of adsorption and the change in the number of carriers in adsorbent. Similar dependencies are given for the case of adsorption of acceptor particles, as well. These are such particles as methyl and ethyl radicals, hydroxyls and nitrogen-containing radicals. Using examples of adsorption of above particles we have shown the applicability of simplest relationships between their concentration and the change of dope electroconductivity of adsorbent.

We outline experimental results and provide theoretical interpretation of effect of adsorption of molecular oxygen and alkyl radicals in condensed media (water, proton-donor and aproton solvents) having different values of dielectric constant on electric conductivity of sensors. We have established that above parameter substantially affects the reversible changes of electric conductivity of a sensor in above media which are rigorously dependent on concentration of dissolved oxygen.

It is shown that similar effects are observed in case of gas or vapor phase media under condition of availability of saturated vapor in the liquid forming a thin layer on the surface of a semiconductor adsorbent.

Chapter 4 deals with several physical and chemical processes featuring various types of active particles to be detected by semiconductor sensors. The most important of them are recombination of atoms and radicals, pyrolysis of simple molecules on hot filaments, photolysis in gaseous phase and in absorbed layer as well as separate stages of several catalytic heterogeneous processes developing on oxides. In this case semiconductor adsorbents play a two-fold role: they are acting both as catalysts and as sensitive elements, i.e. sensors in respect to intermediate active particles appearing on the surface of catalyst in the course of development of catalytic process.

Apart from above studies we dwell on experimental results of evaporation of superstoichiometric metal atoms from the surface of various oxides as well as from the surface of metal foils in order to determine their evaporation heats. The results obtained are consistent with tabulated values obtained by other experimental techniques.

"Plane" oxide sensors were used to study lateral diffusion of H-atoms and spill-over effect.

The combined use of the method of semiconductor sensors and that of molecular beams enabled us to investigate adsorption of atom, molecular and cluster particles of metals on metal oxides.

We indicate a possible use of semiconductor sensors in the studies of interaction of hydrogen ions and electrons with various energy with a surface of oxides and hydrogen layers absorbed to them. These sensors can also be used to study γ-radiolysis of hydrocarbons in gases and liquid media.

Chapter 5 outlines problems of the use of semiconductor sensors to study heterogeneous processes developing in presence of electrically excited atoms and molecules. We give a short description of physical and chemical properties of electrically excited particles, methods of their detection as well as items dealing with their deactivation on the surface of semiconductor oxides. We provide a scientific basis to apply sensor method for detection of electronically excited particles. The method of application of sensors in the studies of such type is described using specific examples of interaction of singlet oxygen and metastable atoms of inert gases with the surface of semiconductors and dielectrics. It is shown in what manner the sensors can be used to determine the efficiency of heterogeneous quenching of electrically excited particles and to elucidate the channels of dissipation of excitation energy on surfaces of various nature. The mechanisms of the change of electric conductivity in semiconductor oxides with pure and dope modified surface due to effect of electrically excited particles are discussed.

Chapter 6 provides experimental results on detection and studies of a wide scope of phenomena dealing with emission of active particles from the surface of solids due to energy released during annihilation of local excitation in surface-adjacent layers. Similar phenomena are indisputably of interest from the standpoint of understanding the mechanism of nucleation and development in the volume of heterogeneous-homogeneous reactions as well as heterogeneous catalytic processes developing through a chain mechanism.

The same Chapter examines specific methodological techniques applicable to identify atoms of hydrogen, oxygen, silver, alkyl radicals, molecules of singlet oxygen in gaseous phase. These methods can be widely applicable for practical physical and chemical studies.

The text of the book is provided with numerous figures and tables exemplifying the applicability of the method of semiconductor sensors to studies of above processes. This illustration material enables one to compare the results obtained with those obtained by similar measurements using different techniques.

Drawing the bottom line, this book is very useful for scientists in various disciplines and experts in domain of interface physical chemistry interested in development and application of the method of semiconductor sensors as well as for post-graduate and graduate students specialized in above domain of science.

Semiconductor Sensors in Physico-Chemical Studies
L. Yu. Kupriyanov (Editor)

Chapter 1

Physical and chemical basics of the method of semiconductor sensors

1.1. What are the semiconductor chemical sensors?

Let us start with a definition. Semiconductor chemical sensor is an electronic device designed to monitor the content of particles of a certain gas in surrounding medium. The operational principle of this device is based on transformation of the value of adsorption directly into electrical signal. This signal corresponds to amount of particles adsorbed from surrounding medium or deposited on the surface of operational element of the sensor due to heterogeneous chemical reaction.

Starting with this definition the semiconductor chemical sensors can be arbitrary classified with respect to following features: the type of electrophysical characteristics chosen for monitoring, such as electric conductivity, thermal-electromotive force, work function of electron, etc.; type and nature of semiconductor adsorbent used as an operational element of the sensor and, finally, the detection method used for monitoring the adsorption response of electrophysical characteristics of the sensor.

In each specific case the choice of an adsorbent, electrophysical parameters and the method of registration of its change as well as the choice of various pre-adsorption treatment techniques of the surface of adsorbent is dictated by the type and nature of analytical problem to be solved. For instance, if particles active from the standpoint of the change in electrophysical parameters of semiconductor adsorbent occur on the surface of the latter due to development of a chemical reaction involving active particles, it is natural to use either semiconductor material catalyzing the reaction in question or if this is not possible specific surface dopes accelerating the reaction. Above substances are used as operational element of the sensor. If such particles occur as a result of adsorption from adjacent volume, one can use semiconductor materials with maximum adsorption sensitivity to the chosen electrophysical parameter with respect to a specific gas as operational element.

Thus, sensor effect deals with the change of various electrophysical characteristics of semiconductor adsorbent when detected particles occur on its surface irrespective of the mechanism of their creation. This happens because the surface chemical compounds obtained as a result of chemisorption are substantially stable and capable on numerous occasions of exchanging charge with the volume bands of adsorbent or directly interact with electrically active defects of a semiconductor, which leads to direct change in concentration of free carriers and, in several cases, the charge state of the surface.

In the case when semiconductors with a characteristic size (film thickness, microcrystal dimensions, etc.) exceeding the Debye shielding length are used as adsorbent the chemisorption change of the surface charge results in distortion of the diagram of the surface in this area and, therefore, in the change of electrophysical characteristics of the material controlled by transition phenomena. Such a large scale curvature of the energy profile of the bottom of the conductivity zone (and respectively the ceiling of the valence band) accompanying chemisorption of various particles which is inherent to semiconductor materials results in sharp change in concentration of free carriers participating in transition of current carriers. The similar change in concentration of free charge carriers can be observed during direct adsorption-induced interaction resulting in a change of concentration of electrically active defects responsible for dope electric conductivity in broad band semiconductors. On the contrary, in metal adsorbents chemisorption-induced change of the charge state of the surface almost does not effect the concentration of the current carriers. It is, mainly, manifested through the change in carrier mobility due to the change of the surface scattering, which, naturally, has a lesser effect on electrophysical characteristics of the material. It is this high sensitivity of electrophysical characteristics of semiconductors to adsorption of various gases as well as capability to control it that makes these materials attractive to manufacture gas-sensitive electronic transducers.

Semiconductor chemical sensors are characterized by low cost, small size, extra high sensitivity (often unattainable in other analytical techniques) as well as reliability. Moreover, concentration of particles detected is being transformed directly into electrical signal and electronic design of the device is the simplest one which can be arranged for on the active part of the substrate.

The crystalline structure of adsorbents directly influences their gas sensitivity. Depending on the type of the problem to be addressed adsorption-sensitive semiconductors are monocrystals or monocrystal films with a predetermined crystallographic orientation of operational surface; vacuum baked polycrystalline films, which, from their electrical standpoint, are similar to monocrystals but differ from the latter by ultimately developed surface; or polycrystalline semiconductors baked in oxidizing medium. In addition to the degree of stoichiometry of the content the difference between the two above types of polycrystalline adsorbents also deals with the type of contacts between various crystals which is manifested in availability (or unavailability) of intracrystalline energy barriers hindering motion of current carriers.

General requirements put forward to all types of operational elements of sensors are the following: semiconductors should have substantial chemical durability, they should not form stable chemical compounds with the particles absorbed. Moreover, they should possess sufficient thermal and mechanical strength. Metal-oxide semiconductors, such as ZnO, SnO_2, TiO_2 and others suit above requirements the best. Apart from that such broad band dope semiconductors exhibit very high adsorption sensitivity (if contrasted to elementary ones), which is explained by small concentration of free current carriers and therefore by exceptionally high sensitivity of their numerous electrophysical parameters to doping of any type.

Numerous experiments showed that depending on adsorption of active particles the "response" of adsorbent, namely the change in its electric conductivity, electron work function, as well as thermal electromotive force, Hall electromotive force, etc., all active particles can be grouped into three categories – electron acceptors, donors or neutral particles. Using oxide semiconductors of the electron n-type such as ZnO, TiO_2, CdS etc. and hole p-type such as NiO, Cu_2O as an example it was shown that irrespective of chemical composition of oxide belonging to any type of dope semiconductor, adsorption of acceptor active particles, i. e. the particles possessing an affinity to electron, e. g. simplest radicals CH_2, CH_3, C_2H_5, C_3H_7, NH, OH etc. as well as atoms of N, O, Cl and active molecules O_2, Cl_2, Br_2, etc. always decrease the dope electric conductivity of oxides of n-type and increase the electron work function.

We should note that adsorption of acceptor particles on oxide semiconductors of p-type influences their electric conductivity and work function in the opposite way. As for donor particles such as atoms of H, Na, K, Zn, Cd, Pb, Ag, Fe, Ti, Pt, Pd and many others, their adsorption at medium and low temperatures (when there is no notable diffusion of atoms proper into the crystal and, consequently, there is no substitution of atoms created, the latter obeying the Vervey rule) is always accompanied by increase in electric conductivity and decrease in the work function for semiconductor adsorbent of n-type, the opposite being valid in case of p-type adsorbent.

Such particles as stable radicals as well as vast majority of molecules of organic compounds, for instance saturated hydrocarbons CH_4,

8

C_2H_6, C_3H_8, etc., unsaturated hydrocarbons as well as numerous classes of cyclic and aromatic compounds belong to neutral particles whose adsorption does not affect electrophysical properties of semiconductors. Speaking of non-organic simplest compounds, there are all noble gases, molecular nitrogen, and molecular hydrogen (at low temperatures) that belong to this class. We should mention, however, that above particles remain neutral with respect to effect of their adsorption of electric conductivity of semiconductor adsorbents only in absence of any vibrational or electronic excitation. In a separate chapter we discuss the impact of adsorption of electron-excited particles on electric conductivity of semiconductor adsorbents.

Additionally to meeting the requirement of high adsorption sensitivity the semiconductor sensors should be response-selective to a specific gas as well as exhibit high signal stability (i.e. reproducibility) over a long operation time.

The need for high selectivity stems from the fact that adsorption of various particles eventually results in identical changes in electrophysical characteristics of an adsorbent which leads to complexity of detection of a specific component in the active gas mixture. Nevertheless, there are techniques to avoid this problem enabling one to obtain adsorbents sensitive only to specific gases. Among such technique there are: i) the choice of a detection temperature interval enabling one to register specific particles which are active in this very interval whereas other components of the mixture either grow passive or have not been excited yet; ii) application of small (~1%) admixtures of catalytically active metals shifting the maximum of sensitivity towards the gas chosen; and iii) a proper choice of the structure and thickness of oxide layers of sensors enabling one to discriminate the effects of various active gaseous particles. The same objective can be achieved through using various filters capable of letting through the particles of interest. The application of multisensor systems (i.e. adsorbent arrays characterized by different sensitivity to different particles) is highly promising. Such systems lead to generation of "adsorption portraits" of specific gases.

In practice, however, as well as in experimental physical chemistry the problem often gets simplified. Usually one has to monitor the development of the process through continuous analysis of highly small concentration of only one component, for instance traces of oxygen in noble gases, hydrogen or nitrogen or in various organic gaseous substances or liquids. The above technique involving semiconductor sensors can be successfully applied to solve not only fundamental physical and chemical problems but also engineering ones in which the requirement to use gases and liquids with utmost purity is being put forward more often. Such media require monitoring the traces of oxygen or other admixtures at a level smaller than $10^{-6} - 10^{-7}$ %-vol. With the laps of time owing to increasing requirements put forward by science and technology to the purity of substances the upper admissible concentration threshold will decrease down to $10^{-9} - 10^{-10}$ %-vol. This is especially the case when one discusses such energy effective, ecologically advantageous and highly wide-spread fuel as hydrogen. Similar problems can be encountered while solving a few biological problems.

The stability of the sensor signal is also dependent on several reasons. Some of them are caused by partial irreversibility of chemisorption of numerous particles, by presence of non-controlled gaseous mixtures, by development of various side chemical reactions and processes on the surface and in adsorbent-adjacent layers. Periodic regeneration of the surface involving various technological operations such as heating to above operational temperatures, treatment by jets of various gases, irradiation, etc., provide the principal means to maintain the signal stability within required margins.

Derivation of simple and unambiguous quantitative relations between the signal amplitude of a sensor, i.e., the value of the change of electric conductivity, work function, etc. and concentration of detected traces of admixture in the medium under study is also important for successful development of the sensor measuring technique. Theoretical considerations given in this book show that such relations exist in most simple form. The purpose of experiment consists in statistical substantiation that these dependencies rigorously hold at proper conditions.

Even listing all above problems and requirements leading to their solution indicates that development of the method of semiconductor chemical sensors opens a wide research domain. In order to resolve this problems and implement all capabilities of the method of semiconductor sensors there are two ways now : the old "trial and error" approach and approach related to further studies of physical and chemical properties of surface phenomena, reactions and processes underlying this method. It is quite clear that the second approach is more promising in order to obtain semiconductor sensors designed for the use in accurate scientific studies and for practical gas analysis.

Without downplaying the importance and suitability of other experimental techniques we can maintain that such miniature, light and low energy consuming analytical devices as semiconductor sensors featuring unique sensitivity (approaching that of sense of smell in wild animals) will soon have a leading edge in science and technology.

1.2. The role of electronic theory of chemisorption in developing ideas on effects of adsorption on electrical and physical properties of semiconductor adsorbents.

The necessity of the use of electronic notions to resolve several chemical and physical problems stemming from the studies of heterogeneous processes was realized by Pisarjevsky already in early twenties [1]. Several non-trivial ideas concerning the effect of adsorption on electrophysical properties of semiconductor adsorbents were formulated in classical studies of Yoffe [2], Roginsky [3] and others. These theoretical ideas were further developed by Volkenshtein and his colleagues (see book [4] and the reference list therein) as well as in studies by Hauffe [5, 6] and some other authors [7, 8].

The principal idea of Volkenshtein, the founder of electronic theory of chemisorption, was that chemisorbed particle and solid body form a unified quantum mechanical system. During the analysis of such systems one should account for the change in electronic state of both adparticle and the adsorbent itself [9]. In other words, in this case adsorption provides for a chemical binding of molecules with adsorbent.

The consideration of quantum mechanical problem regarding the interaction of single valence atom with ion lattice through application of molecular orbitals [9, 10] made it possible to prove that absorbed particles form local surface energy levels for electrons and holes. The studies of energy spectrum of the adsorbate-adsorbent system suggested a possible existence of different forms of chemisorption: neutral and charged ones. In case of the neutral form the binding of an adparticle with adsorbent is provided through merging its electrons with atoms of a lattice. The resultant local level is positioned in the forbidden band and becomes unpopulated. Such single electron binding of a chemisorbed particle with the semiconductor surface is known to be a "weak" one. To a certain extent the chemisorbed particle in this case becomes polarized and its dipole moment induced by chemisorption is of pure quantum mechanical origin [9].

In case of the charged form of chemisorption a free lattice electron and chemisorbed particles get bound by exchange interaction resulting in localization of a free electron (or a hole) on the surface energy layer of adparticles which results in creation of a "strong" bond. Therefore, in case of adsorption of single valence atom the "strong" bond is formed by two electrons: the valence electron of the atom and the free lattice electron.

Thus, in case of a "strong" binding with chemisorbed particle the surface of a solid body becomes charged with relative to its volume which inevitably leads to band bending in the surface-adjacent domain of adsorbent. This peculiarity was accounted for by Hauffe and Veits [5 – 7] while developing the theory of adsorption "adjacent layer". Later on Bonch-Bruevich [11] used an example of interaction of a single valence atom with positively charged defects formed on the surface of ionic crystal to show that adsorption on the defect is more advantageous from the energy standpoint if compared to adsorption on ideal lattice.

As a result of these quantum mechanics calculations the principle option of considering chemisorbed particle as a certain surface defect was proved. Depending on the origin and the properties of adsorbent this defect can be a surface localization center of either free electrons, or holes acting, thereby, as acceptor or donor of electrons [12]. Under such conditions the population of localized levels at thermodynamic equilibrium unambiguously obeys the Fermi statistics. In other words, electrons and holes localized on adsorption-induced energy levels belong to the common electron system of adsorbent. In particular, these are these localized charged carriers that influence the position of the Fermi level in semiconductors [12 – 14]. We should note that the latter makes it feasible to account for interaction of adparticles in case of extremely small deposition of these particles on the surface, when there is almost no direct lateral interaction between them.

Thus, the major conclusions of the early studies by Volkenshtein and his colleagues applicable to the theory of the method of semiconductor gas sensors are the following: a) chemisorption of particles on a semiconductor surface can be accompanied by a charge transfer between adsorption-induced surface levels and volume bands of adsorbent and b) only a certain fraction of absorbed particles is charged, the fraction being dependent on adsorbate and adsorbent.

It is worth pointing out that the latter conclusion contradicts that of the theory of "boundary layer" [5 – 7, 15, 16] predicting the charge in chemisorbed particles and resulting attraction to the surface charge carriers localized close to it. Neutralization of a charge is considered in this theory as an act of desorption of the particle and vanishing of a respective surface level which renders meaningless the very introduction of the notion of local level capable, by definition, of both excepting and emitting electrons [17].

The conclusion of electron theory of chemisorption regarding existence of both charged and neutral form of chemisorption resolves another contradiction dated to early "theory of boundary layer" regarding substantially large values of adsorption of various particle in experiment and availability of "Weitz limit" [7] indicating on existence of a limiting and relatively small (~ 1%) deposition of charged (!) chemisorbed particles. In more advanced versions of this theory [16, 18] the above contradiction was removed through introduction of a notion of a neutral form of adsorption which was attributed to physadsorption. Chemisorption remains attributed to adsorption of charged particles. From this standpoint localization of a free charge carrier on the surface level indicates a transition of adparticles from physadsorbed state to the chemisorbed one [18].

In reality the issue of rigorous differentiation between various forms of adsorption is very complex and ambiguous [19]. Briefly this aspect will be discussed somewhat later on. Now, it is relevant to mention that in several cases a direct linking of chemisorption with a surface charge is misleading while explaining experimental results. For instance, in several papers the data on pure electric measurements were used to make conclusions on mechanisms an adsorption process. Monitoring a dependence of a certain electrophysical parameter of a semiconductor (usually electroconductivity or work function) on the partial pressure of the gas analyzed the conclusions were drawn regarding the shape of adsorption isotherm, whereas kinetics of the change of the studied parameter provided the basis to deduce the kinetics of chemisorption process. In fact, adsorption is one of the stages of a phenomenon of charging a semiconductor surface featuring a transition of a charge on induced surface levels. The kinetics of the process of charging and accompanying response in the values of electrophysical characteristics of the adsorbent is controlled by a limiting stage of the process. In most cases, for instance in adsorption of acceptors in broad band semiconductors of n-type [18, 20], this stage is provided by transition of the charge on adsorption surface levels. Therefore, in such cases the kinetics of the change of electroconductivity directly reflects the kinetics of exchange by the charge carriers between the volume bands of a semiconductor and the energy levels of the surface states.

While studying the dependence of the value of the stationary level of electroconductivity or work function versus gas pressure one encounters the situation of the same complexity because the parameter monitored is dependent on the concentration of the charged chemisorbed particles. Apart from being sufficiently complex, in numerous cases this dependence is superimposed on relationship linking concentration of the charged particles with the total concentration of chemisorbed particles, dependent on the nature of adsorbate and adsorbent. Only in the most simple cases of absorption equilibrium one can expect the total concentration of chemisorbed particles to be described by the known adsorption isotherms. We should note that in cases of not too small adsorbate population on the surface the situation gets complicated due to the "Weitz limit" making unrealistic the chemisorption of charged particles on the surface over reasonable times of attaining equilibrium [7, 21].

In conclusion we would like to mention that basic ideas which formed a ground-stone of the Volkenshtein electron theory of chemisorption have been confirmed and substantially developed in numerous papers on quantum chemistry (e.g., see [22]) accounting for fine peculiarities of local interactions of chemisorbed particles with the immediate atom surroundings on the surface.

1.3. Various forms of adsorption

As it has been already mentioned in previous section the process of adsorption of gaseous particles on the surface of semiconductor adsorbent is one of the major and in several cases the most important stage controlling the adsorption-induced change in the value of surface charge resulting in alteration of numerous surface and volume electrophysical characteristics of adsorbent.

Not pursuing the objective to describe the adsorption phenomenon in detail, we would like, however, to dwell briefly on its principle ideas as well as on main theoretical and experimental results necessary for further understanding of the concept of the method of semiconductor sensors.

Usually adsorption, i.e. binding of foreign particles to the surface of a solid body, is distinguished as physical and chemical the difference lying in the type of adsorbate − adsorbent interaction. Physical adsorption is assumed to be a surface binding caused by polarization dipole-dipole Van-der-Vaals interaction whereas chemical adsorption, as any chemical interaction, stems from covalent forces with plausible involvement of electrostatic interaction. In contrast to chemisorption in which, as it has been already mentioned, an absorbed particle and adsorbent itself become a unified quantum mechanical system, the physical absorption only leads to a weak perturbation of the lattice of a solid body.

From the stand-point of the problems of absorption-induced change in the charge state of a semiconductor surface considered here, the physical adsorption is interesting since it is often considered as an initial step of binding of gases particles preceding chemisorption [21]. Most clearly the relationship between physical and chemical adsorption can be traced in the curves of Lennard-Johns [23, 24], describing the energy in the adsorbate-adsorbent system as a function of the distance between the particle and the surface (Fig. 1.1). These curves were proposed to explain experimentally observed adsorption isobars. They indicate that with increasing temperature the maximum of adsorption specific for substantially low temperatures and ascribed to physically absorbed particles was followed not by an expected monotonous decrease but by a series of maxima. There are two adsorption curves shown in Fig. 1.1 corresponding to physical (curve 1) and chemical (curve 2) adsorption.

14

Here Q_i is the adsorption heat, r_i – is the equilibrium state determining the maximum of potential energy. In case of chemisorption r is much smaller and Q is much larger than corresponding values typical for physical adsorption. The potential barrier δE formed as a result of intersection of these two adsorption curves is often considered as an activation energy of transition from physically absorbed state into the chemically absorbed one. Thereby, the concept of activated adsorption gets introduced. This phenomenon is often observed in experiments and on numerous occasions considered as chemisorption. Moreover, the availability of the activation energy and irreversibility are often referred to as inherent peculiarities to chemical sorption proper. In reality this is not the case. Chemosorption which is irreversible and activated at a specific temperature becomes reversible and non-activated at a higher temperature. Moreover, it is well known from the theory of chemical reaction that there is a whole class of reactions not requiring an activation energy. Addition reactions, for instance, belong to these class. On the other hand, availability of lateral interactions between physically sorbed particles (a typical example of such a reaction is provided by dipole-dipole interaction, dipoles being of inherent and polarized nature) might also result in formation of the activation energy of adsorption.

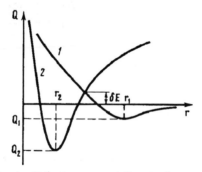

Fig. 1.1. The diagram of potential energy according to Lennard-Johnes.

The fact of a transfer of an electron from an absorbed particle to adsorbent [25] is widely considered as a criterion to differentiate between various forms of adsorption. Yet, as it has been already mentioned in previous section, there is a neutral form of chemisorption, i.e. "weak" binding formed without changing the surface charge state which only affects the dipole component of the work function. On the other hand , in several cases the physical adsorption can result in electron transitions in solids. Indeed, apart from formation of a double layer, changing the work function of adsorbent [26] the formation of surface dipoles accompanying physical adsorption can bring free charge carriers to substan-

tially deep potential wells created by overlapping of random electrostatic fields of the dipoles [17, 27 – 29], provided the surface occupation degree is substantially high.

Distinguishing between physical and chemical adsorption using the
value of adsorption heat cannot lead to unambiguous results, too. An
arbitrary classification of physical adsorption as having small heats
$Q \sim 0.01 \div 0.2$ eV typical for and attributing ~ 1 eV to chemisorption is
often violated. A typical example can be provided by a dissipative
chemisorption characterized by small total heat effect.

Thus, as of today, there is no reliable classification of various types
of adsorption. Presumably, it would be most correct to consider various
types of adsorption interactions consistent with classification of chemical
bonds [19].

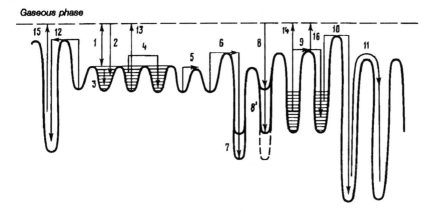

Fig. 1.2. The scheme of processes taking place during interaction of gaseous
phase particles with solid surface.

In most general manner phenomena accompanying collisions of particles of gaseous phase with a semiconductor surface are shown in
Fig. 1.2, borrowed with some alterations from the well known book by
Roberts and McKey [30]. In this figure process *1* describes elastic scattering. As a result of process *2* a fraction of kinetic energy of a particle
dissipates and the particle remains bound to the surface in a state of
weak adsorption moving to the bottom of potential well loosing the
excess of initial oscillation excitation (process *3*). This loss can occur
when the particle penetrates the potential barrier between adjacent potential wells (process *4*). Process *5* corresponds to migration of weakly
bound adsorption particle along the adjacent adsorption centers. In addition to physical adsorption, various molecular complexes formed during interaction of absorbed particles with initially absorbed ones [31]

can act as a weakly bound type of adsorption. Process *6* leads to the neutral form of chemisorption of a particle, the transition into a charged form being described by process *7*. We should mention that states corresponding to physical adsorption and chemical adsorption can develop in the same area which can result in transition of a particle into chemisorbed state without initial migration (processes *8* and *8'*). The case of migration of adparticle through chemisorption centers is described by process *9*. At high temperatures the particle can leave an adsorption state on the surfaces and travel into the semiconductor volume resulting thereby in healing or creation of defects, i.e. process *10*. Process *11* corresponds to the volume diffusion of absorbed particles. Additionally to above processes the absorbed particles can participate in the process of healing or creation of surface defects (process *12*). Simultaneously with considered processes there desorption processes also develop (*13 – 16*). As an example Fig. 1.3 shows results of Gopel [32] obtained in the studies on the energy of interaction of molecular oxygen with thermodynamically stable surfaces of ZnO and TiO$_2$.

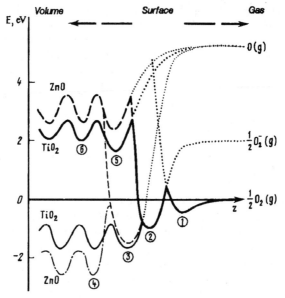

Fig. 1.3. Energy diagram of O$_2$ interactions on ZnO and TiO$_2$ surfaces.
1 – physadsorption: $(1/2)O_2(g) \rightarrow (1/2)O_2(ph)$; *2* – chemisorption: $(1/2)O_2(g) \rightarrow (1/2)O_2^-(ch)$; *3* – healing of surface defects: $V_{o,s} + (1/2)O_2(g) \rightarrow O_{l,s}$; *4* – healing of volume defects: $V_o + (1/2)O_2(g) \rightarrow O_l$; *5* – interstitial surface oxygen: $(1/2)O_2(g) \rightarrow O_{i,s}$; *6* – interstitial volume oxygen: $(1/2)O_2(g) \rightarrow O_i$ [32].

In conclusion one can deduce that adsorption interaction of gas with solids, the type of which is determined by the nature of a specific pair adsorbate-adsorbent, temperature of the system, initial treatment of the surface and, finally, various "sensibilizing" external effects (for instance, photoadsorption) result in formation of various states of particles (which can be both of adsorption and sorption nature) affecting electrophysical characteristics of the semiconductor adsorbent. This effect can be caused by adsorption-induced change of the value of the surface charge of an adsorbent altering the profile of zone diagram in the surface-adjacent domain or by direct interaction of absorbed particles with electrically active defects of a semiconductor. However, to a large extent the dependencies detected for adsorption-caused change in electrophysical characteristics of an adsorbent are dependent on the character of chemosorption process in each case.

1.4. Adsorption isotherms

The classical method to study adsorption deals with establishing a correspondence between thermodynamically equilibrium amount of absorbed gas, the temperature and the partial pressure of the gas in space surrounding adsorbent. Usually, a constant temperature is maintained throughout an adsorption experiment. The shape of isotherms obtained gives information on the character of adsorption.

In the most simple case of ideal, energy homogeneous surface the adsorption equilibrium of non-interacting particles is described by the Langmuir isotherm [33]:

$$N = N_* \left/ \left(1 + \frac{b}{P}\right) \right. , \qquad (1.1)$$

which is applicable to describe non-dissociation adsorption with saturation corresponding to creation of a monomolecular layer. Here N is the concentration of absorbed particles; N_* is the density of adsorption centers; P is the pressure in gaseous phase;

$$b = b_0 \exp\{-Q / kT\} , \qquad (1.2)$$

Q is the adsorption heat; for the simplest case $b_0 \approx 4 \cdot 10^5 (M_A T)^{1/2}$; M_A is the molecular weight of an absorbed particle.

In case of dissociative adsorption when an absorbed particle gets decomposed into n "components", each of which occupies its adsorption site we arrive at the following expression

$$N = N_* \left/ \left[1 + \left(\frac{b}{P}\right)^n \right] \right. , \tag{1.3}$$

For small pressures the Langmuir isotherm becomes the Henry isotherm describing the domain of linear adsorption

$$N = N_* \frac{P}{b} , \tag{1.4}$$

or

$$N = N_* \left(\frac{P}{b}\right)^n , \tag{1.5}$$

for non-dissociation and dissociation adsorption, respectively.

In experiment one can also observe other types of isotherms which do not fit into cases $(1.1) - (1.5)$. In cases of submonomolecular layers the most frequently observed isotherms are:

the power Freundlich isotherm [34]

$$N = CP^\gamma , \tag{1.6}$$

where C and γ are certain constants depending on the adsorbate-adsorbent pair,

logarithmic isotherm of Frumkin [35]

$$N = A + B \ln P , \tag{1.7}$$

where A and B are constants.

Non-compliance with the simple Langmuir adsorption model is indicative of violation under experimental conditions of certain assumptions used to derive the model. Therefore, while developing the theoretical models adequately describing experimental data one usually resorts to one of two approaches: either introduces the notion of a inhomogeneous surface [36, 37] or accounts for various types of interaction developing between the particles absorbed [4, 38].

In case when the notion of inhomogeneous surface is introduced it is assumed that all surface can be represented by areas characterized by various adsorption heats Q_i or, in a more general case, by various inverse adsorption coefficients $b_i = b_{0i} \exp \{-Q_i / kT\}$. Introducing the distribution function of adsorption heats or inverse adsorption coefficients one obtains the following expression for a surface occupation degree

$$\theta = \int\limits_{Q_{min}}^{Q_{max}} \frac{P}{P+b(Q)} f(Q)dQ \equiv \int\limits_{b_{min}}^{b_{max}} \frac{P}{P+b} \Phi(b)db , \tag{1.8}$$

where $P/(P+b(Q))$ is the probability of occupying an area with a given value of $b(Q)$; $f(Q)$, $\Phi(b)$ is the distribution density of adsorption heats and inverse adsorption coefficients. The choice of different types of distribution functions results in different relationships describing adsorption equilibria. In particular, the exponential distribution of areas to be absorbed with respect to adsorption heat results in the Freundlich is isotherm [39], whereas uniform distribution Q leads to derivation of the Frumkin isotherm [40].

Introduction of a inhomogeneous (with respect to the adsorption heat) surface makes it easy to explain the dependence of differential adsorption heat on occupation degree of the surface observed in experiments [36].

Usually the adsorption heat decreases as the occupation degree of the surface rises which from the standpoint of inhomogeneous surface can be easily linked to initial occupation of absorption sites. The latter are characterized by highest Q. As the number of these sites get depleted adsorption occurs in the sites having lower heats. For instance, in case the homogeneous distribution with respect to adsorption heat holds the joint application of equations describing isotherms, isobars and isosteres of adsorption on inhomogeneous surface with exponential Q distribution brings about expression $Q = Q_0 - c \ln\theta$ which is replaced by a linear function with decreasing Q as θ grows.

The key problem of the theory of inhomogeneous surface is to determine the character of inhomogenity on the basis of adsorption data. In general case this inverse problem was resolved by Temkin [41]. It necessitated the development of mathematical apparatus enabling one to use the known type of adsorption isotherm to derive the type of distribution function for adsorption heat. Therefore, the theory of inhomogeneous surface considered the "mapping" of a specific surface as one of the problems to be addressed. However, as it was justifiably mentioned in [4], various adsorbates exhibit different types of adsorption isotherms for the same kind of adsorbent which results in existence of various distribution functions of adsorption heats for the same surface. In best case one can only account for a plausible difference of adsorption centers proper for various gases resulting in proceeding with adsorbent surface "mapping" only relative with a specific given gas.

To a certain extent one can consider an approach to describe non-Langmuir features accounting for probable interaction between adsorption particles as an alternate theory of inhomogeneous surface. Moreover the same properties observed during the studies of adsorption processes can be described using any of above approaches. Up-to-date this fact brings about a certain ambiguity in explanation of experimental data.

Usually the following interactions affecting the adsorption processes are considered between the adsorption particles: dipole-dipole interactions characteristic for physics adsorption [42], Columb interactions with various shielding peculiarities which are classified as chemisorption of charged particles [43], dipole-dipole interactions between neutral chemisorbed complexes, the exchange interaction through the adsorbent lattice [44], exchange interactions resulting in formation of clusters of chemisorbed particles due to expansion of the area where the carrier is localized (the carrier being captured by the energy level of adsorption surface states [45]. Recently, a more correct accounting for interaction present in adsorption layers was provided by considering complex multi-particle arrangements accounting for input of far neighbors [46].

In reality, the adsorption of gas particles on a real surface can be simultaneously influenced by inhomogeneity of the surface and interaction between absorbed particles. Presumably, it is the nature of a specific absorbate-adsorbent pair that controls the major mechanism in each case.

1.5. Kinetics of adsorption

The adsorption rate of a certain substance on a surface of a solid state is described by equation of the type

$$\frac{d\theta}{dt} = AP(1 - \theta)e^{-E_A/kT} - B\theta e^{-E_D/kT}, \qquad (1.9)$$

where θ is the surface coverage degree determined as $\theta = N/N_*$; $N(t)$ is the concentration of particles absorbed to the moment of time t; N_* is the concentration of adsorption sites; E_A and E_D are the activation energies of adsorption and desorption, respectively; A and B are constants containing leading factors, given by the theory of reaction rates and by kinetic theory of gases; P is the pressure in gaseous phase. The leading term in the right-hand side of equation (1.9) describes the number of particles incident from gaseous phase on a unit surface of adsorbent per unit time, second term accounting for amount of particles emitted into the gaseous phase.

According to the Langmuir theory the equation (1.9) can be written as

$$\frac{dN}{dt} = K_0 \frac{SP}{\sqrt{2\pi mkT}} (N_* - N) e^{-E_A/kT} - \nu N e^{-E_D/kT} \qquad (1.10)$$

Here m is the mass of absorbed particles; ν is the frequency of oscillation of absorbed particles; S is the surface area occupied by a single absorbed particles; $\check{K} = K_0 \exp\{-E_A / kT\}$ is the adhesion coefficient; $\tilde{\tau} \approx \nu^{-1} \exp\{+E_D / kT\}$ is the average time of contact of adsorption particle with the surface. Integrating equation (1.10) one gets

$$N(t) = \frac{N_* P}{P + b} (1 - e^{-K_1 t}), \qquad (1.11)$$

where

$$K_1 = \frac{K_0 SP}{\sqrt{2\pi mkT}} e^{-E_A/kT} + e^{-E_D/kT};$$

$$b = \frac{\nu}{K_0 S} \sqrt{2\pi mkT}\, e^{-(E_A - E_D)/kT} \equiv b_0 \exp(-Q / kT),$$

Q being the adsorption heat. For small times ($t \ll K_1^{-1}$) expression (1.11) can be rewritten as

$$N(t) \approx \frac{K_0 S}{\sqrt{2\pi mkT}} e^{-E_A/kT} Pt . \qquad (1.12)$$

We should point out that the Langmuir kinetics given by expressions (1.11) and (1.12) which is often observed in experiment is as often violated. In numerous cases the data on adsorption kinetics follows the well-known Roginsky-Zeldovich-Elovich kinetics isotherm [47, 48]

$$N(t) = A + B \ln\left(1 + \frac{t}{t_*}\right) \qquad (1.13)$$

or that of Bangham-Burt [49]

$$N(t) = At^\gamma - B , \qquad (1.14)$$

where A, B, γ and t_* are the constants dependent on the type of absorbate-adsorbent pair.

Similarly to analysis of adsorption equilibrium in case of investigation of adsorption kinetics the appearance of non-Langmuir features is being described by violation of certain basic assumptions of the Langmuir theory. More often, the deviations detected are ascribed either to inhomogeneity of the surface of adsorbents [36] or to interactions developing in adsorption layers. Either approach describes various experimentally observed types of kinetic isotherms and increase in the adsorption activation energy as the surface occupation degree grows. In case of considering the inhomogeneous properties of the surface, the major role is being played by inhomogeneity of adsorption sites as having different activation energies, which, evidently, leads to close distribution functions for heats and activation energies of adsorption in the Frumkin isotherm and Zeldovich kinetics as well as in the Freundlich isotherm and the Bangham kinetics. These coincidence stems from assumed relationship between kinetic and adsorption characteristics in various areas of inhomogeneous surface (the Temkin-Polyani relationship) [36, 37], which, in several cases, is similar to the Brensted type.

Oxide adsorbents, which are usually polycrystalline samples, exhibit practically the whole array of anomalous properties inherent to disordered systems. Inhomogeneity of the surface which is one of particular manifestations of disorder is, presumably, typical for oxides. This is indirectly proved by the fact that kinetics of adsorption of numerous gases on the surface of various oxides is satisfactorily described by kinetic isotherm of Roginsky-Zeldovich-Elovich. This is relevant, for instance, to adsorption CO on MnO_2 and NiO [36, 47, 50], O_2 on SnO_2 and ZnO [51, 52], H_2O and H_2 on ZnO [53]. Often, the Freundlich isotherm is valid to describe the adsorption equilibrium on oxide surfaces. In particular, this is relevant to adsorption (over specified temperature intervals) of CO on MnO_2 [47, 54], O_2 on ZnO and TiO_2 [55, 56], C_2H_5OH on SnO_2 and ZnO [57], propane on SnO_2 [58]. It is necessary to point out that the Freundlich isotherm is valid to describe the adsorption of O_2 on monocrystal ZnO ($10\overline{1}0$) as well both for sufficiently low temperatures $T \sim 100 \div 200$ K (physical adsorption) [59] and for $T \sim 300 \div 500$ K (chemisorption) [60]. The Freundlich isotherm characterizes the equilibrium in the system $CO_2 -$ ZnO ($10\overline{1}0$) in the temperature range $T \sim 300 \div 400$ K [60], too. At the same time, according to findings [61, 62] adsorption of CO_2 on polycrystalline ZnO and Al_2O_3 satisfies the Langmuir isotherm with adsorption heat independent on the surface occupation degree.

In case when such chemically active particles as atoms and radicals get adsorbed on oxides, the kinetics of the process is characterized by linear dependence over a substantially wide time interval [63]. The transition into the saturation regime in this case is provided by chemical transformations in adsorption layer which are present even at very small surface occupation degrees [64]. The typical example of such transformations is provided by reaction of recombination of absorbed particles of atomic nature or recombination of absorbed atoms with atom particles incident from gaseous phase [65 – 67].

1.6. Characteristic temperature intervals of gas – solid body interaction

As it has been already mentioned in section 1.3 the interaction of gaseous phase with the surface of a solid body may lead to formation of various types of surface and volume states on both absorbed and sorbed particles, the latter being caused by various types of interaction between gas and solid state. It is well-known that chemisorption and physical adsorption of valence-saturated molecules of various gases on oxide adsorbents develop in the range of low and moderately low temperatures: 100 – 300 K for physical adsorption and 300 – 600 K for chemisorption. At high temperatures (starting with T ∼ 700 ÷ 900 K) the interaction of gas with a solid state can lead to formation of various surface and volume defects such as oxygen and metal vacancies or interstitial metal or oxygen atoms in oxides [68 – 70].

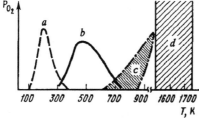

Fig. 1.4. Characteristic temperature intervals of interaction of oxygen with cleaved ZnO ($10\bar{1}0$) [71]

The existence of various temperature intervals characterized by predominant manifestation of one of above interactions can be detected from thermal desorption spectra. For instance, the thermal desorption spectrum obtained in [71] for a cleaved ZnO ($10\bar{1}0$) monocrystal following its interaction with oxygen (Fig. 1.4) indicates the availability of such typical temperature intervals as interval of physical adsorption (*a*), chemisorption (*b*), interval of formation of surface defects (*c*) and, finally, the domain of formation of volume defects (*d*).

The availability of such characteristic temperature intervals makes it possible to distinguish between above processes. Thus, physical adsorption of molecular gases can be easily singled out from chemical sorption at low temperatures due to slow kinetics of the latter. Similar situation can be encountered in case of creation of the surface and volume defects. The energy of creation of the surface point defects is lower than the volume ones therefore the concentration of surface defects is three to four orders of magnitude higher than the volume ones at $T \sim 900$ K in case of ZnO [72]. Hence, one can choose the temperature interval for separate investigation of surface and volume defects created during interaction of gas with a solid body.

The existence of various forms of interaction between solid surface and gaseous phase such as physical adsorption, chemisorption, creation of surface and volume defects may, in certain cases, complicate the interpretation of experimental data. Indeed, we know that chemical sorption of various particles can be accompanied by charge transfer in the absorbate-adsorbent system due to localization of charge carriers on the surface levels of adsorption origin. This process, naturally, affects the electrophysical characteristics of the sorbent. However, in to the case of interaction of gas with solids resulting in formation of surface and volume defects, the change of electrophysical properties of adsorbent is possible. Namely, for a given outer conditions the equilibrium state in the absorbate-adsorbent system is characterized by attaining equilibrium between charged and neutral forms of defects which are generated by ionization and recombination processes. The change in the external conditions, for instance the pressure in gas phase, leads altering the equilibrium concentration of above defects, which, in its tern, shifts the equilibrium between their charged a neutral forms and, thereby, modifies electrophysical characteristics of adsorbent [73]. We should point out that such an interaction may not be accompanied by the change of the surface band bending of adsorbent, i.e. cannot result in the change of the value of the surface charge, all the changes (for instance in electroconductivity) only being caused by the change in concentration of electrically active defects controlling the concentration of free charge carriers.

So, the change in electrophysical characteristics of semiconductor adsorbents due to its interaction with ambient gas phase can stem from various reasons and follow various functions caused by various features of physical and chemical processes governing these changes.

1.7. Effect of adsorption on electrophysical properties of semiconductors

Long before publication of classic papers by Volkenshtein and Hauffe which provided the basis of electron theory of chemisorption the fact that various properties of semiconductor depend on the content of ambient atmosphere was noted. Among these pioneering investigation of the 30-s there were studies by Wagner [74 − 76], Kurchatov [77] and others. In latter studies of Gray and co-authors [78 − 80] it was shown that the change in conductivity of certain oxide adsorbents could be used to investigate the kinetics of adsorption-desorption processes.

However, the principle idea in the studies of that time dealt with assessment of possible changes of inherent properties of a semiconductor caused by its interaction with gaseous phase. In other words, this question was directly linked with problems of quickly developing in that time semiconductor physics. The well known gas cycle of Bardeen-Brattain [81] provides a typical example of the situation of those days. This cycle deals with a opportunity to control the potential of the surface of a semiconductor by adsorption means.

The idea, to a certain extent opposite to above one, deals with an option to control the content of ambient gas atmosphere analyzing the changes in its electrophysical properties. This idea, as far as we are aware, was initially and practically simultaneously put forward by Heiland [82] and Myasnikov [83]. It is this idea which provide the basis for presently widely spread method of semiconductor sensors.

The major prerequisite of above assumption was the following: the detailed data on the change of electrophysical characteristics of a semiconductor adsorbent caused by adsorption of a certain gas over substantially wide pressure range makes it possible to solve the inverse problem concerning determination of the concentration of this gas in ambient atmosphere, due to detected change in electrophysical characteristics of the adsorbent.

Both above ideas dealing with an option to control the surface properties of a semiconductor through adsorption of various gases and a possibility to control gas atmosphere using the value of the change of electrophysical properties of semiconductor could be implemented in practice only after thorough investigation of effect of gas medium on electrophysical properties of semiconductor. The problem became complicated by already mentioned variety of processes inherent to adsorption. These processes might change electrophysical characteristics of adsorbent. Let us dwell in more detail on effects of the change of various electrophysical characteristics of a semiconductor caused both by chemisorption and adsorption induced change in concentration of volume and surface point defects.

1.7.1. Chemisorption induced band bending in semiconductor

The transition of a fraction of chemisorbed particles into the charged form accompanying adsorption results in the change of the value of the surface charge induced by localization of free electrons and holes on the adsorption-related surface states. Creation or change of the value of the charge on the surface of a semiconductor leads to the change of energy arrangement of all surface levels with respect to volume, i.e. in most simple case leads to the change in the bending of the bottom of the conductivity band and the ceiling of the valence band (Fig. 1.5). The zone diagrams are plotted for the cases of initially neutral surface (a) and equilibrium occupation of surface states corresponding to chemisorbed particles (b). Here we used the following designations: E_{vac} is the vacuum level, E_C and E_{CS} are the positions of the bottom of the conductivity band with respect to the vacuum level in volume and on the surface of the semiconductor, E_V and E_{VS} are the ceiling of the valence band in volume and on the surface, respectively, E_F is the Fermi level, E_t is the energy position of the surface level corresponding to chemisorbed particles, qU_S is the value of the surface barrier caused by surface charging, χ is the value of affinity to electron of the semiconductor surface, $q\varphi$ is the electron work function, δ is the dope parameter which gives the depth of the Fermi level below the bottom of the conductivity band inside semiconductor, L_D is the Debye shielding length controlling the degree of penetration of the field of surface charges into the volume of the semiconductor. Diagrams plotted in Fig. 1.5 describe the case of initially neutral surface. In more general case the surface is characterized by a specific charge situated on well-known biographic surface states [4, 84]. In most cases this results in existence of an a priori surface band bending. Moreover, as it has been mentioned above, the adsorbed particles may possess dipole moment of a proper or adsorption-induced nature which can, in several cases, notably change the values of affinity to electron of the adsorbent surface [18, 85, 86]. Thus, in more general case the energy system illustrating the surface-adjacent band bending caused by adsorption (in this case acceptor particles) is shown in Fig. 1.6. In case of adsorption of donor particles the band bending is shown in Fig. 1.5.

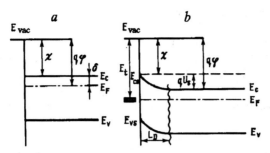

Fig. 1.5. The energy diagram illustrating the surface-adjacent band bending caused by adsorption of acceptor particles on initially neutral surface

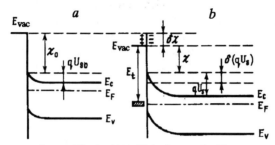

Fig. 1.6. The energy scheme illustrating the change in the zone pattern of the surface-adjacent domain of adsorbent, caused by adsorption of acceptors possessing a specific dipole moment

It should be mentioned that one can detect two types of equilibrium in the model of charge transfer in the absorbate – adsorbent system: (i) complete transition of chemisorbed particles into the charged form and (ii) flattening of Fermi level of adsorbent and energy level of chemisorbed particles. The former type takes place in the case of substantially low concentration of adsorbed particles characterized by high affinity to electron compared to the work function of semiconductor (for acceptor adsorbates) or small value of ionization potential (for donor adsorbates). The latter type can take place for sufficiently large concentration of chemisorbed particles.

The field created by chemisorption induced surface charge results in redistribution of free charge carriers in the surface-adjacent layers of the semiconductor which are situated within the layer determined by the shielding length. The shielding is provided by immobile dope ions and redistributed free electrons and holes. In other words adsorption results in formation of the space-charge region (SCR) affecting the electrophysical characteristics due to transition phenomena in the surface-adjacent layers of the semiconductor.

The theory of the space-charge region in semiconductors was developed by Shottky [87, 88], Mott [89], Davidov [90, 91], Brattain [92] and several other authors. The idea concerning the effect of adsorption on characteristics of SCR of the semiconductor adsorbent was proposed by Hauffe [5, 6]. This theory was developed further by Volkenshtein and his group [4, 93, 94] as well as by Mark [95, 96], Morrison [21] and other scientists. Let us briefly dwell on general features of the model.

Assume, that there are adsorption particles with concentration N_t on the surface of semiconductor which is in adsorption equilibrium with a certain gas. A fraction of adsorption particles is charged with concentration designated as n_t. Apart from them, on the surface there are various biographic surface states with concentration of the charged particles n_B controlling the degree of an a priori band bending qU_{SO}.

The availability of the surface charge results in redistribution of free charge carriers in semiconductor which leads to formation of a compensating space charge and electric field E related to the value of the volume charge through the Poisson equation:

$$\mathrm{div}\, E(r) = \rho(r) / \varepsilon\varepsilon_0 . \tag{1.15}$$

In several particular cases the solution of equation (1.15) enables one to express the value of the volume charge through the degree of the surface band bending qU_S, followed (through applying the condition of electric neutrality to the whole absorbate-adsorbent system) by deduction of adsorbent's SCR characteristics of interest.

Assuming, for the sake of simplicity, that the semiconductor occupies a hemi-space $x \geq 0$, we can rewrite the Poisson equation as

$$\frac{d^2\xi}{dx^2} = \frac{q\rho(x)}{\varepsilon\varepsilon_0 kT} , \tag{1.16}$$

where $\xi(x) = qU(x)/kT$ is the potential energy of the free electron in the field of surface charges expressed in kT units. The expression of electric neutrality in the case under consideration can be written as

$$\pm n_t + n_B = \int_0^\infty \rho(x)dx . \tag{1.17}$$

One obtains from (1.16)

$$\int_0^\infty \rho(x)dx = \frac{kT\varepsilon\varepsilon_0}{q}\frac{d\xi}{dx}\bigg|_0^\infty ,$$

which yields (if one takes into account the condition of neutrality of the volume of adsorbent) that expression

$$\pm n_t + n_B = -\frac{kT\varepsilon\varepsilon_0}{q}\frac{d\xi}{dx}\bigg|_{x=0} , \qquad (1.18)$$

holds under condition that the thickness of the volume is much larger than the shielding length. The relation (1.18) links the density of the surface charge with the value of the strength of electric field directly at the surface (the Gauss theorem). The latter value can be directly obtained through solving the Poisson equation (1.16) by substituting $\rho(x)$ expressed through the value of the band bending $\xi(x)$ in it.

Once the condition of thermodynamic equilibrium is provided by the constant value of the Fermi level through the whole volume of adsorbent, the induced band bending changes the value between the Fermi level and energy zones, which, in its tern, alters the distribution of the charge carriers through the whole energy levels. In general case of a non-degenerate semiconductor we have

$$\rho(x) = q\left\{ n_0\left(1 - e^{-\xi(x)}\right) - p_0\left(1 - e^{\xi(x)}\right) + N_D\left[\frac{1}{2}\exp\left\{\frac{E_D - E_F}{kT}\right\}\right]^{-1} - \right.$$

$$- N_D\left[\frac{1}{2}\exp\left\{\frac{E_D - E_F}{kT} + \xi\right\} + 1\right]^{-1} - N_A\left[\frac{1}{2}\exp\left\{\frac{E_F - E_A}{kT}\right\} + 1\right]^{-1} +$$

$$\left. + N_A\left[\frac{1}{2}\exp\left\{\frac{E_F - E_A}{kT} - \xi\right\} + 1\right]^{-1}\right\}, \qquad (1.19)$$

where terms in the right-hand side describe the change in concentrations of free electrons and holes as well as electrons and holes localized on the volume donor and acceptor levels with energies E_D and E_A and concentration N_D and N_A, respectively. Substituting expression (1.19) in equation (1.16) and integrating the expression obtained one can easily find the first integral of the Poisson equation

$$\pm\left|\frac{d\xi}{dx}\right| = \left\{\frac{2q^2}{kT\varepsilon\varepsilon_0}\left[n_0\left(e^{-\xi} + \xi\right) + p_0\left(e^{\xi} - \xi\right)\right] - \frac{2q^2}{\varepsilon\varepsilon_0(kT)^2}\Lambda(x) + C\right\}^{1/2}, \qquad (1.20)$$

where $\Lambda(x)$ is a certain function satisfying the relationship

$$\left(N_D\left[\left\{\frac{1}{2}\exp\left(\frac{E_D - E_F}{kT}\right) + 1\right\}^{-1} - \left\{\frac{1}{2}\exp\left(\frac{E_D - E_F}{kT} + \xi\right) + 1\right\}^{-1}\right] - \right.$$

$$\left. - N_A\left[\left\{\frac{1}{2}\exp\left(\frac{E_F - E_A}{kT}\right) + 1\right\}^{-1} - \left\{\frac{1}{2}\exp\left(\frac{E_F - E_A}{kT} - \xi\right) + 1\right\}^{-1}\right]\right)\frac{d\xi}{dx} =$$

$$= \frac{d\Lambda(x)}{dx},$$

and C is a constant determined through the boundary conditions imposed on ξ and $d\xi/dx$. We should note that in general case equation (1.20) does not have a simple solution, although it can be easily integrated for various particular cases [97].

As a rule, oxides of various metals usually used as sensitive (operational) elements of sensors are provided by doped broad-band semiconductors of n-type, such as SnO_2, ZnO_2, TiO_2, and others with interstitial atoms of metals or oxygen vacancies operating as a principle dope agents. The energy levels corresponding to them are positioned in forbidden band close to the bottom of conductivity band so that even at temperatures close to the room temperature the dope is almost practically ionized. In this case the concentration of the major carriers is determined by concentration of the volume donors:

$$n_0 = N_D^+ \approx N_D, \quad p_0 \approx n_i^2 / N_D, \tag{1.21}$$

where n_i is the proper concentration of the free charge carriers.
Equation (1.20) acquires the shape with boundary conditions $\xi = 0$ and $d\xi/dx = 0$ and when x $\to \infty$ hold

$$\frac{d\xi}{dx} = \pm\frac{\sqrt{2}}{L_D}\left[e^{-\xi} + \xi - 1 + \left(\frac{n_i}{N_D}\right)^2 e^{\xi}\right]^{1/2}, \tag{1.22}$$

where $L_D = (\varepsilon\varepsilon_0 kT / q^2 N_D)^{1/2}$ is the shielding Debye length. Applying the condition of electric neutrality and neglecting the concentration of holes we get

$$\frac{\pm n_t + n_B}{\sqrt{2N_D L_D}} \cong \pm\left[e^{-\xi_S} + \xi_S - 1\right]^{1/2}, \tag{1.23}$$

where $\xi_S = qU_S/kT$ is the value of band bending at the surface of semi-conductor. Obviously, there is an expression similar to (1.23) linking concentration of the charged form of biographic surface states with the value of pre-adsorption band bending ξ_{S0}:

$$\frac{n_B}{\sqrt{2N_D L_D}} \cong \left[e^{\xi_{S0}} + \xi_{S0} - 1\right]^{1/2}. \tag{1.24}$$

Therefore, the value of ξ_S is related to the equilibrium concentration of chemisorbed charged particles through relationship

$$\pm n_t = \sqrt{2N_D L_D}\left[\pm\sqrt{e^{-\xi_S} + \xi_S - 1} - \sqrt{e^{-\xi_{S0}} + \xi_{S0} - 1}\right], \tag{1.25}$$

where the plus sign in the left-hand side stands for the case of adsorption of acceptor particles and the minus depicts the case of the donor ones. Taking into account that in equilibrium state the occupation of the surface levels of adsorption particles is described by the Fermi statistics we obtain the following equation to determine the height of the surface barrier induced by adsorption of acceptor particles characterized by the value of affinity to electron E_{ta} and total concentration N_{ta}

$$\frac{N_{ta}}{1 + \exp\left\{\dfrac{\chi + \delta - E_{ta}}{kT} + \xi_S\right\}} = \sqrt{2N_D L_D}\left[\sqrt{e^{-\xi_S} + \xi_S - 1} - \sqrt{e^{-\xi_{S0}} + \xi_{S0} - 1}\right]. \tag{1.26}$$

In case of adsorption of donor particles whose chemisorbed state is described by the ionization potential I_{td} and concentration $N_{td} < n_B$ we obtain

$$-\frac{N_{td}}{1 + \exp\left\{\dfrac{I_{td} - \chi - \delta}{kT} - \xi_S\right\}} = \sqrt{2N_D L_D}\left[\sqrt{e^{-\xi_S} + \xi_S - 1} - \sqrt{e^{-\xi_{S0}} + \xi_{S0} - 1}\right]. \tag{1.27}$$

Solution of equations (1.26) and (1.27) for above two limiting cases differing in the value of the surface concentration of adsorption particles brings about different dependencies of the value of the surface band bending as a function of parameters of the absorbate-adsorbent system. Thus, in case of adsorption of acceptors we obtain from (1.26) that

$$\xi_S \approx \left(\sqrt{\xi_{S0}} + \frac{N_{ta}}{\sqrt{2}N_D L_D} \right)^2, \tag{1.28}$$

when $N_{ta} < \sqrt{2}N_D L_D \left(\sqrt{\alpha} - \sqrt{\xi_{S0}} \right)$, were $\alpha = (E_{ta} - \chi - \delta) / kT > 0$ and $\xi_{S0} > 1$. When the opposite condition $\left(N_{ta} > \sqrt{2}N_D L_D \left(\sqrt{\alpha} - \sqrt{\xi_{S0}} \right) \right)$ holds we have

$$\xi_S \approx \alpha + \ln \left\{ \frac{N_{ta}}{\sqrt{2}N_D L_D \left(\sqrt{\alpha} - \sqrt{\xi_{S0}} \right)} - 1 \right\}. \tag{1.29}$$

Expression (1.28) gives the equilibrium height of the surface barrier caused by the total transition of chemisorbed particles into the charged form. In case when expression (1.29) is valid the equilibrium height of the barrier is determined by the leveling-off of energy state of adsorption particle with the Fermi level of adsorbent. In case

$$N_{td} < \sqrt{2}N_D L_D \left(\sqrt{\xi_{S0}} - \sqrt{\theta} \right),$$

where $\theta = (I_{td} - \chi - \delta) / kT > 0$, and reasoning along the same lines we obtain for adsorption of donor particles

$$\xi_S \approx \left(\sqrt{\xi_{S0}} - \frac{N_{td}}{\sqrt{2}N_D L_D} \right)^2. \tag{1.30}$$

In the opposite case $N_{td} > \sqrt{2}N_D L_D \left(\sqrt{\xi_{S0}} - \sqrt{\theta} \right)$ we have

$$\xi_S \approx \theta - \ln \left\{ \frac{N_{td}}{\sqrt{2}N_D L_D \left(\sqrt{\xi_{S0}} - \sqrt{\theta} \right)} - 1 \right\}. \tag{1.31}$$

We should note that expressions (1.28) − (1.31) have been obtained without accounting for plausible recharging of biographic surface states [84, 98] which is reasonable in case when above states have fully occupied energy levels positioned deeply inside the forbidden band.

Knowing the value of ξ_S enables one to obtain the profile distribution of the charge carriers through the depth of adsorbent. Thus, in case of adsorption of acceptor particles at small depths when were the condition $\xi(x) > 1$ holds the solution of equation (1.22) yields

$$\xi(x) \approx \xi_S \left(1 - 2^{3/2} \frac{x}{\sqrt{\xi_S} L_D}\right)^2 . \tag{1.32}$$

It stems from this expression valid for $x < 2^{-3/2} L_D \left(\sqrt{\xi_S} - 1\right)$ that electron depleted layer is formed close to the surface

$$n(x) = N_D \, e^{-\xi(x)} = n_S \exp\left\{2^{5/2} \sqrt{\xi_S} \, \frac{x}{L_D}\right\}, \tag{1.33}$$

where $n_S = N_D \exp(-\xi_S)$. At large values of x we have the expression

$$\xi(x) \approx A \exp\{-x / L_D\} \tag{1.34}$$

for the value of the band bending. Here A is the constant whose value is provided by matching expressions (1.32) and (1.34) at depths where $\xi \sim 1$. We have to point out that expression (1.34) is valid for any x in case when $\xi_{S0} < 1$ and concentration of absorbed particles is so small that $\xi_S \approx (\sqrt{\xi_{S0}} + N_{ta} / \sqrt{2} N_D L_D)^2 < 1$. In this case

$$\xi(x) = \xi_S \exp\{-x / L_D\} . \tag{1.35}$$

The adsorption of donor particles can also be accompanied by various situations. In case when $kT\theta = I_{td} - \chi - \delta > 0$, but $\theta < \xi_{S0}$, i.e. the level of chemosorbed particle is situated in the forbidden band below the Fermi level of the neutral surface, the function $\xi(x)$ is described by expressions (1.32) – (1.35) with values of ξ_S given by relations (1.30) and (1.31). If $\theta > 0$, we can encounter the case with the opposite band bending close to the surface of adsorbent resulting in saturation of the surface-adjacent layers by electrons of conductivity.

Thus, when $N_{td} < N_D L_D \sqrt{2\xi_{S0}}$, $\xi_{S0} > 1$, we have for equilibrium band bending caused by complete depletion of donor levels of adsorption nature

$$|\xi_S| \approx \left(\sqrt{2\xi_{S0}} - \frac{N_{td}}{N_D L_D}\right) > 1. \tag{1.36}$$

Under conditions $\sqrt{2}N_D L_D\left(e^{1/2} + \sqrt{\xi_{S0}}\right) < N_{td} < \sqrt{2}N_D L_D\left(e^{|\theta|/2} + \sqrt{\xi_{S0}}\right)$, the value of the surface band bending is equal to

$$|\xi_S| \approx 2\ln\left\{\frac{N_{td}}{\sqrt{2}N_D L_D} - 1\right\} > 1. \tag{1.37}$$

In this case the bending is caused by complete transition of chemisorbed particles into the charged form. In case when $N_{td} > \sqrt{2}N_D L_D \times$ $\times\left(\exp\left(|\theta|/\sqrt{2}\right) + \sqrt{\xi_{S0}}\right)$, $\xi_{S0} > 1$, the value of the surface band bending is caused by flattening of E_{td} and E_F equaling to

$$|\xi_S| = |\theta| + \ln\left\{\frac{N_{td}}{\sqrt{2}N_D L_D\left(\exp\{|\theta|/2\} + \sqrt{\xi_{S0}}\right)}\right\}. \tag{1.38}$$

The band bending in this case for x, under which $-\xi(x) > 1$ is given by expression

$$\xi(x) \approx 2\ln\left\{e^{\xi_s/2} + x/\sqrt{2}L_D\right\}. \tag{1.39}$$

The concentration of conductivity electrons in this case may by expressed as

$$n(x) = N_D\,e^{-\xi(x)} = \frac{N_D}{\left(e^{\xi_s/2} + x/\sqrt{2}L_D\right)^2}, \quad \xi_S < 0. \tag{1.40}$$

This expression takes place for depths complying with condition $\xi(x) < -1$. We can deduce from expression (1.40) that in case of high concentrations of chemisorbed donor particles sufficient to invert the initial band bending the surface-adjacent level is saturated by conductivity electrons. The graphical illustration of all these cases is given in Fig. 1.7. As it follows from the figure at substantially high concentrations of donor particles and low ionization potentials one can observe not only the inversion of the sign of initial band bending, but also creation of the surface-adjacent degeneration domain [86].

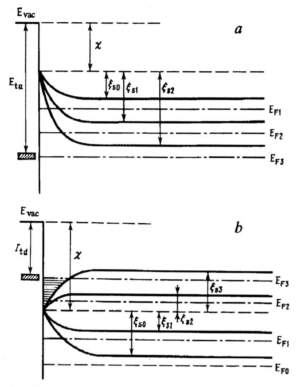

Fig. 1.7. The effect of chemisorption of acceptor (*a*) and donor (*b*) particles on the value of the surface-adjacent band bending (ξ_{S0} is the value of pre-adsorption band bending).

a: ξ_{S1} corresponds to the complete occupation of acceptor levels whose concentration N_{ta} is insufficient to flatten of E_{ta} and E_F; ξ_{S2} is corresponding to the leveling-off E_F and E_{ta}; *b*: ξ_{S1} is caused by charging of all donor adsorption particles, whose concentration N_{td} is insufficient both to provide the inversion in the sign of band bending, and to level-off E_F and I_{td}; ξ_{S2} is the complete occupation resulting in inversion of the band bending; ξ_{S3} corresponds to flattenning of E_F and I_{td}.

1.7.2. The effect of adsorption charging of the surface on the value of conductivity of surface-adjacent layers and semiconductor adsorbent work function

Charging of the surface accompanying adsorption process and resulting in the change of the energy profile of the bottom of the conductivity band and, naturally, the ceiling of the valence band in semiconductors has a substantial impact on transfer phenomena. In the most simple case of monocrystal sample the surface-adjacent band bending accompanying adsorption, or, which is more accurate, the change in pre-adsorption band bending results in the change of concentrations of the current carriers in SCR which directly affects the value of electric conductivity of the surface-adjacent layer:

$$\Delta\sigma \approx q\mu_n(\Gamma_n + b\Gamma_p).$$
(1.41)

Here $b = \mu_p / \mu_n$; μ_n and μ_p are the mobility of electrons and holes; Γ_n and Γ_p are the surface surplus of electrons and holes determined, according to Gibbs [99], as the difference between the total amount of the component in the system which can be attributed to the unit surface area and the amount which might have been in a homogeneous phase

$$\Gamma_n = \int_0^l [n(x) - n_0]\, dx = 2\int_0^{\xi_s} [n(\xi) - n_0]\left(\frac{d\xi}{dx}\right)^{-1} d\xi,$$

$$\Gamma_p = \int_0^l [p(x) - p_0]\, dx = 2\int_0^{\xi_s} [p(\xi) - p_0]\left(\frac{d\xi}{dx}\right)^{-1} d\xi.$$
(1.42)

Here we make use of the same designations $n(x)$ and $p(x)$ for concentrations of electrons and holes in SCR; n_0 and p_0 are their concentrations in volume outside SCR. It is assumed that the thickness and the average size of an adsorbent microcrystal l are much larger than the shielding radius, x is the coordinate plotted from the surface inside the semiconductor.

In the theory of the surface of semiconductor developed by Garret-Brattain [92] it was assumed that mobility of electrons and holes in the surface-adjacent layers of semiconductors is the same as in volume which, however, is true until the thickness of the layer of the volume charge becomes equal to the average free path length of the carriers. The applicability of this approximation was studied by Shrieffer [100]. In that paper the solution of the Boltzmann equation was used to derive the dependence of the value of effective mobility of carriers on the potential value of the surface in case of purely diffuse surface scattering. It was shown that in several cases the effective mobility can be one order of magnitude lower than that in the volume of semiconductor. Thus, to derive an adequate model of adsorption-induced change in conductivity it is necessary to account for correction in the surface component of the scattering.

In case of the doped semiconductor of n-type under consideration the situation gets simple due to large thickness of SCR if compared to the free path length of the carriers. Therefore, substituting expression (1.42) into (1.41) under condition of applicability of the Boltzmann statistics for the free electrons and holes leads to expression

$$\Delta\sigma = \pm\sqrt{2}q\mu_n L_D \int\limits_0^{\xi_s} \frac{N_D\left(e^{-\xi}-1\right)+b\dfrac{n_i^2}{N_D}\left(e^{\xi}-1\right)}{\left[e^{-\xi}+\xi-1+\left(\dfrac{n_i}{N_D}\right)^2 e^{\xi}\right]^{1/2}}\, d\xi, \qquad (1.43)$$

This expression indicates that increase in the surface band bending $(\xi > 0)$ is accompanied by decrease in conductivity, which attains the minimum under value $\xi_S = \xi_* = \ln\{N_D^2 / bn_i^2\}$. Above this value the conductivity starts growing due to availability of surplus holes.

Under conditions of the value of band bending lower than ξ_*, i. e. far from the conductivity inversion domain the expression (1.43) becomes simpler [101]:

$$\Delta\sigma \approx -\sqrt{2}q\mu_n N_D L_D \sqrt{e^{-\xi_s}+\xi_S - 1}, \qquad (1.44)$$

Applying the condition of electric neutrality (1.23) one can easily obtain

$$\Delta\sigma = -q\mu_n n_t, \qquad (1.45)$$

which is valid in case of adsorption of acceptor particles and

$$\Delta\sigma = q\mu_n n_t, \qquad (1.46)$$

the same for adsorption of donors where n_t is the equilibrium concentration on the charged form of chemisorption surface states.

Expressions (1.45) and (1.46) which are valid in case of applicability of above assumptions indicate on availability of direct proportionality between the value of the change of the surface-adjacent conductivity of semiconductor adsorbent and concentration of chemisorbed particles on its surface, the latter being in charged form. This results in the fact that when the surface is covered by adsorbed particles at degrees lower than $N_{ta}^* = \sqrt{2}N_D L_D\left(\sqrt{\alpha} - \sqrt{\xi_{S0}}\right)$ (acceptor particles) and $N_{td}^* =$

$= \sqrt{2}N_D L_D \left(\sqrt{\xi_{S0}} - \sqrt{\theta} \right)$ (donors) the value of the change of electric conductivity is proportional to the total concentration of particles. For
higher concentration of adsorbed particles (if compared to above $N_{ta,d}^{\bullet}$)
the value of the change in electric conductivity during adsorption of
both acceptors and donors is controlled by flattenning of local levels of
chemisorbed particles with the Fermi level of adsorbent and practically
does not depend on concentration of adsorption particles. Thus, in case
of adsorption of acceptors

$$\Delta \sigma \approx -q\mu_n \sqrt{2}N_D L_D \left(\sqrt{\alpha} - \sqrt{\xi_{S0}} \right), \quad (1.47)$$

whereas in case of adsorption of donor particles we have

$$\Delta \sigma \approx \sqrt{2}q\mu_n N_D L_D \left(\sqrt{\xi_{S0}} - \sqrt{\theta} \right). \quad (1.48)$$

Consequently, for high concentration of adsorption particles (which is
directly linked either with their high partial pressure in gaseous phase
or with high value of their adsorption heat) all kinetic curves $\sigma(t)$
whose shape (as it will be showed below) is notably dependent on concentration of adsorption particles N_t tend (at long times) to a specific
value of σ_p dependent on the nature and history of adsorbent and independent on the value of N_t.

We have to note that above reasoning concerning equilibrium states
linked either with complete occupation of adsorption surface states or
with flattenning of the local levels with the Fermi level of adsorbent are
valid in case when respective values of equilibrium barriers do not exceed the height of the Weits barriers mentioned in section 1.2. In the
opposite case (which can be realized at high concentrations of adsorption particles or when the energy levels are situated deeply under the
bottom of the conductivity band) the function $\sigma(t)$ comes to a "kinetic
stop" when the limiting heights in barriers of the order 1 eV are attained. These stops are practically independent of the nature of adsorbent and the stationary value σ becomes insensitive to further increase in
the partial pressure of adsorbed gas.

Adsorption related charging of surface naturally affects the value of
the thermoelectron work function of semiconductor [4, 92]. According to
definition the thermoelectron work function is equal to the difference in
energy of a free (on the vacuum level) electron and electron in the volume of the solid state having the Fermi energy (see Fig. 1.5). In this
case the calculation of adsorption change in the work function $\Delta(q\varphi)$ in

the "first approximation" results in determination of the value of the change in the height of the surface barrier caused by adsorption, which can be determined through solving the Poisson equation.

In large amount of experimental and theoretical studies [4, 102, 103] it was established that the change in electric conductivity and work function observed during adsorption of various particles on different types of adsorbents correlate. Namely, the adsorption of acceptors on semiconductors of n-type leading to depletion of conductivity electrons in the surface-adjacent area causes decrease in σ and increase in $q\varphi$. Adsorption of particles of the donor type on such adsorbents results in the opposite change both for σ and $q\varphi$. In case of semiconductors with the hole conductivity the effect of adsorption of acceptors and donors on the values of σ and $q\varphi$ gets changed to the opposite one, namely, acceptor adsorption leads to increase in σ and decrease in $q\varphi$, adsorption of donors results in decrease in σ and increase in $q\varphi$. Expressions (1.28) − (1.31) and (1.36) − (1.38) indicate that the type of dependence of the value of equilibrium work function of semiconductor adsorbent on concentration of chemisorbed particles is controlled by numerous factors characterizing the absorbate-adsorbent system varying from the power to the logarithmic function.

As it has been already mentioned in numerous cases one should take into account the dipole component in the adsorption − caused change in the work function stemming from availability of dipole moment in adsorption particles. This component alters the value of micropotential of the surface. As it has been already mentioned in section 1.2, the origin of these dipoles can differ ranging from inherent to quantum mechanical one [4].

We should note that recent years have been marked with publishing of several theoretical papers accounting for both discrete character and random distribution of adsorption particles. This analysis was conducted to assess the permeability of the surface barrier, i.e. to examine the thermoemission properties of semiconductor. Thus, paper [104] considered the case of small concentrations of randomly positioned adsorption particle possessing a specific dipole moment. The expressions obtained for thermoemission currents indicate the non-Arrenius type of the temperature dependence. Papers [105, 106] studied the change in the work function of semiconductor when randomly distributed dipoles with concentrations differing over wide range were available on its surface. It has been established that when concentration of adsorption particles is higher than a certain one $N_s^\bullet$ the major input into the value of the change of permeability of the barrier to current is provided by random arrays of adsorption particles [106]. It is worth noting that in case $(N_t > N_s^\bullet)$ expression for $\Delta(q\varphi)$ is similar to the well-known Helmholtz expression $\Delta q \varphi = 4 \pi d N_t$ which is valid for the opposite limiting cases: negligibly small concentrations of adsorption particles and, on the opposite, for a monolayer. However, in contrast to the apparent concentration of adsorption particles N_t in the Helmholtz expression the formula obtained accounts for random positioning of adsorption particles and has a certain apparent concentration.

We shall note that there are numerous experimental papers dealing with studies of effect of adsorption on other electrophysical characteristics of oxide adsorbents such as thermal electromotive force [107], Hall effect [108 − 110], volt-ampere [58, 111] and frequency [112] characteristics. The availability of results of these studies makes one expect that in near future the adequate theoretical model describing adsorption-induced changes in electrophysical characteristics of semiconductor adsorbent will be developed.

Sandomirsky was the first to address applicability of adsorption model to describe the volume charge and electric conductivity of monocrystal semiconductor [93, 113] in sufficient detail. Following that the model of SCR in question was significantly developed due to accounting for and rectifying of numerous simplifications incorporated into the model of Garret-Brattain [92]. They include the studies of effect of arbitrary ionization degree of the volume dope [114, 115] on characteristics of SCR and account for redistribution of the dope over the volume of semiconductor [116] resulting, on the one hand, in increase in adsorption capacity of the surface, and on the other − to a certain straightening out in its energy bands which is manifested in decrease in dependence of the value of electric conductivity of the surface-adjacent layer on the value of the charge of the surface. The important input to the development of the theory of adsorption charging of the surface of semiconductor were provided by studies of Kogan [94] and Peshev [117] which dealt with the adsorption capacity of dispersion semiconductor. These papers established an important role of dispersion of semiconductor adsorbent (defined by the ratio of the surface area of microcrystal to its volume) in effects of adsorption charging of the surface. It has been shown that decrease in this ratio to the order and below the shielding radius leads to effect that adsorption surface states start influencing the position of the Fermi level in the volume of adsorbent. In its turn, with decreasing the size of the sample its capability to adsorb charged adparticles would decrease irrespective of the type of particles in case of negligibly small value of the surface charge caused by biographic surface states. It changes in different manner for donor and acceptor particles when biographic charge of the surface is large. This very change in neutrality of the volume of highly disperse adsorbent is responsible for the well known compensation effect predicted by Peshev [118]. Experimentally it was verified in papers [119, 120]. The gist of the effect deals with the opposite change in the value of potential energy of electrons in the volume of semiconductor and the value of the surface band bending, i.e., between concentration of the free charge carriers and capability of their rating the surface for instance to participate in reactions. It should be mentioned that above effect of "leveling-off" in energy bands with decrease in the size of adsorbent was considered already by Loshkarev [121] without accounting for inverse effect of the change in position of the Fermi level in the volume of semiconductor on the value of its surface charge.

In general, the peculiarities of the surface effects in thin semiconductors, for which application of semi-infinite geometry becomes incorrect were examined in numerous papers. As it has been shown in studies [101, 113, 121 – 123] the thickness of semiconductor adsorbent becomes one of important parameters in this case. Thus, in paper [121] the relationship was deduced for the change in conductivity and work function of a thin semiconductor with weakly ionized dopes when the surface charge was available. Paper [122] examined the effect of the charge on the temperature dependence of the work function and conductivity of substantially thin adsorbents. Papers [101, 123] focused on the dependence of the surface conductivity and value of the surface charge as functions of the thickness of semiconductor and value of the surface band bending caused by adsorption and application of external field.

1.8. Role of recharging of biographic surface states during chemisorption charging of a semiconductor surface

Big number of biographic surface states (BSS) available of a real surface of solids provided by inherent surface states [124] as well as various structural defects, admixture and chemisorbed particles, etc., influences (and sometimes controls) both the mechanism and adsorption kinetics of gaseous particles. In addition to already mentioned fact that above defects may be centers of adsorption it is important that adsorption can be accompanied by recharging of BSS resulting in formation of long-term relaxation of such detectable characteristics of adsorbent as conductivity of surface-adjacent layer, Hall characteristics, work function and others related to the value of charge Q_S. Once the character and the value in the change of above properties during adsorption makes it possible to assess the concentrations of detected particles in the volume around the sensor, the change in Q_S due to recharging of BSS, superimposed on pure adsorption-induced change of Q_S can in several cases mask the real kinetics of development of adsorption processes.

To a great extent (this follows from expressions (1.28) – (1.31) and (1.36) – (1.38)) the presence of an a priori surface band bending ξ_{S0} available in semiconductor adsorbent affects the value of ξ_S caused by adsorption even without accounting for plausible recharging of BSS responsible for existence of ξ_{S0}. The value of ξ_{S0} is related to the history of adsorbent including various means to treat its surface. As it has been pointed out by Volkenshtein [4] different cases having their specific values of charge available in adsorption and biographic surface states lead to different types of dependencies of the values of electrophysical characteristics of adsorbent on concentration of adsorption particles.

The importance of accounting for recharging of the surface states (SS) accompanying various perturbations of the surface of semiconductor was already pointed out by Bardeen [125] while explaining a weak dependence of the Shottky barrier formed at the metal-semiconductor interface as a function of the work function of metal. It has been established that leveling-off of Fermi levels of metal and semiconductor is accompanied by the change in degree of occupation of SS of the latter which, to a certain extent, is equivalent to shielding of the surface-adjacent domain of semiconductor.

The change in position of the Fermi level of semiconductor adsorbent with respect to the ends of bands and local levels on the surface caused by transition of a fraction of chemisorbed particles into the charged form during adsorption frequently results in disturbance of equilibrium in occupation of BSS which substantially affects the value of the surface-adjacent band bending. Let us demonstrate this using adsorption of both acceptor and donor particles in a broad band semiconductor of n-type with characteristic linear dimensions exceeding the shielding length as an example. Assume that the energy levels corresponding to BSS densely occupy the forbidden band E_g, i.e. they are characterized by quasi-continuous spectrum with the density of states $D_S(E)$. The validity of the last assumption was discussed in [84] in detail. The prove of existence of quasi-continuous spectrum of SS in zinc oxide is provided in [126].

In order to determine the equilibrium height of the surface energy barrier ξ_{S0} caused by occupation of BSS let us make use of the Fermi-Dirak distribution in approximation of the absolute zero of temperatures which is valid as it was shown in [127] for weakly changing densities of SS. For the sake of clarity let us assume that an empty zone of SS corresponds to neutral state of the surface, whereas BSS correspond to acceptor type.

In the most simple case of the equilibrium distribution of density of BSS with respect to E_g we arrive to the following expression for the height of equilibrium barrier [128]

$$qU_{S0} = (E_g - \delta)\left[\frac{\sqrt{1 + 4N_B^2} - 1}{2N_B}\right]^2 , \qquad (1.49)$$

where $N_B = qD_S\left[(E_g - \delta) / 2\varepsilon\varepsilon_0 N_D\right]^{1/2}$.

If acceptor adsorption particles with concentrations N_{ta} possessing a value of affinity to electron E_{ta} appear on the surface of adsorbent characterized by pre-adsorption band bending (1.49), then the condition of their transition into the charged form is provided by inequality

$$E_{ta} > \chi + \delta + qU_{S0} , \qquad (1.50)$$

i.e. the charging occurs only on the surface characterized by the value of pre-adsorption band bending smaller than $(E_{ta} - \chi - \delta)$. The condition of electric neutrality makes it possible to determine the equilibrium height of the surface barrier induced by adsorption:

$$qU_S = (E_g - \delta)\left[\frac{\sqrt{1 + 4N_B(N_B + \tilde{N}_{ta})} - 1}{2N_B}\right]^2 , \qquad (1.51)$$

where $\tilde{N}_{ta} = qN_{ta} / \left[2\varepsilon\varepsilon_0 N_D(E_g - \delta)\right]^{1/2}$. As expression (1.50) suggests the relationship (1.51) is valid when $N_B < N_{B1} = (\tilde{a}^{1/2} - \tilde{N}_{ta}) / (1 - \tilde{a})$, where $\tilde{a} = (E_{ta} - \chi - \delta) / (E - \delta)$. If the initial height of barrier does not comply with condition (1.50), then adsorption of acceptors does not result in the change of the barrier's height. The above condition is provided by $N_B > N_{B2} = \tilde{a}^{1/2} / (1 - \tilde{a})$. The profile of a dependence of qU_S on N_B for a given N_{ta} is shown in Fig. 1.8. We should mention that the third domain of concentrations $N_{B1} < N_B < N_{B2}$ characterized by the constant height of barrier qU_S due to flattenning of the Fermi level of adsorbent and local level of adsorption particles gets created while approaching the absolute zero in temperatures used to describe the occupation of adsorption SS.

In the domain of interest $N_B < N_{B1}$ where the recharging of BSS takes place we arrive to the following expression for the value of the surface band bending

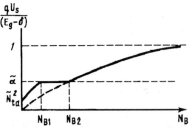

Fig. 1.8. The dependence of the height of the surface barrier caused by adsorption of acceptor particles as a function of the density of BSS.

$$\xi_{SS} = \frac{qU_S}{kT} = \left[\sqrt{\xi_{S0}} + \frac{N_{ta}}{\sqrt{2N_D L_D}} \left(1 - \frac{kT}{E_g - \delta} \xi_{S0} \right) \right]^2, \qquad (1.52)$$

Comparing the above expression with relation (1.28) ($\xi_{S0} = \left(\sqrt{\xi_{S0}} + \right.$

$\left. + N_{ta} / \sqrt{2N_D L_D} \right)^2$) also obtained for the case of complete occupation N_{ta} but without accounting for BSS recharging indicates the shielding effect of BSS on the process of adsorption charging of the surface (Fig. 1.9, *a*).

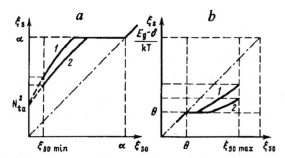

Fig. 1.9. The dependence of barrier height caused by complete occupation of ASS as a function of the initial height. *a* — acceptor adsorption, $\alpha = \tilde{\alpha}(E_g - \delta) / kT$; *b* — donor adsorption, $\theta = \tilde{\theta}(E_g - \delta) / kT$. Curve *1* does not account for BSS charging, curve *2* — the recharging is being taken into account.

In case of adsorption of donor particles with ionization potential I_{td} one can easily obtain the following expression for the equilibrium height of post-adsorption band bending

$$\xi_{SS} = \begin{cases} \left(\sqrt{\xi_{S0}} - N_{td}/\sqrt{2N_D L_D}\right)^2, & N_B < N_{B1} = \sqrt{\tilde{\theta}}/(1-\tilde{\theta}), \\[2mm] \theta = \dfrac{(I_{td} - \chi - \delta)}{kT} = \text{const}, & N_{B1} < N_B < N_{B2} = \dfrac{\left(\sqrt{\tilde{\theta}} + N_{td}\right)}{(1-\tilde{\theta})}, \\[2mm] \left(\sqrt{\xi_{S0}} - \dfrac{N_{td}}{\sqrt{2N_D L_D}}\dfrac{1-\xi_{S0}kT/(E_g-\delta)}{1+\xi_{S0}kT/(E_g-\delta)}\right)^2, & N_B > N_{B2}, \end{cases} \qquad (1.53)$$

This expression indicates on substantial shielding of the charging of the surface caused by BSS recharging in case of adsorption of donors as well (see Fig. 1.9, b). Here $\tilde{\theta} = (I_{td} - \chi - \delta)/(E_g - \delta)$.

Expressions (1.52) and (1.53) prove the dominant role played by BSS recharging in effect of adsorption-induced charging of the surface. However, the direct measurements of equilibrium post-adsorption values of electrophysical characteristics and direct comparison with their pre-adsorption values does not make it possible to establish the role of above recharging, living thereby some uncertainty in singling out adsorption component in the total change in characteristics in question. Resolution of this issue is provided by either availability of an a priori information concerning the pre-adsorption state of the surface (which is possible while proceeding with studies on monocrystal adsorbents [129]) or by examining kinetics of surface charging which in several cases enables one to distinguish between contributions in the change of the value of the surface charge provided by charging of adsorption particles and BSS recharging due to the difference in rates of above processes.

1.9. The kinetic of adsorption charging of the surface of semiconductor under relaxation of biographic surfacing charge

Scrupulous examination of this item is provided in [98] which involved the examination of the system of equations describing the kinetics of charging of adsorption surface states (ASS) and recharging in BSS. This led authors to deriving the kinetic dependencies of the value of the surface barrier during adsorption on semiconductor adsorbent of both acceptor and donor particles in various limiting cases differing in the ratio of characteristic time of the charging of ASS and BSS. We should note that the study of kinetics of the charging of the semiconductor surface while accounting for recharging of BSS was the issue in study [130]. However, this study did not account for effect of the surface barrier on kinetics of transition of the adparticles themselves into the charged form which renders the results obtained as a particular case.

In general case to determine the kinetics of the charging of the surface it is necessary to proceed with self-consistent calculation of kinetics of occupation of ASS and recharging of BSS. One should also use the Poisson equation as a condition linking the height of the surface barrier (band bending) at an arbitrary moment of time t with concentration of the charged SS of both types at a specific moment of time. The capturing of the charge carriers developing during transition of adsorption particles into the charged form and during BSS recharging is considered in adiabatic approximation, i.e. it is assumed that inside the volume zones at each specific moment of time there is an equilibrium and the system is characterized by instant equilibrium values of all SCR characteristics [131].

It is well known [4] that at small surface occupation degrees the concentration of adsorption particles chemisorbed in charged form increases according to the law given by two exponential with different characteristic times corresponding to proper chemisorption and transition of the charge carriers between the energy levels of the formed SS and the volume bands of adsorbent. In case when ASS formation is a limiting stage of surface charging the characteristic time of the whole process is determined by the partial pressure of absorbed gas, by adhesion coefficient of adsorption particles, by the activation energy of chemisorption, by temperature as well as by characteristics of the adsorption centers. In the opposite case when the transition of adsorption particles into the charged form is a limiting stage the characteristic time is dependent on such parameters of absorbate and adsorbent as the value of pre-adsorption surface barrier, the mutual position of the Fermi level of adsorbent and local level of ASS, the capturing cross-section of the free charge carriers on above levels, their concentration, etc.

In numerous cases the experimental data suggests that the limiting stage of the surface charging in semiconductor adsorbent is provided by transition of adsorption particles into the charged form [18]. In case of acceptor adsorption (in addition to a small capturing cross-section of carriers [84]) this, presumably, is mainly caused by availability of sufficiently high energy barriers, whose penetration is linked with transition of the carriers to ASS energy levels. In case of active particle adsorption (such as atoms and radicals) we are dealing with absence or extremely small activation energy of chemisorption of such particles which allows describing the chemisorption through the "adhesion tape" model [132].

It was shown in [98] that during acceptor adsorption in a broad band semiconductor of n-type characterized by availability of an a priori surface-adjacent depletion zone developing depletion of BSS levels slows down the kinetics of surface charging if compared to the case when either BSS or their recharging is absent. The system of equations describing the process of surface charging under assumption of limiting role of transition of adsorption particles into the charged form can be written as

$$\frac{dn_t}{dt} = K_{nt}\left\{\left(N_{ta} - n_t\right)e^{-\xi(t)} - n_t\,e^{-\alpha}\right\},$$ (1.54)

$$\frac{dn_B}{dt} = K_{nB}\left\{\left(N_B - n_B\right)e^{-\xi(t)} - n_B\,e^{-\gamma_B}\right\},$$ (1.55)

$$\sqrt{\xi + e^{-\xi} - 1} = \frac{n_B(t) + n_t(t)}{\sqrt{2}N_D L_D}.$$ (1.56)

As usual here N_{ta} and N_B are total concentrations of ASS and BSS; $n_t(t)$ and $n_B(t)$ are the concentrations of their charged forms at the moment of time t; E_B is the BSS energy level; $\alpha = (E_{ta} - \chi - \delta)/kT$; $\gamma_B = (E_B - \chi - \delta)/kT$; $\xi(t)$ is the value of the surface barrier at moment t; K_{nt} and K_{nB} in the diode approximation applicable to the broad band oxides with high concentration of superstoichiometric metal atoms of [133] are equal to: $K_{nt} = v n_v \sigma_t$ and $K_{nB} = v n_v \sigma_B$, where v is the average heat velocity of free charge carriers, σ_t and σ_B are the capture cross-sections of the carriers on ASS and BSS surface levels, n_v is the volume concentration of free charge carriers which is assumed equal to a concentration of small totally ionized donors: $n_v = N_D$.

The solution of system (1.54) − (1.56) at initial conditions: $n_t(0) = 0$ and $n_B(0) = n_{B0}$, where n_{B0} is determined from equation $(N_B - n_{B0})\exp(\xi_{S0}) = n_{B0}\exp(-\gamma)$ and $\xi_{S0}^{1/2} = n_{B0}/\sqrt{2}N_D L_D$ indicates that when the condition

$$\eta \equiv \left(1 + \frac{\kappa_t}{\kappa_B}\right)^{-1}\exp\xi_{S0} > 1,$$ (1.57)

is met the adsorption-induced change in barrier height is small: $\xi < 2\xi_{S0}$ (Fig. 1.10, curve 1). (The above condition restricts from above the rate of ASS recharging relative to that of BSS). Here $\kappa_t = \bar{v}N_D\sigma_t N_{ta}$, $\kappa_B = \bar{v}N_D\sigma_B N_B$. In other words in this case the biographic defects are a "fast subsystem", and ASS are "slow subsystem". When $\eta \gg 1$ the charging of the surface during adsorption will not practically take place which is caused by fast relaxation of the charge in BSS. However, if such relaxation processes develop sufficiently slow they do not affect the kinetics of the surface charging at initial moment of time. The quantitative expression of this condition is provided by

$$\eta < 1 \quad \text{and} \quad t < t_B = \left[2\kappa_B \sqrt{\xi_{S0}}\, \exp(-\xi_{S0})\right]^{-1}. \qquad (1.58)$$

With the laps of time when $t \gg t_B$ the rate of ASS occupation and BSS depletion become equal to each other and function $\xi(t)$ reaches a kinetic stop. This case is depicted in Fig. 1.10 (curve 2).

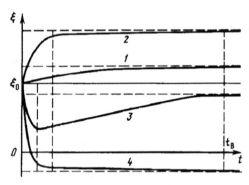

Fig. 1.10. Kinetic curves of surface charging of semiconductor of n-type during adsorption. Acceptors: $1 - \eta > 1$; $2 - \eta < 1$. Donors: $3 - N_t \ll N_B$; $4 - N_t > N_B$.

In case of small density of adsorbed particles if contrasted to the density of charged BSS the adsorption of donors can be accompanied by non-monotonous kinetics change in $\xi_S(t)$ which is caused by fast ASS depletion with subsequent slow BSS recharging (see Fig. 1.10, curve 3). The use of typical values of parameters in absorbate-adsorbent systems shows that depletion of donor levels is characterized by the times of the order of seconds whereas the relaxation of charge in BSS takes hours.

In case when concentration of donor adsorption particles is sufficient for complete removal of the surface barrier caused by BSS charge, the extremum behavior in $\xi_S(t)$ gets replaced by monotonous decrease in the value of ξ (see Fig. 1.10, curve 4). The leveling in the rate of ASS charging and additional BSS charging occurring after inversion of the sign of the band bending results in saturation of $\xi_S(t)$. This is accompanied by the system attaining the regime of lasting ASS and BSS charging. Finally BSS get charged completely (if we consider the case when $N_{td} > N_B$) and ASS get the equilibrium charge accounting for a compensating role of BSS.

Summing up the above reasoning we may conclude that the availability of an a priori energy barrier in adsorbent impeding the transition of free charge carriers on the adsorption-induced surface levels is equivalent to a substantial viscosity in an arbitrary dynamic system, this viscosity being caused by monotonous character of kinetic curves. On the contrary, in case of small concentration of adsorption particles the availability of an a priori band bending facilitating the charging of absorbate results in the change of electrophysical parameters of adsorbent as non-monotonous features in kinetic curves. The removal of this barrier, i.e. the inversion of the sign of the band bending which is possible at high concentration of adsorbed particles is equivalent to creation of viscosity in the system. This viscosity is larger as the value of ξ_S grows up.

It is quite obvious that adsorption-caused changes of various electrophysical characteristics of adsorbent such as electric conductivity, work function, thermal electromotive force, etc. can be related to an adsorption-caused change in the value of the surface barrier, which, in its tern, is directly linked with the change in the value of the surface charge. Therefore, apart from direct experiments on the field effects controlling the value of the surface charge Q_S the intensity, monotonicity or non-monotonicity in kinetics of the change in ξ_S can be traced by the change in kinetics of the alteration of other electrophysical characteristics of adsorbent, too.

Bearing this in mind we should point out that conclusions made in [98] have been confirmed in large number of experimental papers dealing with examination of adsorption of molecular and atom gases on semiconductor oxides. In most of these papers it has been shown that adsorption of such active particles as atoms of hydrogen, oxygen, nitrogen, atoms of various metals as well as simplest radicals on thin sintered films and monocrystals is usually accompanied by monotonous change in their electrophysical characteristics over the wide range of concentrations of absorbate and biographic states. In these experiments the concentration of the latter was often changed through doping the surface of oxides by atoms of various metals. For instance in [134] the series of monotonous kinetic curves of the change of electric conductivity $\sigma(t)$ during adsorption of H-atoms on sintered polycrystalline ZnO films, pure or doped by atoms of Zn, Pb, Na, In was obtained. The stationary volume concentration of H-atoms was about 10^7 cm^{-3} in these experiments. Similar monotonous dependencies were obtained for $\sigma(t)$ during adsorption of H-atoms on monocrystals of ZnO. These results were published in [82, 129], as well as in [135] for the case of adsorption of H-atoms at low temperatures on various sintered oxides. Similar results were also obtained in case of adsorption on above oxides of atoms of numerous metals [136].

We should mention, however, that there are papers in which the non-monotonous change in electric conductivity with time was observed for adsorption of H-atoms. Thus, it was shown in paper [137] which dealt with adsorption of H-atoms on CdS in particular, and which was performed under reliable experimental conditions, that non-monotonous dependence of $\sigma(t)$ was observed only at low temperatures (77 K) and for small concentrations of hydrogen in the reaction cell (10^{-5} Torr). The increase of the hydrogen pressure up to 10^{-4} Torr lead to leveling-off of curves $\sigma(t)$ with transition to monotonous dependence. We should note that at room temperature adsorption of H-atoms in the same CdS samples was also accompanied by monotonous dependence in $\sigma(t)$.

So far, we have focused our attention on adsorption of donor particles on semiconductor oxides. As for the effect of adsorption of acceptor particles on electrophysical characteristics, in concurrence with conclusions made none of adsorption phenomenon involving such characteristic acceptor particles as molecular and atom oxygen on n-semiconductor, atoms of nitrogen and simplest alkyl and amine radicals brought about a non-monotonous change in characteristics of adsorbents, despite the fact that experiments had been conducted at various conditions.

For instance in [67] the function $\sigma(t)$ is given for the case of adsorption on films of ZnO for oxygen molecules and atoms. Monotonous kinetic curves of the change in $\sigma(t)$ have been obtained in [138]. Work function taking place during adsorption of O-atoms on sintered samples of ZnO was also determined in above paper. A vast experimental material of the change in $\sigma(t)$ was gathered in course if investigations [139, 140]. These data were obtained for doped and pure films of ZnO, CdO, NiO during adsorption of nitrogen atoms. None of above experiments showed a non-monotonous character in dependence $\sigma(t)$. In [141] it was shown that in case of adsorption on films of various semiconductor oxides of simplest alkyl radicals (typical acceptors of electrons) the monotonous kinetic dependence of function $\sigma(t)$ was always observed. As for the rare data on non-monotonous kinetic in the change of electric conductivity and the value of the surface charge observed (for instance, during adsorption of NO_2 on PbS, paper [142]), it could, in our opinion, be caused by irreversible chemical interaction of acceptor particles NO_2 (forming nitric acid vapor in presence of water) with the surface of sulfuric lead resulting in oxidation of its surface and, consequently, in agreement with findings of classical study [143] in formation of surface-adjacent level of p-type, resulting in non-monotonous character in function $\sigma(t)$ which was proved by experiment.

Therefore, the theoretical analysis and numerous experimental data enable one to assume that charging of the surface of adsorbent occurring during adsorption can result in relaxation of the charge state of an a priori existing BSS and influence the kinetics of transition of adsorption particles into the charged form. However, during chemisorption of active particles characterized by small charging rate if compared to BSS recharging the kinetics and the value of the change in electrophysical characteristics of adsorbent is determined by kinetics of charging of absorbate at initial stages. The notable manifestation of effects relating to BSS recharging should be detected over substantially large times often not attainable under experimental conditions. These effects are mainly dealing with decrease in the stationary values of adsorption-induced changes in electrophysical parameters if contrasted to an a priori clean surfaces. Thus, this is the data on the initial rates of the change in electrophysical parameters of adsorbent during adsorption caused by surface charging that enables us to assess the concentration of particles of the surface of adsorbent as well as in the gaseous phase.

1.10. The effect of adsorption surface charging on electrophysical characteristics of polycrystalline semiconductor adsorbents

In sections 1.7 − 1.9 we have examined effects of surface charging in semiconductor adsorbent on electrophysical characteristics of the adsorbent. Although we did not go into details with respect to the crystalline origin of adsorbent, the consideration of effect of adsorption on electric conductivity of surface-adjacent layer led to conclusion that we considered monocrystalline samples.

Adsorbents involving oxides of various metals to be used as operational elements of chemical sensors are usually polycrystalline materials possessing the whole set of anomalous properties which drastically differ them from monocrystalline samples. First of all this deals with anomalously high sensitivity of various characteristics of these materials to external effects [144]. For instance, if we consider electrophysical characteristics of materials such as electric conductivity, thermal electromotive force, the Hall effect the high sensitivity of above characteristics to illumination, application of external electro-magnetic fields, thermal effects, etc. is substantially related with the change of current permeability through intra-domain barriers existing in the material, whose penetration is related with the transfer of free charge carriers. To the full extent this is relevant to adsorption of various particles which in numerous cases results in the change of the height of inter-crystalline barriers due to chemisorption-induced change in the value of the surface charge of microcrystal samples.

However, the role of availability of the broad distribution in height of these barriers [145] caused by variation in the size and shape of microcrystals, in defect and dope statistical distribution at the surface and throughout the volume of microcrystals and finally the random crystallographic orientation of facets at crystallite interface is also important. For instance, it was established in [146] that the amplitude of the spread in barriers with respect to height in polycrystalline zinc oxide is about 0.3 – 0.7 eV. Such randomly inhomogeneous character of the energy profile of the bottom of the conductivity band results in appearance of a percolation structure in this material caused by its splitting into classical domains where motion of the free charge carriers is either allowed or forbidden. Owing to this not all barriers take part in the process of conductivity of the constant current. This process is supported only by those barriers which are included into the well known infinite cluster responsible for ohmic conductivity of the material [147, 148].

Existence of such barrier disarrangement is responsible for availability of such phenomena as anomalously delayed kinetics decay in photoconductivity [149], anomalously early onset of non-ohmicity phase [150], adsorption change in the shape of V-A characteristics (VAC) in polycrystal semiconductor materials. It is obvious to suggest that above disarrangement should underlie the effects of adsorption-catalytic processes on electrophysical characteristics of polycrystalline semiconductor adsorbents [151-153]. These processes involve the particles chemisorbed in charged form. In so doing it is obvious that above effect cannot be adequately described by any of developed models describing effects of adsorption on electrophysical characteristics of monocrystals [4-6], nor by the model of chemisorption response of electric conductivity in polycrystalline semiconductor [4, 154, 155] provided by the Volger-Petritz theory of biocrystals [156, 157], modeling a polycrystalline adsorbent by an array of inter-crystalline barriers with identical height and properties linked through relatively low ohmic conductivity domains. The presence of a wide spread in height of these barriers as well as existence of obvious correlation between the rate of adsorption-induced change in the height of a barrier and its initial height [158] result in the fact that adsorption changes the barrier disarrangement in adsorbent determined by the type and interval in distribution of height of the barriers.

The developed theory of effect of adsorption on electric conductivity in polycrystalline adsorbents in weak and high electric fields was the purpose of the numerous papers [151, 158 – 161]. The choice of ohmic electroconductivity as an object of studies was initially dictated by a traditional investigation adsorption-catalytic processes through analyzing respective changes in electric conductivity of adsorbent, which led to development of the method of semiconductor sensors [83, 162]. As for electric conductivity in high electric fields, namely in differential coefficients in V-A characteristics of polycrystalline adsorbent, it is these characteristics as a whole that are determined by degree of barrier disarrangement of adsorbent. We will show this latter on. Moreover, its ad-

sorption-caused change is linked not to the change in the height of separate barriers but with collective behavior of the whole system of inter-
crystalline barriers present in adsorbent. Note, that here we are referring
to the well known pre-relaxation VAC, i.e. the dependence of initial
values of current going through the system plotted as a function of applied unit voltage [146, 163] measured prior to development of relaxation processes caused by transition of the system to the new equilibrium
state characterized by the new degree of occupation of surface states
corresponding to the value of applied voltage [164].

Consistent theory of dark conductivity and the model of disordered
polycrystalline semiconductor were developed in studies [146, 150].
Within the framework of this model the height of inter-crystalline barriers ξ (measured in kT units) is provided by independently measured
statistical values with dispersion much larger than a unity, the polycrystal being represented as a statistical electric lattice with exponentially
broad resistance spectrum $R = R_0 \exp\xi$, whose electric conductivity is
calculated through the known methods of the percolation theory [147,
148]. In the ohmic regime the main term of the logarithm of electric
conductivity is determined by the height of critical barriers ξ_c
"responsible" for percolation level of a statistical potential profile of the
bottom of the conductivity band in the polycrystalline semiconductor.
The conductivity is affected by impedance included into infinite cluster
(IC) which is a critical subgrid of the carrying grid whose characteristic
size is provided by the correlation length L (the average unit size of
critical subgrid controlling the average distance between critical barriers
and satisfying the condition $L \gg l \gg L_D$, where l is the average size of
microcrystal, L_D is the Debye shielding length). The density of distribution of heights in inter-crystalline barriers $f(\xi)$ gives both the value
of the percolation level, i.e. the height of critical barriers

$$x_C = \int_{\xi_{min}}^{\xi_C} f(\xi)\, d\xi \ , \qquad (1.59)$$

and the value of the correlation length

$$L = l\left|x(\xi_C + 1) - x_C\right|^{-\nu} \approx l\, f^{-\nu}(\xi_C) \ , \qquad (1.60)$$

controlling the activation energy and the value of leading factor in effective electric conductivity of material [146, 148, 159], respectively

$$\sigma = \sigma_v f^\nu(\xi_C)\exp\{-\xi_C\} \ , \qquad (1.61)$$

where σ_v is the value dependent on the volume conductivity of microcrystals, their average size and certain general characteristics of the contacts between microcrystals; v is the critical index of the correlation length; $f(\xi_C)$ is the density of critical barriers; $x(\xi)$ is the fraction of contacts, containing barriers with height not exceeding ξ; $x(\xi_C) \equiv x_C$ is the percolation threshold in the contact problem, i.e. the value which is dependent on the density of packing in microcrystals.

Apart from purely field decrease in the height of the barriers using application of electric field with value $E > kT/ql$, where q is the electron charge, the change in electric conductivity of barrier-disarranged system can be caused by IC restructuring responsible for the current [150]. In this case one can deduce for VAC [150, 146, 159]

$$J = J_0 \exp\left\{\left[c\,\frac{q|E|l}{kT}f^{-v}(\xi_C)\right]^{\frac{1}{1+v}}\right\} , \qquad (1.62)$$

i.e. the tangent of the VAC inclination angel plotted in coordinates $\ln(J\,/\,J_0) - |E|^{1/2}$ ($v_3 \approx 0.88$ in three dimensional case) is given by expression

$$\beta \approx \left[\frac{Cql}{kT}\right]^{\frac{1}{1+v}} f^{-\frac{v}{1+v}}(\xi_C) , \qquad (1.63)$$

where C is a constant factor; J_0 is the value weakly dependent on E. It follows from expressions $(1.59) - (1.63)$ that both in ohmic (1.61) and in non-ohmic (1.62) regimes the electric conductivity of the material is dependent on concentration of critical barriers $f(\xi_C)$ determined both by the shape of function $f(\xi)$ and the interval of distribution of heights of inter-crystalline barriers $\Delta\xi$). It is the tangent of the angle of the slope in VAC of adsorbent that is wholly determined by the value of $f(\xi_C)$, whereas the ohmic conductivity contains this characteristics in the leading factor.

Charging the surface of crystallines accompanying adsorption due to transition of a portion of chemisorbed particles into the charged form [4] changes the height of inter-crystalline barriers. In such a case the change in barrier height results in the change of the rate of the charge transfer to the not yet occupied surface energy levels of the adsorption type. Consequently, the rate of change ξ is dependent on the initial height of barrier ξ_0 and changes together with occupation of the adsorption surface states, i.e. there is a time-dependent correlation between adsorption-caused change in the height of barrier and its initial height which, naturally, leads to adsorption-induced transformation of initial distribution of heights of inter-crystallines barriers. In its turn, this results in the change of the height and concentration of critical barriers which is manifested in the change of both electric conductivity and the tangent of VAC inclination angle of adsorbent. Therefore, to determine the value and kinetics of adsorption-induced response $\sigma(t)$ and $\beta(t)$ one needs accessing information regarding the behavior of the density of distribution of heights of barriers $f(\xi,t)$ during adsorption of both acceptor and donor particles.

To determine correlation between $\xi(t)$ and ξ_0 and, therefore, to find out the type of dependence $f(\xi,t)$ let us consider the occupation kinetics for ASS levels by free charge carriers. The capturing of charge carriers occurring during transition of adsorption particles into the charged form will be considered, as usual, in adiabatic approximation, i.e. assuming that at any moment of time there is a quasi-equilibrium and the system of crystallites is characterized by immediate equilibrium values ξ_C and L inside the conduction (valence) band.

As it has been mentioned in section 1.9, depending on what is a limiting process in formation of the charged form of chemisorbed particles (establishing an electron equilibrium between ASS and the volume of semiconductor or the proper process of formation of ASS) the kinetics of the change of the height of surface barriers is determined by either rate of exchange by charges between the bands of semiconductor and surface levels, or by chemisorption rate. Thereby, in general case one has to solve the kinetic equation describing occupation of ASS with time dependent concentration of the latter to determine function $\xi(t,\xi_0)$. As usual, the position of energy levels in ASS with respect to the edges of bands is controlled by the value of affinity to electron E_a or by ionization potential I_d for adsorption of acceptors and donors, respectively. In what follows we will consider the broad band doped n-type semiconductor with the volume concentration of doping donors N_D and doping parameter δ characterizing the depth of the Fermi level beneath the bottom of the conductivity band in the crystalline volume.

1.10.1. Adsorption of acceptor particles

In case of adsorption of acceptor particles the kinetic equation describing the rate of change in density of ASS occupied by electrons can be written as

$$\frac{dn_t}{dt} = K_{nt}\left\{\left(N_t - n_t\right)e^{-\xi(t)} - n_t\,e^{-\alpha}\right\}\,,\tag{1.64}$$

where all notations correspond to those adopted in section 1.9. The height of barrier $\xi(t)$ at a moment of time t is determined from solving the Poisson equation under assumption: 1) the condition of electric neutrality in the absorbate-adsorbent system is observed ; 2) the effects of BSS recharging do not influence the surface charging kinetics. As a result we obtain the expression linking the value $\xi(t)$ with current magnitude in $n_t(t)$ (see formula (1.25)):

$$\sqrt{\xi + e^{-\xi} - 1} - \frac{n_t(t)}{N_D L_D} = \sqrt{\xi_0 + e^{-\xi_0} - 1}\,.\tag{1.65}$$

Accounting for the fact that according to the definition of barrier $\xi_{0_{min}} > 1$ for $\xi(t)$ we have

$$\xi(t) \approx \left[\sqrt{\xi_0} + n_t(t)\right]^2\,.\tag{1.66}$$

Here and in what follows N_t and n_t stand for the values reduced to the product $N_D L_D$.

Substituting (1.66) into expression (1.64) leads to the fact that far from the equilibrium position owing to either full occupation of adsorption surface states or with leveling off in E_a with the Fermi level of adsorbent E_F solution of the latter can be written as

$$\xi(t) \approx \xi_0 + \ln\left\{1 + 2K_{nt}\sqrt{\xi_0}\,e^{-\xi_0}\int_0^t N_t(\tau)\,d\tau\right\}\,.\tag{1.67}$$

Note that the large amount of experimental data makes it possible to assume that processes related to the transfer of the charge to the surface states formed during adsorption of acceptors on oxidated oxides develop much slower than the process of formation of the proper adsorption surface states and, therefore, they are the limiting stage of the process of charging of the surface [18, 20]. Thus, in this case one can consider that $N_t(t) = N_t = $ const and expression (1.67) can be written as

$$\xi(t) \approx \xi_0 + \ln\left\{1 + \kappa T\sqrt{\xi_0}\,e^{-\xi_0}\right\}\,,\tag{1.68}$$

where $\kappa = 2K_{nt}N_t$. The analysis of the expression obtained yields that to an arbitrary moment of time t the initially lowest barriers with height ξ_0 satisfying the condition $\exp(\xi_0)/\sqrt{\xi_0} < \kappa T$ are almost leveled-off and following that they increase with the similar, not depending on ξ, rate, whereas barriers with high ξ_0 are changing weakly (Fig. 1.11). Therefore, expression (1.68) is satisfactorily approximated by two asymptotes:

$$\xi(t) \approx \begin{cases} \ln\{\kappa t \sqrt{\ln \kappa t}\}, & \xi_0 < \ln\{\kappa t \sqrt{\ln \kappa t}\}, \quad (1.69) \\ \xi_0 + \kappa t \sqrt{\xi_0}\, e^{-\xi_0}, & \xi_0 > \ln\{\kappa t \sqrt{\ln \kappa t}\}. \quad (1.70) \end{cases}$$

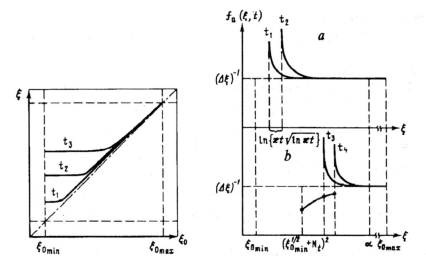

Fig. 1.11. The dependence of the height of barrier ξ caused by adsorption of acceptors as a function of initial height ξ_0 at various moments of time $t_1 < t_2 < t_3$.

Fig. 1.12. The density of distribution of the heights of barriers during different moments of time during adsorption of acceptors $a - t_1 < t_2 < t_{ex_{min}}$; $b - t_{ex_{min}} < t_3 < t_4$.

Now it is easy to trace the effect of adsorption of acceptors on degree of barrier disordering in polycrystalline adsorbent.

For the sake of simplicity we can assume that initially the heights of barriers are distributed homogeneously over the interval $\Delta\xi = \xi_{0_{max}} - \xi_{0_{min}}$, i.e. $f(\xi_0) = (\Delta\xi)^{-1}$. Using (1.68) we arrive to expression

$$f_a(\xi,t) = (\Delta\xi)^{-1}\left[1 - \kappa t\sqrt{\xi}\,e^{-\xi}\right]^{-1} ,$$ (1.71)

for the density of distribution of heights at moment of time t. This expression underlines the existence of the "group of leveled-off barriers" (Fig. 1.12, a).

However the obtained functions $\xi(t,\xi_0)$ and $f_a(\xi,t)$ are valid unless there is no (if allowed) depletion of non-occupied ASS. The above depletion regime for a crystalline with a fixed value of ξ_0 is possible only for a concentrations of adsorption particles N_t, insufficient (even in case of a complete transition to the charged form) to level-off E_a and E_F, or, in other words, ξ and α. In case of acceptor adsorption under consideration such smallness condition providing the complete occupation of adsorption surface states (at least on the smallest barriers at initial times is $N_t < \sqrt{\alpha} - \sqrt{\xi_{0_{min}}}$). The inequality $t < t_{ex}(\xi_0) = \kappa^{-1}\exp\{(N_t + \sqrt{\xi_0})^2\}/\sqrt{\xi_0}$ provides the ground for applicability of expressions (1.68) – (1.70) for a barrier with a specific ξ. When the opposite condition holds $N_t > \sqrt{\alpha} - \sqrt{\xi_{0_{min}}}$ the equilibrium state provided by full fill in of N_t cannot be obtained for any ξ_0 and the transformation functions (1.68) – (1.70) are valid up to times at which the leveling-off of ξ and α develops, i.e. of the order $t_\alpha = \kappa^{-1}\exp(\alpha)/\sqrt{\xi_0}$.

Thus, for a concentration of acceptor particles $N_t < \sqrt{\alpha} - \sqrt{\xi_{0_{min}}}$ starting with $t \approx t_{ex_{min}} = t_{ex}(\xi_{0_{min}})$ the function $f_a(\xi,t)$ alters due to the change in function $\xi(t,\xi_0)$ for barriers with smallest ξ_0. For these concentrations the kinetic equation (1.64) acquires the shape

$$\frac{dn_t}{dt} \approx K_{nt}(N_t - n_t)\exp\left\{-\left(\sqrt{\xi_0} + N_t\right)^2\right\} ,$$ (1.72)

with solution given by expression

$$\xi(t,\xi_0) = \left[\sqrt{\xi_0} + N_t\left(1 - \exp\left\{-\frac{t}{t_f}\right\}\right)\right]^2 ,$$ (1.73)

where $t_f = 2N_t\xi_0^{1/2}t_{ex}$. Consequently, for small N_t at an arbitrary moment of time $t > t_{ex_{min}}$ all barriers present in the system are broken down into groups differing in shape of $f_a(\xi,t)$:

a) barriers with $\left(\xi_{0_{\min}}^{1/2} + N_t\right)^2 < \xi < \ln\{K_{nt}t\}$ do not change and are characterized by complete occupation N_t;

b) barriers with $\ln(K_{nt}) < \xi < \ln\{\kappa t\,(\ln\kappa t)^{1/2}\}$ change very slowly according to expression (1.73):

c) barriers with $\ln\{\kappa t\,(\ln\kappa t)^{1/2}\} < \xi < \alpha$ increase according to formula (1.68);

d) barriers with $\xi \geq \alpha$ do not change at all during adsorption (we are considering the case when $\xi_{0_{\max}} > \alpha$).

Consequently, for density of distribution ξ at an arbitrary moment of time $t > t_{ex_{\min}}$ one can write down an approximate expression

$$
f_a(\xi,t) \approx
\begin{cases}
(\Delta\xi)^{-1}\left(1 - \dfrac{N_t}{\sqrt{\xi}}\right), & \left(\sqrt{\xi_{0_{\min}}} + N_t\right)^2 < \xi < \ln\{\kappa t\sqrt{\ln\kappa t}\}, & (1.74) \\[2ex]
(\Delta\xi)^{-1}\left[1 - \kappa t\sqrt{\xi}\,e^{-\xi}\right]^{-1}, & \ln\{\kappa t\sqrt{\ln\kappa t}\} < \xi < \alpha, & (1.75) \\[2ex]
(\Delta\xi)^{-1}, & \alpha < \xi < \xi_{0_{\max}}, & (1.76)
\end{cases}
$$

illustrated by Fig. 1.12, b. Thereby, the type of function $f_a(\xi,t)$ indicates on complex features of adsorption-related change in density of distribution ξ dealing with the initial decrease in the interval of distribution ξ which is caused by superseding rate of increase in smaller barriers if compared to large ones resulting in formation of the "group of leveled-off barriers". In case of small concentration of adsorption particles (small partial pressures of absorbed gas) the lowest barriers start "falling down" from the "leveled-off group" due to full occupation of N_t.

As it has been already shown the adsorption-induced change in the value of percolation electric conductivity of adsorbent and tangent of inclination angle in its pre-relaxation VAC are determined by adsorption-related change in the value of the percolation level and correlation length of the system, i.e. by functions $\xi_C(t)$ and $f(\xi_C,t)$. One can readily conclude for $\xi_C(t)$ that when $N_t < \sqrt{\alpha} - \sqrt{\xi_{C0}}$, i.e. at small N_t and sufficiently high initial conductivity of adsorbent ($\ln\sigma_0 \sim -\xi_{C0}$ is valid) we have

$$\xi_C(t) \approx \begin{cases} \xi_{C0} + \ln\left\{1 + \kappa t \sqrt{\xi_{C0}} \ e^{-\xi_{C0}}\right\}, & 0 < t < t_{exc}, \qquad (1.77) \\ \left(N_t + \sqrt{\xi_{C0}}\right)^2 = \text{const}, & t > t_{exc} \qquad\qquad (1.78) \end{cases}$$

where $t_{exc} = \kappa^{-1} \exp\left\{\left(N_t + \sqrt{\xi_{C0}}\right)^2\right\} / \sqrt{\xi_{C0}}$ is the characteristic time over which the "falling down" of critical barriers develops from the "leveled-off group". When the opposite inequality holds $N_t > \sqrt{\alpha} - \sqrt{\xi_{C0}}$ the kinetics of change in $\xi_C(t)$ is described by expression (1.77) under condition $\xi_{C0} < \alpha$ up to times of the order of t_α. When $\xi_{C0} > \alpha$ we have $\xi_C(t) = \xi_{C0} = \text{const}$, i. e. it is not changed during adsorption.

The adsorption-caused change in concentration of critical barriers $f(\xi_C, t)$ is also different for various relationships between parameters N_t, ξ_{C0} and α. Thus, in case of small concentration of adsorption particles $N_t < \sqrt{\alpha} - \sqrt{\xi_{C0}}$

$$f\left(\xi_C(t)\right) \approx \begin{cases} (\Delta\xi)^{-1}\left[1 + \dfrac{e-1}{e}\right], & t < t_{exc}, \qquad (1.79) \\[2ex] (\Delta\xi)^{-1}\left[1 + \dfrac{N_t}{\sqrt{\xi_{C0}}}\right]^{-1}, & t > t_{exc}, \qquad (1.80) \end{cases}$$

where e is the base of the natural logarithm; $t_{0C} = \kappa^{-1} e^{\xi_{C0}} / \sqrt{\xi_{C0}}$ is the time over which capturing of critical barriers by the "leveled-off group" occurs. When the opposite inequality holds $N_t > \sqrt{\alpha} - \sqrt{\xi_{C0}}$ the behavior of $f(\xi_C, t)$ is described by expression (1.79) up to times of the order t_α.

Thus, having examined expressions (1.61) – (1.63) and (1.77) – (1.80) we have for the case of small concentration of acceptors $N_t < \sqrt{\alpha} - \sqrt{\xi_{C0}}$ the following expressions describing the kinetics of the change in $\sigma(t)$ and $\beta(t)$

$$\frac{\sigma(t)}{\sigma_0} \approx \begin{cases} \left(1 + t/t_{0C}\right)^{-m}, & t < t_{exc}, \qquad (1.81) \\[2ex] \left(1 + N_t/\sqrt{\xi_{C0}}\right)^{-\nu} \dfrac{t_{0C}}{t_{exc}}, & t > t_{exc}, \qquad (1.82) \end{cases}$$

$$\frac{\beta(t)}{\beta_0} \approx \begin{cases} \left(1 + t / t_{0C}\right)^{-\frac{m}{1+v}}, & t < t_{exc}, & (1.83) \\[3mm] \left(1 + N_t / \sqrt{\xi_{C0}}\right)^{-\frac{v}{1+v}}, & t > t_{exc}, & (1.84) \end{cases}$$

where $m \approx 0.46$ is valid for the samples with thickness $h > L$ ($v_3 \approx 0.88$) for the three-dimensional case, and $m \approx 0.15$ stands for the two-dimensional case which is applicable for samples with $h < L$ ($v_2 \approx 1.35$). For thickness of adsorbents of the order of correlation length ($h - L$) both σ and β become dependent on h [148, 165].

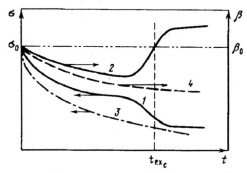

Fig. 1.13. The kinetics of the change in $\sigma(t)$ (curves 1, 3) and $\beta(t)$ (curves 2, 4) during acceptor adsorption 1,2 – $N_t < \sqrt{a} - \sqrt{\xi_{C0}}$; 3,4 – $N_t > \sqrt{a} - \sqrt{\xi_{C0}}$.

When concentration of adsorption particles exceeds the limiting one $N_{t^*} = \sqrt{a} - \sqrt{\xi_{C0}}$ (which is proportional to the value of electric conductivity of adsorbent), the kinetics of change in $\sigma(t)$ and $\beta(t)$ is described by expressions (1.81) and (1.83) for all times $t < t_a$, respectively. Graphically obtained dependencies of $\sigma(t)$ and $\beta(t)$ are given in Fig. 1.13.

One can deduce from these profiles that in case of small concentrations of adsorption particles that an additional notable change in ohmic electric conductivity takes place for the case of small concentration of adsorption particles during change of t in vicinity of t_{exc}. This happens due to decrease in the value of leading factor from $(1 + t_{exc} / t_{0C})^v > 1$ to $(1 + N_t / \sqrt{\xi_{C0}})^{-v} < 1$ the value of the power index of exponent $\xi_C = (N_t + \sqrt{\xi_{C0}})^2$ being practically the same. The tangent of the angle of inclination in VAC over this interval of times increases from

$(1 + t_{ex_C} / t_{0C})^{-m/(1+\nu)} < 1$ to $(1 + N_t / \sqrt{\xi_{C0}})^{\nu/(1+\nu)} > 1$. Such anomaly in adsorption kinetics of $\sigma(t)$ and $\beta(t)$ over large times $t \approx t_{ex_C}$ is distinctly shown in experiment [159]. This is caused by effect of adsorption on concentration of critical barriers controlling the value of leading factor of electric conductivity and VAC inclination angle. The rise in $f(\xi_C)$ due to narrowing of distribution interval for ξ at $t < t_{0C}$ and increase of the fraction of critical barriers for $t_{0C} < t < t_{ex_C}$ caused by expansion of "leveled-off barrier group" absorbing new barriers with $\xi_0 > \xi_{C0}$ which initially did not participate in conductivity gets replaced by substantially abrupt decrease in $f(\xi_C)$ for $t \approx t_{ex_C}$ due to "falling down" of ξ_C from above group at complete occupation of N_t.

We should note that availability of the power index $m < 1$ in expression (1.81) can provide the delay in kinetics of the change in $\sigma(t)$ during adsorption of acceptors if contrasted to the case of theoretical predictions based on biocrystal model [84] observed in experiment [51, 84, 166]. In our case the smallness of the power index and degree of delay of kinetics in $\sigma(t)$ related to it are caused by manifestation of a specific compensation effect dealing with substantial rise in the value of leading factor during increase in the conductivity activation energy which is related to simultaneous increase of both the height of critical barriers and the density of current channels.

It should be noted, however, that all above reasoning dealing with equilibrium state touching on either complete occupation of surface adsorption states or with leveling-off in E_a and F are valid in case when respective values of equilibrium barriers do not exceed the height of the Weisz barrier [7]. The opposite case realized at a high concentration of adsorption particles and at deep positioning of their energy levels beneath the bottom of conductivity band $\sigma(t)$ results in "kinetics shut down" when the barriers reach 1 eV and the stationary value of σ becomes insensitive to further increase in the partial pressure of gas of acceptor particles.

Commenting on above we should mention that initial expressions (1.59) – (1.63) are valid for disordered systems with exponentially broad spectrum of local values of electric conductivity. Due to existing dependence of $\xi(t)$ on ξ_0 over long times in our case the broad pre-adsorption spread in ξ can grow narrow. At specific ratios between parameters of the absorbate-adsorbent system it can either vanish at all or there is a notable concentration of leveled-off barriers being formed with the fraction higher than the threshold one x_C. The straightforward analysis of each specific case characterized by a certain relationships between parameters of the system enables one easily obtain conditions and time intervals of applicability of expressions obtained within the frame work of the percolation theory.

As for equilibrium values of σ_S and β_S they are mainly dependent on relations between such parameters of the systems as initial electric conductivity of adsorbent, concentration of chemisorbed particles, reciprocal position of the energy levels of absorbate and adsorbent. Thus, during acceptor adsorption in case of small concentration of adsorption particles one can use (1.82) and (1.84) to arrive to expressions for equilibrium values of ohmic electric conductivity and the tangent of inclination angle of VAC:

$$\sigma_S = \sigma_0 \left(1 + \frac{N_t}{\sqrt{\xi_{C0}}}\right)^{-\nu} \exp\left\{-\left(N_t^2 + 2N_t\sqrt{\xi_{C0}}\right)\right\} , \tag{1.85}$$

$$\beta_S = \beta_0 \left(1 + \frac{N_t}{\sqrt{\xi_{C0}}}\right)^{\frac{\nu}{1+\nu}} , \tag{1.86}$$

from which we readily obtain for the case $N_t < \sqrt{\xi_{C0}}$

$$\sigma_S \approx \sigma_0 \left(1 - \nu \frac{N_t}{\sqrt{\xi_{C0}}}\right) \exp\left\{-2N_t\sqrt{\xi_{C0}}\right\} , \tag{1.85'}$$

$$\beta_S \approx \beta_0 \left(1 + \frac{\nu}{1+\nu} \frac{N_t}{\sqrt{\xi_{C0}}}\right) , \tag{1.86'}$$

for the case $\sqrt{\xi_{C0}} < N_t < \sqrt{\alpha} - \sqrt{\xi_{C0}}$ one gets

$$\sigma_S \approx \sigma_0 \left(\frac{N_t}{\sqrt{\xi_{C0}}}\right)^{-\nu} \exp\left\{-N_t^2\right\} , \tag{1.85''}$$

$$\beta_S \approx \beta_0 \left(\frac{N_t}{\sqrt{\xi_{C0}}}\right)^{\frac{\nu}{1+\nu}} , \tag{1.86''}$$

When the concentration of acceptors increases up to values $\sqrt{\alpha} - \sqrt{\xi_{C0}} < N_t < \sqrt{\alpha} - \sqrt{\alpha - \xi_{C0}}$ we have

$$\sigma_S \approx \sigma_0 \left(2\sqrt{\alpha}N_t\right)^{-\nu} \exp\{-(\alpha - \xi_{C0})\} \ . \tag{1.87}$$

The pre-relaxation VAC acquire the following shape for above relationship between the parameters of the system

$$\ln\frac{J}{J_0} \approx \begin{cases} \dfrac{\beta_0^{(1+\nu)}\,|E|}{2\sqrt{\alpha}N_t} \,, & |E| < |E_\bullet| \,, \tag{1.88}\\[2ex] \beta_0\sqrt{|E|} - 2\sqrt{\alpha}N_t \,, & |E| > |E_\bullet| \,, \tag{1.89} \end{cases}$$

where $|E_\bullet| = 4\alpha N_t^2 \beta_0^{-(1+\nu)}$. One can deduce from the last expression that when $|E| < |E_\bullet|$ the tangent in VAC inclination angle decreases with the rise in N_t, but VAC can be linearized in coordinates $\ln(J/J_0) - |E|$. When $E > E_\bullet$ the value of β_S almost returns to its initial value β_0 and the rise in N_t only leads to the parallel shift in VAC for a constant β. When $N_t > \sqrt{\alpha} - \sqrt{\alpha - \xi_{C0}}$ the fraction of the leveled-off barriers exceeds x_C and expressions (1.59) – (1.63) are inapplicable. The qualitative shape of functions $\sigma_S(N_t)$ and $\beta_S(N_t)$ is given in Fig. 1.14.

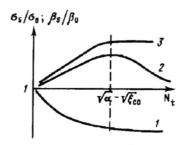

Fig. 1.14. The qualitative profile of dependencies of equilibrium values of electric conductivity σ_S (curve *1*) and the tangent of inclination angle of VAC β_S (curves *2, 3*) as a function of concentration of chemisorbed acceptors *2* – $\ln\sigma_0 \approx \xi_{C0} > \alpha/2$; *3* – $\ln\sigma_0 \approx \xi_{C0} < \alpha/2$

Note that expressions (1.85′) – (1.86″) are valid for substantially low ohmic samples with $\xi_{C0} = \ln(\sigma_0/\sigma_v) < \alpha/4$. Thus, in case of adsorption of O_2 on ZnO at temperatures $T < 200°C$ $kT\alpha \approx 0.9$ eV [20] and for ξ_{C0} we obtain $kT\xi_{C0} < 0.23$ eV. With increase in concentration of adsorption particles above $N_{t^*} = \sqrt{\alpha} - \sqrt{\xi_{C0}}$ expressions (1.59) – (1.63) for such low ohmic adsorbent do not hold and percolation transition becomes inapplicable. Thus, for $kT\alpha \approx 0.9$ eV, $kT\xi_{C0} \approx 0.2$ eV, $N_D \approx 10^{17}$ cm^{-3}, $L_D \approx 10^{-5}$ cm and $T \approx 300$ K we get $N_{t^*} \approx 3 \cdot 10^{12}$ cm^{-2}.

On the contrary, formulas (1.87) – (1.89) are applicable for high ohmic samples with $\xi_{C0} > \alpha/2$. Thus, when $kT\alpha \approx 0.9$ eV and $kT\xi_{C0} \approx 0.6$ eV the domain of applicability of expressions (1.87) – (1.89) with respect to concentration of adsorption particles amounts to 10^{12} cm$^{-2} < N_t < 3 \cdot 10^{12}$ cm^{-2}, and expressions (1.85') – (1.86'), to $N_t < 10^{12}$ cm^{-2}, respectively.

For different acceptor particle adsorption isotherms expressions (1.85) – (1.89) provide various dependencies of equilibrium values of σ_S for a partial pressure P (ranging from power indexes up to exponential). Thus, in case when the logarithmic isotherm $N_t \sim \ln P$ is valid the expression (1.85') leads to dependence $\sigma_S \sim P^{-k}$ often observed in experiments [20, 83, 155]. In case of the Freundlich isotherm we arrive to the same type of dependence of $\sigma_S \sim P^{-k}$ observed in the limit case described by expression (1.87).

It should be noted, however, that due to slow kinetics in change of $\sigma(t)$ observed during adsorption of numerous acceptors one cannot rule out a possibility of detecting of a "quasi-equilibrium" dependence of electric conductivity on pressure in experiment. In this case (which is more characteristic for high ohmic adsorbents with large t_{0C}) the use of expressions (1.77) – (1.81) yields

$$\hat{\xi}_{CS} \approx \xi_{C0} + \ln(1 + \lambda t \cdot N_t) , \qquad (1.90)$$

$$\hat{\sigma}_S \approx \sigma_0 (1 + \lambda t \cdot N_t)^{-m}, \quad \lambda = 2K_{nt}\sqrt{\xi_{C0}}\, e^{-\xi_{C0}} , \qquad (1.91)$$

Therefore, the activation energy of "quasi-equilibrium" conductivity changes as a logarithm of concentration of adsorption particles which, when the linear dependence between N_t and P is available, corresponds to situation observed in experiment [155]. We should note that due to small value m function (1.91) satisfactorily approximates the kinetics $\sigma(t) \sim A - B\ln(1 + t/t_*)$ observed in experiments [51, 167, 168]. Moreover, substantially high partial pressures of acceptor gas, i.e. at high concentrations of N_t expression (1.81) acquires the shape $\sigma(t) \approx \sigma_0(t/t_{0C})^{-m} \sim (t, N_t)^{-m}$ when $t > t_{0C} > t_{ex_{min}}$. This suggests that for large times characterized by slow kinetics in $\sigma(t)$ there is a power function of "quasi-equilibrium" electric conductivity plotted against concentration of adsorption particles and, consequently, against their pressure in gaseous phase when adsorption is described by a linear function. Considering samples with $h > L$ $m \approx 0.5$ we arrive at the root function $\hat{\sigma}_S \sim P^{-1/2}$. These "quasi-equilibrium" profiles of β plotted versus P described by expression $\beta_S \sim (1 + \text{const } tN_t)^{-\frac{m}{1+\nu}}$ can explain the decrease in VAC inclination angle for ZnO with increase in P_{O_2} observed in study [111].

BSS recharging of which may accompany adsorption influences the adsorption-caused transformation of the type of barrier height distribution function. In this case, similarly to the situation which was addressed in section 1.8 one can easily obtain

$$f_{aS}(\xi) = (\Delta\xi)^{-1}\left[\left(1 - \frac{N_t}{\sqrt{\xi}}\right) + \tilde{\mu}\frac{N_t}{\sqrt{\xi}}\left(\sqrt{\xi} - N_t\right)^2\right], \quad \tilde{\mu} \equiv \frac{kT}{E_g - \delta}. \quad (1.92)$$

Comparing expressions (1.92) and (1.74) obtained accounting and without accounting for BSS recharging shows that the latter slows down the decrease in value $f(\xi_S)$ occurring at complete occupation N_t and, therefore, hinders the rise in β and decrease of factor σ.

In case of plausible BSS recharging the equilibrium values of σ and β caused by acceptor adsorption can be described by the following expressions

$$\sigma_{SS} \approx \sigma_0\left[1 + \frac{N_t}{\sqrt{\xi}}\left(1 - \tilde{\mu}\xi_{C0}\right)\right]^{-\nu} \exp\left\{-\left[N_t^2\left(1 - \tilde{\mu}\xi_{C0}\right) + 2N_t\sqrt{\xi_{C0}}\left(1 - \tilde{\mu}\xi_{C0}\right)\right]\right\}, \quad (1.93)$$

$$\beta_{SS} \approx \beta_0\left[1 + \frac{N_t}{\sqrt{\xi_{C0}}}\left(1 - \tilde{\mu}\xi_{C0}\right)\right]^{\frac{\nu}{1+\nu}}. \quad (1.94)$$

Hence, BSS recharging substantially hinders the adsorption-induced change in degree of barrier disarrangement of adsorbent and in several cases can bring about small sensitivity of electrophysical characteristics of adsorbent to the adsorption process. We should note that above BSS recharging can be one of reasons of notable effect of adsorbent manufacturing prehistory of and the methods to treat its surface on the value and kinetics of chemisorption response.

1.10.2. Adsorption of donor particles

The kinetics of depletion of the adsorption surface states corresponding to chemisorbed donor particles characterized by ionization potential I_d is described by equation

$$\frac{dn_t}{dt} = k_n\left[\left(N_t(t) - n_t\right)e^{-\theta} - n_t e^{-\xi}\right] , \qquad (1.95)$$

where $\theta = (I_d - \chi - \delta)/kT$. Solving the Poisson equation with respective boundary conditions under assumption that $\left(\sqrt{\xi_{0_{min}}} - N_{ts}\right) > 1$ where N_{ts} is the equilibrium concentration of chemisorbed particles both in charged and neutral form leads us to relationship between the value $\xi(t)$ and $n_t(t)$. In our case this expression is the following

$$\xi(t) = \left[\sqrt{\xi_0} - n_t(t)\right]^2 . \qquad (1.96)$$

The solution of equation (1.95) far from the equilibrium given either by leveling-off in I_d and E_F, or with complete depletion of donor adsorption surface states makes is feasible to determine the time dependence of concentration of the charged form of adsorption surface states:

$$n_t(t) = t_d^{-1}\int_0^t N_t(\tau)\, e^{-\frac{t-\tau}{t_d}}\, d\tau , \qquad (1.97)$$

where $t_d = k_n^{-1}\exp\theta$; function $N_t(t)$ describing the kinetics of chemisorption of donor particles. Therefore, the kinetics of the change in the height of barrier caused by adsorption of donor particles is described by expression

$$\xi(t) \approx \left[\sqrt{\xi_0} - t_d^{-1}\int_0^t N_t(\tau)\, e^{-\frac{t-\tau}{t_d}}\, d\tau\right]^2 , \qquad (1.98)$$

valid for $t < \min\{t_*, t_\theta\}$, where t_* and t_θ stand for characteristic times of obtaining the equilibrium caused by either complete depletion of N_t or leveling-off in E_F and I_d, respectively. t_* and t_θ can be determined through equations

$$t_d^{-1} \int_0^t {}^*N_t(\tau) \exp\left\{-\frac{t_* - \tau}{t_d}\right\} d\tau = N_t(t_*), \tag{1.99}$$

$$t_d^{-1} \int_0^t {}^\theta N_t(\tau) \exp\left\{-\frac{t_\theta - \tau}{t_d}\right\} d\tau = \sqrt{\xi_0} - \sqrt{\theta}. \tag{1.100}$$

In contrast to the case of acceptors these expressions differ by the time of chemisorption kinetics $N_t(t)$.

Similar to our previous reasoning, we can assume that ξ_0 are distributed homogeneously over interval $\Delta\xi = \xi_{0_{max}} - \xi_{0_{min}}$. We obtain for the donor case the following expression for adsorption caused transformation of function $f(\xi)$

$$f_d(\xi, t) = (\Delta\xi)^{-1}\left(1 + \frac{n_t(t)}{\sqrt{\xi}}\right), \quad \xi < \theta, \tag{1.101}$$

valid for domain far from the equilibrium; n_t being determined by expression (1.97).

When ξ approaches the equilibrium value ξ_s caused by complete depletion in N_t the equation (1.95) acquires the shape

$$\frac{dn_t}{dt} = k_n\left[N_t(t)e^{-\theta} - n_t\left(e^{-\theta} + e^{-\left(\sqrt{\xi} - N_t(t)\right)^2}\right)\right], \tag{1.102}$$

whose solution is provided by

$$n_t(t) = t_d^{-1} \int_0^t N_t(\tau)e^{-\frac{t-\tau}{t_{ds}}} d\tau, \tag{1.103}$$

where

$$t_{ds} = t_d\left[1 + \exp\left\{\theta - \left(\sqrt{\xi_0} - N_t(t)\right)^2\right\}\right]^{-1}. \tag{1.104}$$

Recalling condition $\lim_{t\to\infty}\left(\sqrt{\xi_0} - N_t(t)\right)^2 > \theta$ which assumes that even complete depletion of donor levels of adsorption surface states for major part of barriers does not provide the leveling-off in F and I_d one can derive from (1.104) that $t_{ds} \approx t_d$ and that $n_t(t)$ is weakly dependent on ξ_0 and almost does not influence the shape of $f_d(\xi,t)$ at large t. Thus, for the donor case when $\xi_{0_{min}} < \theta < \xi_{0_{max}}$ the time dependence of the distribution function of the heights of barriers is

$$f_d(\xi,t) \approx \begin{cases} (\Delta\xi)^{-1}, & \xi_{0_{min}} \leq \xi < \theta, & (1.105) \\ (\Delta\xi)^{-1}\left[n_t^2(t) + 2\sqrt{\theta}n_t(t)\right], & \xi = \theta, & (1.106) \\ (\Delta\xi)^{-1}\left(1 + n_t(t)/\sqrt{\xi}\right), & \theta < \xi < \left(\sqrt{\xi_{0_{max}}} - n_t(t)\right)^2, & (1.107) \end{cases}$$

They are shown in Fig. 1.15.

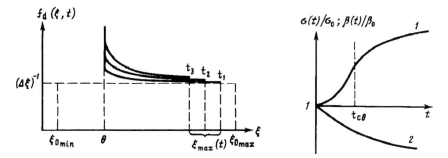

Fig. 1.15. The density of distribution of barriers with respect to heights at various moments of time for adsorption of donors ($t_1 < t_2 < t_3$)

Fig. 1.16. The qualitative type of dependencies $\sigma(t)$ (curve 1) and $\beta(t)$ (curve 2) in case of linear kinetics of chemisorption of donors.

Similar to the acceptor case, information regarding the peculiarities in change of functions $\xi(t)$ and $f_d(\xi,t)$ enables one to determine the kinetics of the change of values $\xi_C(t)$ and $f_d(\xi_C,t)$ and, thereby, using expressions (1.59) − (1.63) establish the functions $\sigma(t)$ and $\beta(t)$ which differ for various values of parameters in the absorbate-adsorbent system.

Thus, when $N_t < \sqrt{\xi_{C0}} - \sqrt{\theta}$ for any t the following expressions are valid

$$\sigma(t) \approx \sigma_0\left(1 + \frac{n_t(t)}{\sqrt{\xi_{C0}}}\right)^\nu \exp\left\{2\sqrt{\xi_{C0}}n_t(t)\right\}, \qquad (1.108)$$

$$\beta(t) \approx \beta_0 \left(1 + \frac{n_t(t)}{\sqrt{\xi_{C0}}}\right)^{-\frac{v}{1+v}}. \tag{1.109}$$

In case when the concentration of adsorption particles increases $\left(N_t > \sqrt{\xi_{C0}} - \sqrt{\theta}\right)$ formulas (1.108) and (1.109) become valid only for times $t < t_{C\theta}$, where $t_{C\theta}$ is the characteristic time determined from equation $\sqrt{\xi_{C0}} - n_t(t_{C\theta}) = \sqrt{\theta}$, which is the condition of leveling-off of critical barriers ξ_C with θ.

When $t > t_{C\theta}$ we have

$$\sigma(t) \approx \sigma_0 \left[\left(\sqrt{\theta} + n_t(t)\right)^2 - \xi_{C0}\right]^v \exp\{\xi_{C0} - \theta\}, \tag{1.110}$$

$$\beta(t) \approx \beta_0 \left(1 + \frac{n_t(t)}{\sqrt{\theta}}\right)^{-1/2} \tag{1.111}$$

when $|E| > |E_*(t)|$, where $|E_*(t)|$ is a certain function of time and parameters of the system. When the strength of field E is smaller than $E_*(t)$ VAC inclination angle also decrease with time, although VAC themselves in this case get linearized in coordinates $\ln(J / J_0) - |E|$. We should note that functions (1.110) and (1.111) are valid only when inequality $n_t(t) < \sqrt{\xi_{C0} + \theta} - \sqrt{\theta}$ is fulfilled, which provides the condition of absence of infinite claster dealing with leveling-off of barriers, i.e. the condition of applicability of percolation transition. The qualitative shape of dependencies $\sigma(t)$ and $\beta(t)$ is provided by Fig. 1.16.

For equilibrium values of σ_s and β_s in case when $N_{ts} < \sqrt{\xi_{C0}} - \sqrt{\theta}$ the kinetic formulas (1.108) and (1.109) yield

$$\sigma_s \approx \sigma_0 \left(1 + \frac{N_{ts}}{\sqrt{\xi_{C0}}}\right)^v \exp\{2N_{ts}\sqrt{\xi_{C0}}\}, \tag{1.112}$$

$$\beta_s \approx \beta_0 \left(1 + \frac{N_{ts}}{\sqrt{\xi_{C0}}}\right)^{-\frac{v}{1+v}}. \tag{1.113}$$

When $N_{ts} > \sqrt{\xi_{C0}} - \sqrt{\theta}$ relations (1.110) and (1.111) provide that

$$\sigma_s \approx \sigma_0 N_{ts}^{2v} \exp\{\xi_{C0} - \theta\} ,$$ (1.114)

$$\beta_s \approx \beta_0 \left(1 + \frac{N_{ts}}{\sqrt{\theta}}\right)^{-\frac{v}{1+v}}$$ (1.115)

At different types of adsorption isotherms plotted for adsorption of donor particles on oxides (see section 1.5) expressions (1.112) − (1.115) provide the rise in σ_s and decrease in β_s with the growth of partial pressure of gas P, the functions themselves being different. Thus, in case of applicability of the Henry isotherm at small P we have the function $\sigma_s \sim$ $\sim \exp\{const\cdot P\}$ becoming a power function $\sigma_s \sim P^k$ with the rise in P which is often observed in experiments [154, 155, 169].

Accounting for BSS recharging having high importance for adsorption of the donor particles proper [84, 98], which can be done similarly to acceptor case yields

$$f_{dS}(\xi_C) \approx \left[1 + \frac{N_{ts}}{\sqrt{\xi_C}} - 6\tilde{\mu}N_{ts}\sqrt{\xi_C}\right] ,$$ (1.116)

where, as above, $\tilde{\mu} = kT / (E_g - \delta)$. From this expression we can obtain the equilibrium values for ohmic electric conductivity and the tangent of VAC inclination angle in case of small concentration of adsorption particles $N_{ts} < \sqrt{\xi_{C0}} - \sqrt{\theta}$

$$\sigma_{sS} \approx \sigma_0 \left[1 + \frac{N_{ts}}{\sqrt{\xi_{C0}}} - 6\tilde{\mu}N_{ts}\sqrt{\xi_{C0}}\right]^v \exp\left\{2N_{ts}\sqrt{\xi_{C0}} \frac{1 - \tilde{\mu}\xi_{C0}}{1 + \tilde{\mu}\xi_{C0}}\right\} ,$$ (1.117)

$$\beta_{sS} \approx \beta_0 \left[1 + \frac{N_{ts}}{\sqrt{\xi_{C0}}} - 6\tilde{\mu}N_{ts}\sqrt{\xi_{C0}}\right]^{-\frac{v}{1+v}} .$$ (1.118)

Comparing expressions (1.117) and (1.118) with respective expressions (1.112) and (1.113) obtained without accounting for BSS recharging indicates that apart from decrease in the value of adsorption response of σ and β BSS recharging is responsible for the fact that adsorption of donors developing at substantially high-ohmic samples may result in increase in β up to values higher than the initial one β_0. In case of low-ohmic samples the adsorption of donors brings about the decrease in β and the input of BSS recharging is small. This conclusion is in agreement with theoretical findings of paper [98] which (as it was mentioned in section 1.9) established the notable effects of above recharging in the case of donor adsorption. In compliance with the comments made in section 1.9 in several cases this recharging might result in formation of extrema on kinetic curves of surface charging during donor adsorption or weak manifestation of recharging effects of during adsorption of acceptors. This was often observed in experiments.

1.10.3. Adsorption response of electrophysical characteristics of semiconductor sensors made of barrier-disordered polycrystalline oxides

Above theoretical analysis of adsorption effects on electric conductivity and VAC profiles in polycrystalline semiconductor adsorbent with accounting for its barrier disorder indicate that the value and kinetics of change in $\sigma(t)$ and $\beta(t)$ during adsorption of both acceptors and donors sharply differ from those predicted by theory both for the case of ideal monocrystal and for polycrystal considered from the standpoint of bi-crystal model.

Let us dwell briefly on experimental situation encountered during investigation of adsorption response of electrophysical parameters in polycrystalline semiconductor adsorbents having high inter-crystalline barriers.

The percolation model of adsorption response outlined in this section is based on assumption of existence of a broad spread between heights of inter-crystalline energy barriers in polycrystals. This assumption is valid for numerous polycrystalline semiconductors [145, 146] and for oxides of various metals in particular. The latter are characterized by practically stoichiometric content of surface-adjacent layers. It will be shown in the next chapter that these are these oxides that are characterized by chemisorption-caused response in their electrophysical parameters mainly generated by adsorption charging of adsorbent surface [32, 52, 155]. The availability of broad spread in heights of inter-crystalline barriers in above polycrystallites was experimentally proved by various techniques. These are direct measurements of the drop of potentials on probe contacts during mapping microcrystal pattern [145] and the studies of the value of exponential factor of ohmic electric conductivity of the material which was L/l times lower than the expected one in case of identical barriers [146], the studies of anomalously early manifestation of non-ohmicity [146, 163] and, finally, direct experimental determination of the function of distribution of heights of inter-crystalline barriers [170].

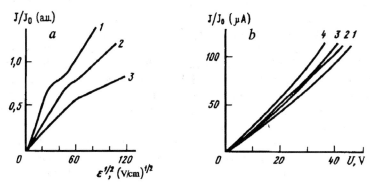

Fig. 1.17. Pre-relaxation VAC of ZnO slabs at different pressures of O_2 (Torr) (*a*) and steady-state VAC of SnO_2 films doped by 15 wt.% Nb at different concentration of propane in air (vol.%) (*b*). *a*: 1 – $5 \cdot 10^{-2}$; 2 – $5 \cdot 10^{-1}$; 3 – 50 [111]; *b*: 1 – 0; 2 – 0.1; 3 – 0.4; 4 – 0.6 [58].

The adsorption-induced charging of the barriers of such adsorbents results in the change in heights of inter-crystalline barriers and transforms the profile of their distribution function. As the model suggests, it is this change in the distribution function of the heights of barriers, that is responsible for adsorption-induced change of such an important characteristic of polycrystal as the differential coefficients of its volt-ampere characteristics or, which is more convenient for our studies, VAC inclination angle in coordinates $\ln(J / J_0) - |E|^{1/2}$. Note that from the standpoint of the theory of bicrystal one should have expected the change in position of the curve of VAC as a whole into the domain of smaller or larger values of current while retaining their shape during adsorption. In particular, the inclination angle should have been the same. Nevertheless, the experiments indicate that adsorption of various gaseous particles brings about the change in ohmic electric conductivity of adsorbent resulting in transformation of the shape of VAC, as well. Figure 1.17 shows the effect of transformation of VAC caused by adsorption of oxygen on polycrystalline zinc oxide (*a*) and adsorption of propane on polycrystalline tin dioxide doped by transitional metals (*b*) [58, 111]. We should note that in the first case the experiments dealt with monitoring the pre-relaxation VAC, the second case dealt with adsorption-caused change in the shape of stationary VAC, i.e. the effect of adsorption on equilibrium value of the current corresponding to the potential difference applied to the sample and monitored after a completion of the relaxation process.

74

In our opinion the method of "pre-relaxation VAC" is the most informative from the standpoint of the studies of effects of adsorption on the degree of barrier disordering in polycrystalline adsorbent. This method provides an information on adsorption-caused change in current percolation through inter-crystalline barriers. In case when the effect of adsorption on the shape of stationary (after-relaxation) VAC is in question, one cannot expect to obtain unambiguous information regarding the development of adsorption process. This deals with the masking effect of relaxation processes caused by disarrangement of equilibrium condition between the free charge carriers in the conductivity band and those localized on energy levels of SS of both adsorption and biographic origin when the external potential difference is applied. Therefore, the stationary values of current corresponding to the value of the voltage applied are dependent not only on characteristics of adsorption SS but also on such characteristics of biographic SS as density of their states in forbidden band, the value of the capturing cross-section of free charge carriers, etc. [171, 172]. Moreover in this case it is absolutely impossible to study kinetics of adsorption change in the inclination angle of VAC which provides information on kinetics of charging of the surface. Apart from that the adsorption caused change in the density of states in forbidden band can also lead to the change of the shape of steady-state VAC and a separate inter-crystalline barrier [171].

Summing up the above said we examined in our experiments the adsorption-caused change in ohmic electric conductivity and the inclination angle of pre-relaxation VAC of polycrystalline oxide adsorbents.

We used polycrystalline films of ZnO and SnO$_2$ as adsorbents. The films were deposited from the water suspension of respective oxides on quartz substrates. These substrates contained initially sintered contacts made of platinum paste. The gap between contacts was of about 10^{-2} cm. All samples were initially heated in air during one hour at T ~ 500°C. We used purified molecular oxygen an acceptor particle gas. H and Zn atoms as well as molecules of CO were used as donor particles. We monitored both the kinetics of the change of ohmic electric conductivity and the tangent of inclination angle of pre-relaxation VAC caused by adsorption of above gases and the dependence of stationary values of characteristics in question as functions of concentrations of active particles.

For instance, in case of acceptors the experiments dealt with monitoring the kinetic changes in σ (the voltage applied was ~ 1 V) and β during letting inside the volume surrounding adsorbent of various amounts of molecular oxygen. (Initially this volume was kept at pressures down to 10^{-7} Torr). The method of monitoring the function $\beta(t)$ involved multiple recording of pre-relaxation VAC of adsorbent, accomplished in certain moments of time during the process of observed adsorption-caused change in $\sigma(t)$. VAC profiles were obtained as experimental points (when $\sigma(t)$ got saturated), i. e. when the step-like voltage of various amplitude was applied [146, 163] and while applying triangle-shaped pulses [173] with amplitude 20 V, the span duration being ~ 7 s providing the recording of the dependence of current on the voltage applied over the field strength range $0 - 2 \cdot 10^3$ V/cm. Due to weak manifestation of relaxation processes observed at small pressures of O_2 [111] the chosen span time enabled us to remain in the domain of pre-relaxation VAC. The experiments were conducted both for one adsorbent for various pressures of O_2 and for different adsorbents changing in pre-adsorption values of electric conductivity. In case when one adsorbent was used introduction of various portions of O_2 was preceded by evacuation of the experimental cell down to pressures of residual gases of 10^{-7} Torr. The heating of the whole cell did not increase 150°C in order to avoid the process of reduction of oxide adsorbent.

Figure 1.18 shows profiles of functions $\sigma(t)$ and $\beta(t)$ observed during inputting of various amounts of oxygen into the volumes surrounding the same adsorbent characterized by initial resistance ~ 10^8 Ohm. In case of small pressures of O_2 (which were small for the adsorbent chosen) we observed anomalous kinetics (see curves *1, 2*) over long times which differed from those observed at large pressures (see curves *3, 4*). The shape of curves *1* and *2* indicates that at specific, usually notably large times, the decrease in the VAC inclination angle gets replaced by notably abrupt rise with simultaneous decrease in ohmic electric conductivity of adsorbent. Note that time over which this "switching" is observed increases with the rise of initial resistance in the sample. Above anomalous changes in σ and β finally led to attaining the plateau value, the stationary values of β for all adsorbents for which the "switching" in VAC exceeded their pre-adsorption values.

Therefore, the kinetic studies enable us to conclude that functions $\sigma(t)$ and $\beta(t)$ have different shapes when concentrations of acceptor particles in the volume surrounding adsorbent differ. For small pressures of O_2 (with respect to a given adsorbent) over large times the "switching" in VAC accompanied by notable change in electric conductivity is distinctly visible in compliance with predictions of percolation model. This "switching" does not occur at large (note again that we refer to a given adsorbent) pressures of oxygen. We should mention that in agreement with the model the initial rate of adsorption-caused change in σ observed in experiments exceeds the similar characteristics in β.

Having acknowledged a qualitative agreement in experimentally determined change in σ and β during adsorption of acceptor particles with theoretical predictions we should comment the correct choice of parameter t_{0C} in expressions (1.81) and (1.83) makes it feasible to obtain a satisfactory quantitative description of kinetics of $\sigma(t)$ and $\beta(t)$ (Fig. 1.19). Plausible discrepancies can be related to possible dependence of the power index m on the depth of the adsorption layer which might be the case for depths of the order of the correlation length (in our case this is tens of microns) used in experiments. On the other hand, this may be caused by diffusion hindrance occasionally resulting in violation of the condition incorporated in the model regarding the free gas penetration through the adsorbent layer. We should point out that rigorous accounting for two above conditions may led to a time dependence in the power index in expressions (1.81) and (1.83).

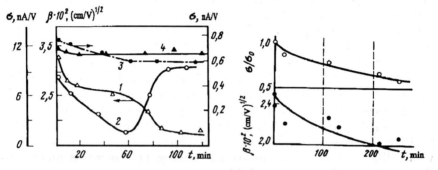

Fig. 1.18. The kinetics of the change in $\sigma(t)$ (curves *1* and *3*) and $\beta(t)$ (curves *2* and *4*) during adsorption of O_2 on ZnO ($1,2 - P_{O_2} = 5 \cdot 10^{-3}$ Torr; $3,4 - P_{O_2} = 7 \cdot 10^{-2}$ Torr; $R_0 = 10^8$ Ohm [159])

Fig. 1.19. The comparison of theoretical predictions (1.81) and (1.83) describing kinetics of $\sigma(t)$ and $\beta(t)$ in case of large pressures of gas of acceptors with experimental results ($P_{O_2} = 300$ Torr; $\sigma_0 = 0.16$ nA/V; $t_{0C} \approx 120$ min)

As for the relation between the stationary values of σ_s and β_s and the oxygen pressure the experiments proved the availability of the power dependence $\sigma_s \sim P_{O_2}^{-m}$ with $m \approx 0.4 \div 0.5$ (Fig. 1.20) as well as complex dependence of β_s on P_{O_2}, which for majority of samples was bell-shaped (Fig. 1.21, curve *1*). This agrees with predictions of the percolation model. For several samples with low value of initial resistance the dependence of β_s on P_{O_2} was characterized by monotonous increase in β_s with the rise in P_{O_2} followed by reaching the saturation domain (Fig. 1.22, curve *2*). This, in our opinion, may be explained by creation of a large number of leveled-off barriers and, therefore, violation of condition of applicability of percolation transition. In this case, as it has been mentioned above, the value of the inclination angle of pre-relaxation VAC should remain insensitive to further change in P_{O_2}, which was confirmed by experiment.

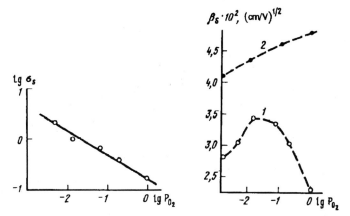

Fig. 1.20. The dependence of equilibrium electric conductivity of ZnO sample on oxygen pressure

Fig. 1.21. The dependence of equilibrium value of the tangent of inclination angle of pre-relaxation VAC on oxygen pressure ($1 - \sigma_0 = 11.8$ nA/V; $2 - \sigma_0 = 2.25$ μA/V)

During the studies of adsorption-caused response in σ and β by donor particles we monitored both kinetics of the change in above characteristics and dependence of their stationary values on concentration of adsorption particles. The partial pressure in gaseous phase (and in case of atom particles the time of treatment of the surface of adsorbent by the flux of above particles) used to be the measure of the quantity of absorbed particles during adsorption of molecular particles.

In our experiments the adsorption of CO was developed from atmosphere of analytically pure nitrogen and dry air kept at atmospheric pressure and $T \sim 300°C$. The volume concentration of CO varied over the range 10 ppm – 1.5 vol-%.

The titan cup was the source of zinc atoms. It contained the metal initially melted in hydrogen atmosphere and distilled. The operational temperature of the cup was maintained at the level of 185°C which provided the high percentage of the content of single zinc atoms in the flux.

The hydrogen was introduced into the system through a nickel membrane. The atoms were obtained by platinum pyrolysis filament having a required temperature to provide necessary concentration of H-atoms. The values of hydrogen pressures for a given cell geometry were chosen accounting for the Smith conditions [174] so that concentration of H-atoms decreased exponentially with P_{H_2} - insensitive decrement while increasing the distance from the source.

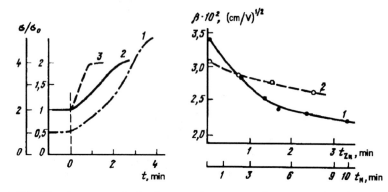

Fig. 1.22. The kinetics of change in electric conductivity of polycrystalline oxide during adsorption of donor particles: 1 – CO on SnO_2, 0.11 vol.-% CO in atmosphere, $T = 300°C$; 2 – H atoms on ZnO, $P_{H_2} = 5·10^{-2}$ Torr, $T_{Pt} = 1400°C$, $T_{ZnO} = 20°C$; 3 – Zn atoms on ZnO, $T_{Ti} = 185°C$, $T_{ZnO} = 20°C$

Fig. 1.23. The kinetics of the change of tangent of inclination angle of pre-relaxation VAC during adsorption of donors: 1 – H atoms on ZnO; 2 – Zn atoms on ZnO

Typical curves for $\sigma(t)$ kinetics caused by adsorption of H and Zn atoms are shown in (Fig. 1.22). The kinetics of change of $\sigma(t)$ in case of CO adsorption has the similar shape, however, in contrast to two above cases the signal gets reversible during removal of CO. The availability of quasi-exponential sections in kinetic curves $\sigma(t)$ for all three types of adsorbents provides yet another indirect prove of existence of barrier mechanism of adsorption-caused change in electric conductivity of poly-crystalline films of ZnO and SnO_2. The adsorption-related change in $\beta(t)$ (Fig. 1.23) suggests the validity of the percolation model. Comparing the obtained kinetic results with results of the percolation model (see Fig. 1.16) proves their qualitative agreement and possibility to model the process of adsorption-caused change in σ and β provided the parameters of the system are properly selected.

As for stationary values of the signal, as the experiment indicates the dependence of σ_s on concentration of CO for different adsorbents based on SnO_2 has a linear shape with increase in inclination with growing temperature over the temperature interval $100 – 400°C$, concentration of CO being lower than 150 pmm. In high concentration domain

(> 150 pmm) one observes the power law in the change of $\sigma - \sigma_* =$ $= \{[CO] - [CO]_*\}^m$ with m $\approx$ 0.68 ÷ 0.77 which qualitatively agrees with predictions of the percolation model. Here [CO] is the value of the volume content of CO expressed in percent; σ_* and $[CO]_*$ are constants depending on characteristics of a specific adsorbent.

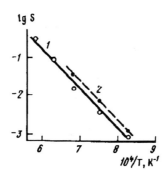

Fig. 1.24. The dependence of the value of signal as a functions of the temperature of filament (source of H atoms): 1 – P_{H_2} = 0.035 Torr; 2 – P_{H_2} = 0.05 Torr

In conclusion we should mention that experimental points obtained while measuring the value of signal $S = [(d\sigma / dt) \sigma^{-1}]_{t\to 0}$ as a function of temperature of the filament (the source of H atoms), fit satisfactorily into the linear functions plotted in coordinates $\ln S - T^{-1}$ for different pressures of molecular hydrogen (Fig. 1.24). The angle of above straight lines determining the value of the dissociation energy of molecular hydrogen yields the value of 104 ± 3 kcal/mole. Its closeness to the tabulated value (102.9 kcal/mole) is indicative of a linear response of detector based on polycrystalline zinc oxide to effects of hydrogen atoms in domain of their relative concentrations differing by more than three orders of magnitude.

Thus, as it can be concluded from above, we obtained a satisfactory agreement between results predicted by the percolation model of adsorption response of electrophysical parameters of polycrystalline semiconductor adsorbent with experimental data. This clarified several cases which have not been recently satisfactorily explained.

The method proposed to deal with polycrystalline adsorbent is highly promising while considering polycrystalline composites, the latter being mixtures of various oxides used as adsorbents who made chemical sensors [175, 177]. It was established in above papers that mixtures of various oxides exhibit the highest sensitivity to specific gases. Moreover, in several cases the adsorption-caused response in electrophysical parameters of such composites exhibit anomalous behavior [58, 176, 177],

80

inherent only to specific content of composite (Fig. 1.25). The availability of such anomalies is very interesting not only from the fundamental standpoint but from the application aspect as well. These anomalies make it feasible to manufacture detectors for specific concentration ranges of a given gas. The theoretical analysis accomplished in [161, 162, 178] indicate that the transformation of the type of the function of distribution of heights of inter-crystalline barriers accompanying the change of its content plays a substantial role in the change of electric conductivity of composite. Therefore, it was first shown in these papers that both adsorption and the change of the content of composite influence its electric conductivity through the change of the degree of barrier disorder. The availability of such correlation made it possible to formulate the principle of choosing the content of composite exhibiting the highest sensitivity to adsorption. Namely, the above property is inherent to composites with contents determined through examining the sections of electric conductivity-content profiles characterized by the highest rate of change in electric conductivity during changing the composite content.

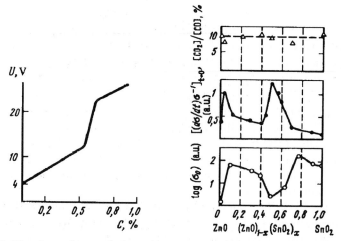

Fig. 1.25. The dependence of the voltage response of detection U on concentration of propane C for SnO_2 sample doped with Nb (9 weight-%) [58]

Fig. 1.26. The dependence of initial electric conductivity (σ_0), sensitivity to adsorption of CO ($S = [(d\sigma / dt)\sigma^{-1}]_{t\to0}$) and catalytic activity ($[CO_2]/[CO]$) in oxidation reaction of CO to CO_2 for adsorbent made of the mixture ZnO + SnO_2 as a function of the content

Figure 1.26 provides an experimental proof to above through profiles of initial conductivity σ_0, sensitivity to CO $S = [(d\sigma / dt) \sigma^{-1}]_{t \to 0}$ and catalytic activity in oxidation reaction of CO to CO_2 plotted as function of the content of adsorbent (the mixture of polycrystalline ZnO and SnO_2). It was established long ago that adsorbents made of the mixture of above oxides possess high sensitivity to ethanol in air over specific content ranges [175].

The analysis of results shown in Fig. 1.26 makes it possible to conclude the availability of above correlation between the rate of change in electric conductivity of adsorbent during changing of its content and the value of sensitivity to adsorption. We should mention that small changes in catalytic activity observed during the change of the content of composite do not correlate with the change in its sensitivity. Moreover, conducted X-ray analysis of the composites used showed that within the accuracy of measurements there is no new phase which might be responsible for the change in electric conductivity caused by adsorption of CO. Presumably, all composites used were mechanical mixtures of initial oxides substantially differing in the type of distribution function of the heights of inter-crystalline barriers. It is this feature that enables one, from our standpoint, to consistently explain the anomalous dependencies of sensitivity of detector on the content of composite adsorbent observed in experiment.

Thus, all above experimental examples indicate that additionally to effect of disordering of the surface of adsorbent on its adsorption characteristics it is necessary to account for the effect of the adsorption itself on degree of adsorbent disordering controlling both its own electrophysical characteristics and the adsorption-caused change while deriving the theory of adsorption response of polycrystalline semiconductor adsorbent.

1.11. The effect of adsorption on concentration of the surface and volume interstitial defects in semiconductor adsorbent and resultant change in electric conductivity

In above sections the main attention has been paid to adsorption-caused change in electrophysical characteristics of semiconductor adsorbent caused by surface charging effects. However, as it was mentioned in section 1.6, the change in electrophysical characteristics of such adsorbents can be caused by other mechanisms, e.g. by direct interaction of absorbate with the surface defects provided (as in the case of oxide adsorbents) by superstoichiometric atoms of metals and oxygen vacancies. Above processes provide the main input into the change in electrophysical characteristics of adsorbent. According to experimental data their role is dependent on conditions of adsorbent manufacturing and prehistory. As it will be shown in Chapter 2 it is assumed today that effects of charging of the surface are principal in case of adsorbents characterized by almost stoichiometric content of their surface-adjacent layers, whereas for partially reduced oxides the major input into the change of their electrophysical characteristics is provided by direct interaction of absorbate with the surface defects which almost does not lead to formation of the surface band bending [32]. In this section we consider the case of effect of gaseous phase on electrophysical characteristics of adsorbent induced by adsorption-caused change in equilibrium concentration of interstitial defects positioned both in the volume and on the surface of semiconductor.

There are several types of various equilibria available in the gas-solid body system if a pure substance, for instance metal oxide MeO is in equilibrium state with surrounding medium. They are: equilibrium both between atoms of metals and between atoms of oxygen in various states, namely in outer phase, in lattice nodes, in interstitial points and in substitution nodes of the lattice; equilibrium between atoms of Me and O as well as equilibrium between Me and O having different charge forms. The existence of the latter is provided by processes of ionization and recombination. For each process leading to the change in concentration of any component one can write down an equilibrium equation using the law of acting masses. The balance condition would determine the equilibrium content of the system for a given condition. For each specific system the main role in attaining the equilibrium content is only provided by specific processes and defects. Examples of most widely spread defects in oxides of various metals are provided by vacancies of oxygen (V_O^0, V_O^+, V_O^{++}), metal vacancies (V_{Me}^0, V_{Me}^0, V_{Me}^0), interstitial metal atoms ($Me_{\bullet}^0$, $Me_{\bullet}^+$, $Me_{\bullet}^{++}$) and, finally, interstitial oxygen atoms ($O_{\bullet}^0$, $O_{\bullet}^-$, $O_{\bullet}^{--}$). Here ^{0}, $^+$ and $^-$ denote neutral and single and double charged form of a defect.

In case of equilibrium the balance between various atoms, defects and various charged forms is maintained due to reactions [73, 69]

$$MeO(v) \underset{\leftarrow}{\rightarrow} Me(g) + \frac{1}{2}O_2(g),$$

$$MeO(v) \underset{\leftarrow}{\rightarrow} Me(g) + V_{Me}^0,$$

$$O(v) \underset{\leftarrow}{\rightarrow} O_2(g) + V_O^0,$$

$$V_{Me}^0 \underset{\leftarrow}{\rightarrow} V_{Me}^- + p,$$

$$V_{Me}^{-} \rightleftarrows V_{Me}^{--} + p,$$

$$V_{O}^{0} \rightleftarrows V_{O}^{+} + e,$$

$$V_{O}^{+} \rightleftarrows V_{O}^{++} + e,$$

where v and g designate that a respective atom belongs to the lattice of a solid state or to gaseous phase; p and e stand for free hole and electron, respectively. Owing to the law of acting masses one can write down the equilibrium equation for each process:

$$K_{MeO} = P_{Me} \cdot P_{O_2}^{0.5}, \tag{1.119}$$

$$K_{V_{Me}^{0}} = P_{Me} \cdot [V_{Me}^{0}], \tag{1.120}$$

$$K_{V_{O}^{0}} = P_{O_2}^{0.5} \cdot [V_{O}^{0}], \tag{1.121}$$

$$K_{V_{Me}^{-}} = [V_{Me}^{-}] \cdot [p] / [V_{Me}^{0}], \tag{1.122}$$

$$K_{V_{Me}^{--}} = [V_{Me}^{--}] \cdot [p] / [V_{Me}^{-}], \tag{1.123}$$

$$K_{V_{O}^{+}} = [V_{O}^{+}] \cdot [e] / [V_{O}^{0}], \tag{1.124}$$

$$K_{V_{O}^{++}} = [V_{O}^{++}] \cdot [e] / [V_{O}^{+}], \tag{1.125}$$

where P_{Me} and P_{O_2} are respective pressures and angular brackets designate, as usual, respective concentrations. Apart from equilibrium conditions (1.119) − (1.125) there is a condition of electric neutrality which always holds in equilibrium case. In our situation it can be written as

$$[e]+[V_{Me}^{-}]+2[V_{Me}^{--}]+[N_A^{-}]+2[N_A^{--}]=[p]+[V_{O}^{+}]+2[V_{O}^{++}]+[N_D^{+}]+2[N_D^{++}], \tag{1.126}$$

where $[N_A^{-}]$, $[N_A^{--}]$, $[N_D^{+}]$ and $[N_D^{++}]$ are the concentrations of single and twice ionized dope acceptors and donors.

One can readily conclude from expression (1.119) − (1.126) that conditions in gaseous phase affect the values of equilibrium concentrations of all point defects and, therefore, the concentration of free charge carriers.

If the interstitial metal atom is the major donor of electrons one should include to consideration the following reactions

$$Me(g) \overset{\leftarrow}{\rightarrow} Me_\bullet^0 \,,$$

$$Me_\bullet^0 \overset{\leftarrow}{\rightarrow} Me_\bullet^+ + e \,,$$

whose conditions of equilibrium are described by expression

$$[Me_\bullet^0] = K_{Me_\bullet^0} / P_{Me} \,, \tag{1.127}$$

$$[Me_\bullet^+] = K_{Me_\bullet^+} [Me_\bullet^0] / [e] \,, \tag{1.128}$$

Thus, for a broad-band doped semiconductor of n-type which stays in equilibrium with outside medium containing oxygen with partial pressure P_{O_2} the neutrality equation acquires the shape

$$[e] = [V_O^+] + 2 \, [V_O^{++}] + [Me] \,. \tag{1.129}$$

In case when the major input into the change of concentration of free carriers is provided by double-ionized oxygen vacancies which is valid at high temperatures the concentration of conductivity electrons is $[e] \approx$ $\approx 2 \, [V_O^{++}]$, which, recalling expressions (1.121), (1.124) and (1.125) brings us to formula

$$[e] = (2K_{V_O^{++}} \, K_{V_O^+} \, K_{V_O^0})^{1/3} \, P_{O_2}^{-1/6} \,, \tag{1.130}$$

and further to the well-known dependence of equilibrium electric conductivity of adsorbent on oxygen pressure:

$$\sigma \sim P_{O_2}^{-\tilde{m}} \,, \tag{1.131}$$

where $\tilde{m} = 1/6$. We should note that for ZnO at temperatures higher than 1100 K one has $\tilde{m} \approx 1/5.5$ [170] and for SnO_2 at $T > 1400$ K $\tilde{m} \approx 1/6.5$ [180]. When concentration of free charge carriers is provided by $[V_O^+]$ or by concentration of interstitial metal atoms the dependence of concentration of conductivity electrons on P_{O_2} can be written as $\sigma \sim [e] \sim P_{O_2}^{-1/4}$ [181]. When the dominant defects are provided by interstitial oxygen atoms it can be easily shown that an adsorbent featuring p-type conductivity exhibits the following dependence of electric conductivity on oxygen pressure

$$\sigma \sim P_{O_2}^{1/6} \;, \qquad\qquad\qquad\qquad (1.132)$$

which is characteristic for such oxides as SrO, ThO$_2$, ZrO, BaO [180]. Note that issues dealing with creation of volume point defects, their stabilization during cooling of material and their influence on concentration and mobility of free charge carriers in monocrystalline semiconductors have been scrutinized in review [32].

At temperatures of the order 700 – 900 K the surface point defects play the dominant role in controlling the various electrophysical parameters of adsorbent on the content of ambient medium [32]. As it has been mentioned in section 1.6, these defects are being formed in the temperature domain in which the respective concentration of volume defects is very small. In fact, cooling an adsorbent down to room temperature results violation of uniform distribution due to redistribution of defects. The availability of non-homogeneous defect distribution led to creation of a new model of depleted surface layer based on the phenomenon of oxidation of surface defects [182] which is an alternative to existing model of the Shottky barrier [183].

The basis of this model is provided by assumption that decrease in the diffusion coefficient of defects accompanying cooling of the sample from the healing temperature establishes a non-homogeneous distribution of frozen defects with decrease in concentration towards the surface. This results in decrease in chemical potential which, naturally, causes the band bending in surface-adjacent domain providing the constant value of the Fermi level over the whole volume of adsorbent. Einzinger [182] made a conclusion that such a homotransition corresponds to a higher degree to the surface-adjacent depletion domain which is available, for instance, in ZnO in contrast to SCR caused by the Shottky barrier. However, in study [69] the solution of diffusion equation with time-dependent value of the diffusion coefficient was used to calculate the Aizinger model. It was established that indeed such effect exists, however, its value is small. Thus, for ZnO the value of the band bending caused by this effect is only one tenth of the experimentally observed height of the surface-adjacent barrier. This led the authors to conclusion that the major reason of the surface-adjacent band bending is provided by the surface charging of adsorbent resulting in formation of depletion described by the Shottky model. However, the last conclusion of study [69] dealing with the dominant role of the Shottky band bending profiles questions the result indicating the presence of depletion of the surface domain in defects. Indeed, the presence of an a priori band bending caused by existence of charged surface substantially influences the distribution of ionized defects resulting, as it has been shown in [4, 116], in saturation of surface-adjacent layer by positively charged defects of Me_i^+ and V_O^+ type and depletion of negatively charged O_i^- and V_{Me}^- defects. This, in its turn, changes the value of band bending. Assessment of the joint effect of both an a priori band bending and the process of surface oxidation of defects on concentration of the surface defects was given in [70]. It was established that at high temperatures, when oxygen vacancies are in equilibrium with the gaseous phase surrounding adsorbent their surface-adjacent concentration may be higher by three orders of magnitude if compared to homogeneous distribution.

The high concentration of surface defects in adsorbent results in substantially higher change in the value of electrophysical characteristics of adsorbent when it is introduced into the reaction medium if compared to the case of ideal surface. As it has been mentioned in [70] experiments performed at room temperature involving ZnO healed at 810°C show the dependence of $\sigma \sim P_{O_2}^{-m}$, where $m \approx 3$.

Therefore, the detailed analysis of concentration of defects in surface-adjacent layer and in the volume of adsorbent as well as assessment of the values of diffusion coefficients of defects and particles of various gases in material of adsorbent are very important for understanding the processes of both reversible and irreversible change in electrophysical characteristics of semiconductor during low temperature (if compared to the temperature of creation of defects) interaction with gaseous phase.

As an another example let us consider results of study [184], dealing with the change of electric conductivity and work functions of monocrystal adsorbent TiO_2 (110) with defects caused by its interaction with oxygen (Fig. 1.27). For ~ 300 K the changes in σ and $q\varphi$ are mainly linked to effects of chemisorption of O_2 on the value of the surface charge. At higher temperatures the effects of chemisorption-caused change in the value of the surface charge are negligible if compared to those caused by penetration of oxygen into the surface-adjacent domain resulting in removal of the inherent electrically active defects (oxygen vacancies). This causes substantial decrease in σ at practically absent change in the value of band bending which is proven by constant value in work function. In this case the approach to determine $\Delta\sigma$ based on assessment of the change in concentration of free charge carriers caused by band bending in the surface area becomes inapplicable. We should note that formation of point defects was provided by heating and fast cooling of samples from $T \approx 1200$ K at $P_{O_2} \sim 10^{-9}$ mBar. It was established that heating up to ~ 350 K is sufficient to activate energy barriers by diffusing oxygen resulting in irreversible (at room temperatures) changes in electric conductivity of adsorbent.

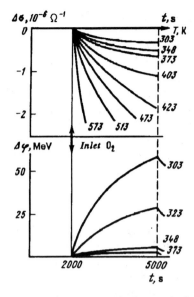

Fig. 1.27. The temperature dependence of observed changes in surface-adjacent area of conductivity $\Delta\sigma$ and work functions $\Delta\varphi$ for TiO_2 (110) with defects caused by interaction with molecular oxygen [184]

1.12. The effect of the nature of adsorbent on adsorption-caused change in its electrophysical characteristics. The nature of adsorption centers and their effect on the process of charging of the surface

All major characteristics of chemisorption response of electrophysical parameters of semiconductor adsorbents such as sensitivity, selectivity, inertia, reversibility are naturally dependent both on the nature of adsorbent and on chemical activity of absorbate with respect to adsorbent chosen.

The comparison of experimental data on adsorption of various particles on different adsorbents indicate that absorbate reaction capacity plays a substantial role in effects of influence of adsorption on electric conductivity of oxide semiconductors. For instance, the activation energy of adsorption of molecular oxygen on ZnO is about 8 kcal/mole [83] and molecular hydrogen – 30 kcal/mole [185]. Due to such high activation energy of adsorption of molecular hydrogen at temperatures of adsorbent lower than 100°C (in contrast to O_2) practically does not influence the electric conductivity of oxides. The molecular nitrogen and many simplest hydrocarbons (CH_4, C_2H_6, C_3H_8, etc.) remain inert over broad temperature domain from that standpoint [64]. Atoms of inert gases are absolutely non-active. On the contrary, the chemisorption of atoms and radicals develop with substantially low activation energy than chemisorption of molecules.

In paper [141] it was shown that the change in electric conductivity in case of chemisorption of various alkyl radicals on the same oxide is notably dependent on chemical nature of free radicals. In this case the arrangement of simplest radicals in the order of decreasing degree of effect of electric conductivity of ZnO given in [132, 186] will be the following:

$$CH_2 > CH_3 > C_2H_5 > C_3H_7 \, ,$$

which is in agreement with decrease in their chemical activity [187], however, this contradicts their ordering with respect to decrease in affinity to electron [188], required to explain above effect from the standpoint of collective electron factors. In same paper [144] it was shown that the value of effect (change in conductivity) is dependent on chemical character of adsorbent during chemisorption of simplest radicals on oxides of n-type under fixed conditions. Thus, the ordering of oxides with respect to decrease in the value of the change of electric conductivity during chemisorption of CH_3-radicals yields

$$ZnO > TiO_2 > CdO > WO_3 > MoO_2 \, ,$$

which corresponds to decrease in capacity of metal to form the surface metal-organic compounds.

As it has been mention in preceding section, the vast effect of the mechanism of adsorption-caused change in electrophysical characteristics of adsorbent is provided by availability of defects [32]. However, various admixtures play similarly important role on effects of properties of oxides including the sensitivity of their electrophysical properties to adsorption [4, 5]. Small amounts of admixtures (of the order of 0.5 – 1 mol.-%) can both increase the sensitivity of oxide for instance to oxygen (addition of Y_2O_3 to calcium oxide over pressure interval 10^{-2} – 10^{-6} Torr [189]) and decrease it (addition of Ga_2O_3 to ZnO [190]), or can result in insensitivity of electric conductivity on the pressure of the gas in question (as it is the case with respect to O_2 while adding 0.5 – 1 mol.-% of lithium to NiO [190]).

Adopting the current standpoint that the process of chemisorption can be treated as chemical reaction of adsorption particle with adsorption center accounting for effect of both the reaction on the whole adsorbent and adsorbent on the reaction proper, i.e. accounting for the fact that the system adsorption particle – adsorption center – adsorbent is a unified quantum mechanical system it is not allowable to ignore the local factor of chemisorption, i.e. the issue related to the nature of one of the components of reaction, namely the center of adsorption and the mechanism of interaction of components.

Usually, during examination of electron theory of chemisorption it is assumed that all information concerning adsorption centers and mechanisms of interaction of adsorption particles with the former deals with expression for the total equilibrium concentration of chemisorbed particles both in charged and neutral form, the value of the concentration being dependent on characteristics of adsorbent and absorbate. Due to the lack of reliable information the nature of the surface compound being formed during chemisorption and its parameters are not being specified. However, as the experiment indicates [64] a priori information concerning the centers of adsorption of particular particles is very informative and often is necessary to control various processes and reactions and to create selective semiconductor sensors [141].

From the theoretical standpoint the above issues are addressed by quantum chemistry. On the basis of calculations of various cluster models [191] the properties of surfaces of solid body are being studied as well as issues dealing with interaction of gas with the surface of adsorbent. However, fairly good results have been obtained in this area only to calculate adsorption on metals. The necessity to account for more complex structure of the adsorption value as well as availability of various functional groups on the surface of adsorbent in case of adsorption on semiconductors geometrically complicates such calculations.

On experimental level the question regarding the centers of adsorption was addressed in numerous papers. For instance, in [66] the experimental data were used to show that in case of adsorption of hydrogen atoms on the surface of zinc oxide the centers of chemisorption can be provided by regular oxygen ions of the lattice, i.e. the process of chemisorption of H-atoms can be shown as the following sequence of reactions:

$$H_g + O_s^{2-} \ (HO^{2-})_s \rightleftarrows (HO)_s^- + e \,,$$

where g and s designate the free and surface-bound states, respectively. This conclusion is supported by the fact that under effect of chemisorption of H-atoms the electric conductivity of oxides of various metals (ZnO, TiO$_2$, CdO) changes in approximately the same manner irrespective of the metal itself. The creation of surface OH-groups during chemisorption of H-atoms is confirmed by increase in intensity of respective bounds in IR-spectra [192]. In paper [136] it was established that adsorption of atoms of various metals such as Na, Zn, In, Cd, Pb as well as of H atoms dramatically increases the conductivity of oxide films of n-type, developing on the same adsorption centers, namely, oxygen ions of the lattice.

There are some data in the literature regarding the centers of adsorption of acceptor particles on oxide adsorbents. The vast experimental material on effect of adsorption of simplest radicals on electrophysical properties of oxide semiconductors obtained in [132, 186] led to conclusion that centers of adsorption in this case are provided by admixture atoms of metal and interaction of radicals with these centers result in formation of unstable surface metal-organic compounds. It was mentioned that in case of such interaction the amount of donor centers provided by atoms of superstoichiometric metal decreases which, in opinion of the authors, is the major reason of decreasing the electric conductivity of thin oxide films [141]. The conclusion regarding the origin of centers of adsorption of above particles is confirmed by results of the studies aimed at establishing a relationship between the surface concentration of initially deposited atoms of metal and the value of the change of electric conductivity of activated film during adsorption of CH_3-radicals. Similar results were obtained during studies of adsorption of other acceptor particles (O_2, O, OH, N, etc.) on metal oxide films whose surface was doped with Zn, Ti, Fe atoms [139, 193] and on the surface of monocrystal oxides doped by atoms of respective metals. For instance, in paper [194] in which adsorption of O_2 on the surface of TiO_2 (110) doped with Ti atoms was studied it was established that in majority of cases the adsorption centers of such particles are provided by admixture atoms of metal responsible for conductivity of adsorbent.

In several investigations dealing with the studies of oxygen adsorption on monocrystal and powder zinc oxide it was established that the centers of adsorption of O_2 are provided by either admixture carbon atoms [195, 196] or active surface complexes containing carbon similar to Zn–COO– [197]. Such a conclusion was made in papers [195, 196] on the basis of results of mass – spectroscopic studies of the products of photodesorption from the surface of ZnO. These investigations were conducted parallel with monitoring the kinetics of the change of the surface conductivity of adsorbent. On the basis of results of above experiments the conclusion was made that the photogeneration of holes results in neutralization and desorption of the surface complexes of CO_2. The results of Auger-analysis [195] show the direct correlation between the degree of carbon deposition on the surface and the value of the change of surface conductivity of ZnO during oxygen chemisorption as well as photodesorption-caused activity of adsorbent.

On the contrary, in paper [197] the investigation of adsorption of O_2 on the surface of monocrystal zinc oxide by methods of contact potential, surface photospectroscopy and Auger-spectroscopy the conclusion was made that without special treatment dealing with its irradiation in presence of mixture of gases CO + O_2 the surface of ZnO does not exhibit correlation between concentration of the surface-deposited carbon and the value of chemisorption of oxygen. However, it was pointed out that the value of O_2 chemisorption increases by three orders of magnitude on the surface of ZnO preliminary treated in such a manner. The conclusion was made that such an increase during practically intact value of the surface band bending was caused by creation of active surface complexes including carbon, oxygen and ions of zinc which provide the active centers of oxygen chemisorption.

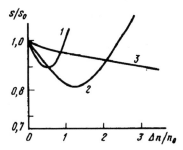

Fig. 1.28. The change in sticking coefficient of atoms of hydrogen during doping of film of ZnO by atoms of Ag (*1*), Zn (*2*) and applying the transverse electric field to the film (*3*) [198, 199]

The examples provided indicate the importance of accounting for local mechanism of chemisorption, i.e. accounting for local binding of adsorbent on certain adsorption centers during which the effect of the band bending in solid state can be consider as a weak perturbation.

The direct prove of the local mechanism of chemisorption of active particles is provided by conclusion of experiments outlined in [198, 199] in which the measurement of scattering of atom beams by targets made of various oxides was used to show that during increase in electric conductivity of the target − film of TiO_2 by seven times provided by applying to the film of transverse electric field the sticking coefficient of oxygen atoms increased only by 1.2 times and decrease for hydrogen atoms by 1.3 times. Similar conclusion is supported by a complex profile of dependence of sticking coefficient of H atoms as a function of degree of doping of the surface of ZnO film by atoms of various metals [199] (Fig. 1.28). In opinion of numerous authors this profile can be explained by simultaneous effect of both local and collective factors.

An a priori information on the centers of adsorption of various particles on oxide adsorbents is important from the standpoint of having an option of separate detection of various particles in their mixtures [200].

One of options of selective detection is provided by introduction on to the surface of adsorbent of an admixture enabling to control the adsorption capability of the surface with respect to the particles of various origin. Thus it was shown in [134] that metal atoms absorbed on the surface of ZnO result in decreasing sensitivity to atoms of hydrogen due to blocking of major centers of adsorption of H atoms. As it was shown in [201] this feature explains the fact that starting with a specific degree of doping the character of effect of H atoms on electric conductivity of n-semiconductor under consideration gets changed from the donor to acceptor one, which is caused by manifestation of adsorption of H atoms on excessive atoms of metal [202]. On the other hand, adsorption on the surface of ZnO of atoms of Zn which are the centers of chemisorption of oxygen atom activates detectors with respect to the latter [67]. One can arrive to similar results while examining the initial adsorption of metal atoms with respect to nitrogen adsorption [139], CH_3- [64] and OH-radicals [193].

Summing up we can conclude that up-to-date there is a reliable physical and chemical basis to build up a method of semiconductor chemical sensors. The notions on effect of adsorption on electrophysics of semiconductor as a thermal perturbation caused by local heating of surface caused with heat released during adsorption became history. On the contrary, now the phenomena linked with transition of the charge between absorbate and adsorbent resulting in the change of the value of surface charge or local chemical interaction of adsorbent with electrically active defects of semiconductor manifested in the change of experimentally controlled electrophysical characteristics are adopted by all scientists.

It has been proven by experiment that there are donor acceptor atoms and molecules of absorbate and their classification as belonging to one or another type is controlled not only by their chemical nature but by the nature of adsorbent as well (see, for instance [18, 21, 203-205]). From the standpoint of the electron theory of chemisorption it became possible to explain the effect of electron adsorption [206] as well as phenomenon of luminescence of radical recombination during chemisorption [207]. The experimental proof was given to the capability of changing of one form of chemisorption into another during change in the value of the Fermi level in adsorbent [208].

The theoretical models of effects of recharging of the surface on the band diagram in the surface-adjacent domain of semiconductor adsorbent accompanying adsorption have been developed. The effect of the surface band bending in semiconductor adsorbent on its electrophysical characteristics caused by transition phenomena have been studied. The theories of adsorption-caused response of above characteristics were derived for both ideal monocrystalline adsorbent [4] and monocrystal with

93

inhomogeneous surface [154], for polycrystalline adsorbent, characterized by availability of an a priori barrier disorder, i.e. broad spread in heights of inter-crystalline energy barriers [158, 160].

On the basis of advances of the chemistry of imperfect crystals [209] the role of point defects existing in the volume and on the surface of oxide adsorbents and their effect on electrophysical characteristic during adsorption-caused change is being elucidated [32].

In conjunction with latest progress in quantum chemistry the availability of vast experimental data makes it possible to analyze the character of possible centers of adsorption of particles of various gases as well as type, chemical and electron properties of surface compounds formed during interaction of adsorption particles with adsorption centers.

The experimental studies of the surface properties of monocrystals of oxides of various metals recently conducted at well-controlled conditions [32, 210] enable one to proceed with detailed analysis of separate effects of various factors on characteristics of semiconductor gas sensors. In this direction numerous interesting results have been obtained regarding the fact of various electrophysical characteristics of monocrystalline adsorbents on the value of adsorption-related response. Among these characteristics there are crystallographic orientation of facets [211], availability of structural defects, the disorder in stoichiometry [32], application of metal additives, etc. These results are very useful while manufacturing sensors for specific gases with required characteristics.

However, it should be mentioned that as of today the majority of industrially produced sensors or those made in laboratories are being designed using empirical ideas. Naturally, this is explained by complexity of the object of studies, which can be described in quantitative terms only in most simple cases. Nevertheless, the thorough information regarding the behavior of detected particles on the gas – solid state interface can and should be the ground stone during designing and manufacturing the devices of such type.

References

1 L.V. Pisarzhevski, Selected Papers, Ukrainian Ac. of Sci. Publ., Kiev, 1936
2 A.F. Ioffe, Reports on Republican Scientific-Engineering Activity: Catalysis, NKhTI Publ., Leningrad, 1930
3 S.Z. Roginski and E.I. Shults, Ukr. Khim. Zhurn., 3 (1928) 177 - 182

4   F.F. Volkenshtein, Electron Processes on the Surface of Semiconductors during Chemisorption, Nauka Publ., Moscow, 1987

5   K. Hauffe, Reactions in Solids and their Surface, Foreign Literature Publ., Moscow, 1963

6   H.J. Engell and K. Hauffe, Ztschr. Electrochem., 56 (1952) 336 - 343

7   P.B. Weisz, J. Chem. Phys., 21 (1953) 1931 - 1939

8   M. Boudart, J. Amer. Chem. Soc., 74 (1952) 1531 - 1540

9   F.F. Volkenshtein, Zhurn. Fiz. Khimii, 21 (1947) 1317 - 1322

10   F.F. Volkenshtein, Zhurn. Fiz. Khimii, 26 (1952) 1462 - 1470

11   V.L. Bonch-Bruevich, Zhurn. Fiz. Khimii, 27 (1953) 960 - 968

12   F.F. Volkenshtein and S.Z. Roginski, Zhurn. Fiz. Khimii, 29 (1955) 458 - 491

13   F.F. Volkenshtein, Zhurn. Fiz. Khimii, 32 (1958) 2383 - 2390

14   F.F. Volkenshtein and Sh.M. Kogan, Ztschr. f. Phys. Chem., 211 (1959) 282 - 289

15   I.E. Zhermen, Heterogenous Catalysis, Foreign Literature Publ., Moscow, 1962

16   K. Hauffe, Reaktionen in und an festen Stoffen. 2. Aufl.: Springer, 1966

17   V.L. Bonch-Bruevich and S.G. Kalashnikov, Physics of Semiconductors, Nauka Publ., Moscow, 1977

18   A. Many, Advances in Studies of Surface of Solids, T. Jayadevaya and R. Vanselov (eds.), Mir Publ., Moscow, 1977

19   V.F. Kiselev and O.V. Krylov, Adsorption Processes on the Surface of Semiconductors and Dielectrics, Nauka Publ., Moscow, 1978

20   U.Kh. Nymm, Physics and Technology of Semiconductors, 8 (1974) 2111 - 2117

21   S.R. Morrison, Physical Chemistry of the Surface of Solids, Mir Publ., Moscow, 1982

22   N.D. Chuvylkin and G.M. Zhidomirov, Zhurn. Fiz. Khimii, 55 (1981) 1 - 27

23   J.E. Lennard-Jones, Trans. Faraday Soc., 28 (1932) 333 - 342

24   J.E. Lennard-Jones, Proc. Roy. Soc. A, 106 (1924) 463 - 474

25   B. Trepnel, Chemisorption, Foreign Literature Publ., Moscow, 1958

26   J.C.P. Mignolet, Chemisorption. L.; Butterworths, Wash. (DC), 1957

27   V.L. Bonch-Bruevich, Proceedings in Electron Theory of Highly Doped Semiconductors, VINITI Publ., Moscow, 1965

28   N. Mott and E. Davis, Electron Processes in Non-Crystalline Substances, Mir Publ., Moscow, 1982

29   E.N. Figurovskaya and V.F. Kiselev, DAN SSSR, 182 (1968) 1365 - 1369

30 M. Roberts and C. MacKey, Chemistry of Metal-Gas Interface, Mir Publ., Moscow, 1981
31 K.N. Spiridonov and O.V. Krylov, Problemy Kinetiki i Kataliza, 16 (1975) 7 - 18
32 W. Gopel, Progr. Surface Sci., 20 (1985) 9 - 18
33 I. Langmuir, J. Amer. Chem. Soc., 38 (1916) 2217 - 2235
34 H. Freundlich, Kapillarchemie. 3. durches und erw. Aufl. Leipzig: Acad. Verl. Ges., 1923, 1225 - 1237
35 A. Frumkin and M. Chlyguin, Acta Physicochim. USSR, 3 (1935) 791 - 803
36 S.Z. Roginski, Adsorption and Catalysis on Inhomogeneous Surfaces, USSR Ac. of Sci. Publ., Moscow-Leningrad, 1948
37 M.I. Temkin, Problemy Kinetiki i Kataliza, 6 (1948) 54 - 62
38 A.R. Miller, The Adsorption of Gases on Solids, Cambridge Univ. Press, 1949
39 Ya.B. Zeldovich, Acta Physicochim. USSR, 1 (1935) 961 - 975
40 M.I. Temkin, Zhurn. Fiz. Khimii, 15 (1941) 296 - 303
41 M.I. Temkin and V.L. Levich, Zhurn. Fiz. Khimii, 20 (1946) 1441 - 1453
42 I. Langmuir, J. Amer. Chem. Soc., 54 (1932) 2798 - 2811
43 B.Yu. Aibinder and E.Kh. Enikeev, VINITI Abstracts, 30 (1973)
44 J. Koutecky, Trans. Faraday Soc., 54 (1958) 1038 - 1042
45 N.S. Lidorenko and E.L. Nagaev, ZhETF Pisma, 31 (1980) 505 - 508
46 Yu.K. Tovbin, Zhurn. Fiz. Khimii, 62 (1988) 2728 - 2738
47 S. Roginski and Ya. Zeldovich, Acta Physicochim., 1 (1934) 554 - 561; 595 - 607
48 S.Yu. Elovich and F.F. Kharakhorin, Problemy Kinetiki i Kataliza, 3 (1937) 322 - 331
49 D.H. Banham and F.P. Burt, Proc. Roy. Soc. A, 105 (1924) 481 - 490
50 S.Yu. Elovich and B.A. Koridorf, Zhurn. Organ. Khimii, 9 (1939) 673 - 680
51 H.E. Matthews and E.E. Kohnke, J. Phys. and Chem. Solids, 29 (1968) 653 - 659
52 T.I. Barry and F.S. Stone, Proc. Roy. Soc. A, 255 (1960) 124 - 128
53 C. Aharony and F.S. Tomkins, Adv. Catal., D. Eley, H. Pines and P.B. Weisz (eds.), Acad. Press, N.Y., 21 (1970) 108 - 115
54 Bray Haskins, J. Amer. Chem. Soc., 48 (1928) 1473 - 1481
55 L.E. Drain and J.A. Morrison, Trans. Faraday Soc., 49 (1953) 654 - 662
56 J.M. Honig and L.H. Reyerson, J. Phys. Chem., 56 (1952) 140 - 148

57 H. Pink, L. Treitinger and L. Vite, Jap. J. Appl. Phys, 19 (1980) 513 - 521

58 M. Nitta, S. Kanefusa and L. Haradome, J. Electrochem. Soc., 125 (1978) 1676 - 1683

59 P. Esser and W. Gopel, Surface Sci., 97 (1980) 309 - 315

60 W. Hotan, W. Gopel and R. Haul, Surface Sci., 83 (1979) 162 - 168

61 E. Garrone, G. Ghiotti, E. Giamello et al., J. Chem. Soc., Faraday Trans. Pt.1, 77 (1981) 2613 - 2622

62 E. Giamello and B. Fubini, J. Chem. Soc., Faraday Trans. Pt.1, 79 (1983) 1995 - 2009

63 I.A. Myasnikov, E.V. Bolshun and E.E. Gutman, Kinetika i Kataliz, 4 (1963) 867 - 872

64 I.A. Myasnikov, Mendeleev ZhVKhO, 20 (1975) 19 - 29

65 I.N. Pospelova and I.A. Myasnikov, DAN SSSR, 167 (1966) 625 - 629

66 I.N. Pospelova and I.A. Myasnikov, Kinetika i Kataliz, 10 (1969) 1097 - 2003

67 G.V. Malinova and I.A. Myasnikov, Kinetika i Kataliz, 10 (1969) 328 - 332

68 K.I. Hagemark, J. Solid State Chem., 16 (1976) 293 - 298

69 G.D. Mahan, J. Appl. Phys., 54 (1983) 3825 - 3831

70 S. Strassler, A. Reis and D. Wieser, Plycryst. Semicond.: Phys. Prop. and Applications. Intern. Sch. Mater. Sci. and Technol, G. Harbeke (ed.), Verlag, Berlin, 1985

71 W. Gopel, Surface Sci., 62 (1977) 165 - 171

72 W. Gopel and U. Lampe, Phys. Rev. B, 22 (1980) 6447 - 6454

73 N. Henney, Chemistry of Solids, Mir Publ., Moscow, 1971

74 C. Wagner, Ztschr. Phys. Chem., 22 (1933) 181 - 190

75 H. Dunwald and C. Wagner, Ztschr. Phys. Chem., 22 (1933) 212 - 225

76 C. Wagner and H. Hammen, Ztschr. Phys. Chem., 40 (1983) 197 - 202

77 V.P. Zhuze and I.V. Kurchatov, Phys. Ztschr. Sovjetunion, 21 (1932) 453 - 460

78 T.J. Gray, Nature, 162 (1948) 260 - 273

79 T.J. Gray, Proc. Roy. Soc. A, 197 (1949) 314 - 321

80 W.E. Garner, T.J. Gray and F.S. Stone, Trans. Faraday Soc. Heterogen. Catal., 1950

81 W.N. Brattain and J. Bardeen, Bell Syst. Techn. J., 32 (1953) 1 - 15

82 G. Heinland, Ztschr. Phys., 148 (1957) 15 - 26

83 I.A. Myasnikov, Zhurn. Fiz. Khimii, 31 (1957) 1721 - 1728

84 V.F. Kiselev and O.V. Krylov, Electron Phenomena in Adsorption and Catalysis on Semiconductors and Dielectrics, Nauka Publ., Moscow, 1979
85 B.Ya. Moizhes, Physical Processes in Oxide Cathode, Nauka Publ., Moscow, 1968
86 J.D. Levine, Surface Sci., 34 (1973) 90 - 98
87 W. Schottky, Ztschr. Phys., 113 (1939) 367 - 374
88 W. Schottky and E. Spenke, Wiss. Veroffentl. Siemens-Werkon., 18 (1939) 3 - 12
89 N.F. Mott, Proc. Roy. Soc. A, 171 (1939) 27 - 43
90 B.I. Davydov, ZhETF, 9 (1939) 451 - 456; 10 (1940) 1342 - 1351
91 B.I. Davydov and S.I. Pekar, ZhETF, 9 (1939) 534 - 543
92 C.G. Garret and W.N. Brattain, Phys. Rev., 99 (1955) 376 - 382
93 V.B. Sandomirski, Izv. AN SSSR Sr. Fiz., 21 (1957) 211 - 225
94 Sh.M. Kogan, Problemy Kinetiki i Kataliza, 10 (1960) 52 - 66
95 T.A. Goodwin and P. Mark, Progr. Surface Sci., S.G. Davison (ed.), Pergamon Press, N.Y. - London, 1 (1972) 1 - 19
96 S. Baidyarov and P. Mark, Surface Sci., 30 (1972) 53 - 60
97 P.S. Kireev, Semiconductor Physics, Vys. Shkola Publ., Moscow, 1975
98 V.Ya. Sukharev and I.A. Myasnikov, Zhurn. Fiz. Khimii, 60 (1986) 3016 - 3031
99 I. Gybbs, Thermodynamic Studies, Foreign Literature Publ., Moscow, 1950
100 J.R. Schrieffer, Phys. Rev., 97 (1955) 641 - 647
101 V.Ya. Sukharev and I.A. Myasnikov, Zhurn. Fiz. Khimii, 58 (1984) 697 - 705
102 Electron Phenomena on Semiconductor Surface, V.I. Lyashenko (ed.), Nauk. Dumka Publ., Kiev, 1968
103 E.E. Gutman and I.A. Myasnikov, Zhurn. Fiz. Khimii, 46 (1972) 131 - 134
104 S.F. Timashev, Elektrokhimiya, 15 (1979) 333 - 340
105 S.G. Dmitriev and Sh.M. Kogan, FTT, 21 (1979) 29 - 35
106 V.Ya. Sukharev and I.A. Myasnikov, Double Layer and Adsorption on Solid Electrodes, Tartu Univ. Publ., Tartu, 1985
107 J.F. McAleer, P.T. Moseley, J.O.W. Norris et al., Proc. 2 Meet. Chem. Sensors. Bordeaux, 1 (1986) 201 - 208
108 B.S. Agayan, I.A. Myasnikov and V.I. Tsivenko, Zhurn. Fiz. Khimii, 47 (1973) 1292 - 1294
109 Z. Bastl, Ya. Adamek and V. Ponets, Problemy Kinetiki i Kataliza, 14 (1970) 88 - 96
110 H. Chon and J. Pajares, J. Catal., 14 (1969) 257 - 268
111 B.V. Chistyakov, V.Ya. Sukharev and I.A. Myasnikov, FTT, 29 (1987) 2305 - 2313

112 C. Pijolat and R. Lalauze, Sensors and Actuators, 14 (1988) 27 - 35

113 F.F. Volkenshtein and V.B. Sandomirski, Problemy Kinetiki i Kataliza, 8 (1955) 184 - 193

114 R. Seiwatz and M. Green, J. Appl. Phys., 29 (1958) 1034 - 1040

115 Yu.I. Gorkun, FTT, 3 (1961) 1061 - 1071

116 V.S. Kuznetsov and V.B. Sandomirski, Kinetika i Kataliz, 3 (1962) 724 - 732

117 O. Peshev, Zhurn. Fiz. Khimii, 52 (1978) 3015 - 3019

118 O. Peshev, J. Chem. Phys., 14 (1977) 183 - 201

119 I.M. Herrman, P. Vergnon and S.I. Teichner, J. Catal., 37 (1975) 57 - 69

120 I.M. Herrman, P. Vergnon and S.I. Teichner, React.. Kinet. and Catal. Lett., 2 (1975) 199 - 207

121 V.E. Lashkarev, Izv. AN SSSR, ser. fiz. 16 (1952) 203 - 209

122 G.E. Pikus, ZhETF, 21 (1951) 1227 - 1241

123 L.S. Gasanov, Fiz. i Tekhn. Poluprovodnikov, 1 (1967) 809 - 812

124 S. Davison and G. Levin, Surface (Tamm) States, Mir Publ., Moscow, 1973

125 J. Bardeen, Phys. Rev., 71 (1947) 717 - 722

126 K.B. Demidov and I.A. Akimov, Zhur. Nauch. i Prikl. Fotogr. i Kinematogr., 22 (1977) 66 - 77

127 C.R. Crowell and G.J. Roberts, J. App. Phys., 40 (1969) 3726 - 3735

128 V.Ya. Sukharev, Zhurn. Fiz. Khimii, 62 (1988) 2429 - 2440

129 W. Gopel and G. Rocker, J. Vac. Sci. Technol., 21 (1982) 389 - 393

130 S.N. Kozlov, Izv. Vuzov. Fizika (1975) 116 - 128

131 A.I. Anselm, Introduction into Semiconductor Physics, Nauka Publ., Moscow, 1978

132 I.A. Myasnikov, DAN SSSR, 120 (1958) 1298 - 1304

133 Kh.S. Valeev and V.B. Kvaskov, Nonlinear Metal-Oxide Semiconductors, Energoizdat Publ., Moscow, 1983

134 I.N. Pospelova and I.A. Myasnikov, DAN SSSR, 170 (1966) 1372 - 1376

135 I.N. Pospelova and I.A. Myasnikov, Zhurn. Fiz. Khimii, 41 (1967) 1990 - 1997

136 I.A. Myasnikov and E.V. Bolshun, Kinetika i Kataliz, 8 (1967) 182 - 187

137 G. South and D.M. Hughes, Thin Solid Films, 20 (1974) 135 - 139

138 G.V. Malinova, E.E. Gutman and I.A.Myasnikov, Zhurn. Fiz. Khimii, 46 (1973) 131 - 136

139 V.I. Tsivenko and I.A.Myasnikov, Kinetika i Kataliz, 17 (1976) 522 - 527

140 V.I. Tsivenko, I.A.Myasnikov and E.S. Shmukler, Kinetika i Kataliz, 17 (1976) 454 - 458

141 I.A.Myasnikov, Electron Phenomena during Adsorption and Catalysis on Semiconductors, F.F. Volkenshtein (ed.), Mir Publ., Moscow, 1969

142 A.E. Bazhanova, Yu.A. Zarifyants, V.F. Kiselev et al., DAN SSSR, 217 (1974) 1099 - 1103

143 J.C. Slater, Phys. Rev., 103 (1956) 1631 - 1646

144 Polycrystalline and Amorphous Thin Films and Devices, L. Kazmerski (ed.) Acad. Press, N. Y., 1980

145 O.L. Krivanek, P. Williams and Yi-Ching Lin, Appl. Phys. Lett., 34 (1979) 805 - 810

146 A.Ya. Vinnikov, A.M. Meshkov and V.N. Savvushkin, FTT, 22 (1980) 2989 - 3001

147 S. Kirkpatrick, Rev. Mod. Phys., 45 (1973) 574 - 582

148 B.I. Shklovski and A.L. Efros, Electron Properties of Doped Semiconductors, Nauka Publ., Moscow, 1979

149 M.K. Sheikman and A.Ya. Shik, Fiz. i Tekhn. Poluprorvodnikov, 10 (1976) 209 - 213

150 B.I. Shklovski, Fiz. i Tekhn. Poluprorvodnikov, 13 (1979) 93 - 115

151 V.Ya. Sukharev and I.A.Myasnikov, FTT, 27 (1985) 705 - -712

152 J. Gutierrez, F. Cebollada, C. Elvira et al., Mater. Chem. and Phys., 18 (1987) 265 - 278

153 V. Lantto and P. Romppainen, Surface Sci., 192 (1987) 243 - 251

154 P.K. Clifford, Chemical Sensors: Anal. Chem. Symp. Ser., T. Seiyama et al. (eds.), Elsevier, Amsterdam, 17 (1983) 135 - 138

155 J.F. McAleer, P.T. Moseley, J.O.W. Norris et al., J. Chem. Soc. Faraday Trans. Pt. 1, 84 (1988) 441 - 458

156 R.L. Petritz, Phys. Rev., 104 (1956) 1508 - 1517

157 W.E. Taylor, N.H. Odell and H.T. Fan, , Phys. Rev., 88 (1952) 867 - 881

158 V.Ya. Sukharev and I.A.Myasnikov, Phys. Status Solidi (a), 100 (1987) 277 - 284

159 V.Ya. Sukharev and V.V. Chistyakov, FTT, 31 (1989) 264 - 269

160 V.Ya. Sukharev, Zhurn. Fiz. Khimii, 63 (1989) 674 - 684

161 V.Ya. Sukharev, V.V. Chistyakov and I.A.Myasnikov, J. Phys. Chem., 40 (1988) 333 - 341

162 V.Ya. Sukharev and I.A.Myasnikov, Zhurn. Fiz. Khimii, 60 (1986) 2385 - 2392; 61 (1987) 302 - 309

163 V.Ya. Sukharev, N.E. Lobashina, N.N. Savvin et al., Zhurn. Fiz. Khimii, 57 (1983) 405 - 410

164 A.Ya. Vinnikov, A.M. Meshkov and V.N. Savushkin, FTT, 24 (1982) 1352 - 1357

165 N.M. Beekmans, J. Chem. Soc. Faraday Trans. Pt. 1, 74 (1978) 31 - 42

166 G. Heiland, E. Mollwo and F. Stockmann, Solid State Physics, F. Seitz and D. Turnbull (eds.), Acad. Press, N.Y., 8 (1959) 191 - 194

167 E.Kh. Enikeev, L.Ya. Margolis and S.Z. Roginski, DAN SSSR, 129 (1959) 372 - 376

168 D.A. Melnik, J. Chem. Phys., 21 (1953) 1531 - 1541

169 S.R. Morrison, Sensors and Actuat., 11 (1987) 283 - 298

170 A.Ya. Vinnikov, A.M. Meshkov and V.N. Savushkin, FTT, 283 (1986) 1229 - 1234

171 E.Kh. Roderik, Contacts of the Metal-Semiconductor Type, Radio i Svyaz Publ., Moscow, 1982

172 E.I. Goldman and A.G. Zhdan, Fiz. i Tekhn. Poluprovodnikov, 10 (1976) 1839 - 1849

173 B.Sh. Galyamov, Pisma v ZhETF, 31 (1980) 539 - 542

174 W.V. Smith, J. Chem. Phys., 2 (1943) 110 - 120

175 T. Seiyama and S. Kagawa, Anal. Chem., 38 (1966) 1069 - 1074

176 S. Kanefusa, N. Nitta and M. Haradome, J. Appl. Phys., 50 (1979) 1145 - 1151

177 V.Ya. Sukharev and I.A.Myasnikov, Zhurn. Fiz. Khimii, 60 (1986) 755 - 762

178 J.S. Cauhape, C. Lucat, C. Bayle et al., Proc. 2 Meet. Chem. Sensors. Bordeaux, 1 (1988) 258 - 266

179 J. Rudolph, Ztschr. Naturforsch. A, 14 (1959) 727 - 732

180 S. Samson and C.G. Fonstad, J. Appl. Phys., 44 (1973) 4418 - 4423

181 D.J.M. Bevan and J.S. Anderson, Discuss. Faraday Soc., 8 (1950) 238 - 243

182 R. Einzinger, Appl. Phys. Surface Sci., 1 (1978) 329 - 339

183 M. Matsuoka, Jap. J. Appl. Phys., 10 (1971) 763 - 770

184 W. Gopel, G. Rocker and R. Feierabend, Phys. Rev. B, 28 (1983) 3427 - 3432

185 I.A. Myasnikov, Zhurn. Fiz. Khimii, 32 (1958) 841 - 846

186 I.A. Myasnikov and E.V. Bolshun, DAN SSSR, 135 (1960) 1164 - 1168

187 E.W.R. Steaclie, Atomic and Free Radical Reactions, Acad. Press, N.Y., 1946

188 Handbook: Chemical Bonds, Ionization Potentials and Affinity to Electron, Nauka Publ., Moscow, 1964

189 K. Hauffe and G. Traukler, Ztschr. Phys., 136 (1953) 166 - 171

190 K. Hauffe and J. Block, Ztschr. Phys. Chem., 196 (1951) 438 - 443

191 Kh. Dunken and V. Lygin, Quantum Chemistry of Adsorption on the Surface of Solids, Mir Publ., Moscow, 1980

192 E.E. Gutman, N.N. Savvin, I.A. Myasnikov et al., Application of Optical Spectroscopy in Adsorption and Catalysis, Nauka Publ., Alma-Ata, 1980

193 N.N. Savvin, I.A. Myasnikov and E.E. Gutman, Zhurn. Fiz. Khimii, 48 (1974) 1262 - 1256

194 G. Rocker, PhD Thes., Montana State Univ., Bozeman, 1985

195 Y. Shapira and S.M. Cox, Surface Sci., 54 (1976) 43 - 56

196 Y. Shapira, R.B. McQuistan and D. Lichtman, Phys. Rev. B, 15 (1977) 2163 - 2174

197 J. Lagowski, H.C. Gatos, R. Holmstrom et al., Surface Sci., 76 (1978) 575 - 585

198 I.A. Myasnikov, E.E. Gutman, S.A. Zavyalov et al., 5 USSR-Japan Seminar on Catalysis, Tashkent, 1979

199 N.V. Ryltsev, E.E. Gutman and I.A. Myasnikov, Zhurn. Fiz. Khimii, 57 (1983) 381 - 384

200 A.I. Lifshits, E.E. Gutman and I.A. Myasnikov, Zhurn. Fiz. Khimii, 52 (1978) 2953 - 2957

201 E.E. Gutman, A.I. Lifshits and I.A. Myasnikov, FTT, 19 (1977) 3146 - 3153

202 R.P. Eischens, W.A. Pliskin and J. Low, J. Catal., 1 (1962) 180-191

203 I.A. Myasnikov and S.Ya. Pshezhetski, Problemy Kinetiki i Kataliza, USSR Ac. of Sci. Publ., Moscow, 8 (1955) 175 - 184

204 V.I. Lyashenko, I.I. Stepko and O.I. Serba, DAN SSSR, 3 (1962) 350 - 354

205 I.A. Karpovich, A.N. Kalinin and B.I. Bednyi, Fiz. i Tekhn. Poluprovodnikov, 10 (1976) 1402 - 1410

206 V.I. Lyashenko and I.I. Stepko, Problemy Kinetiki i Kataliza, USSR Ac. of Sci. Publ., Moscow, 8 (1955) 180 - 191

207 F.F. Volkenshtein, A.N. Gorban and V.A. Sokolov, Radical-Combination Luminescence in Semiconductors, Nauka Publ., Moscow, 1976

208 G.M. Kozub, Zhurn. Fiz. Khimii, 52 (1878) 3025 - 3030

209 F.A. Kroger, Nonstoichiometric Oxides, O. Toft (ed.), Acad. Press, N.Y., 1981

210 G. Heiland and D. Kohl, Chemical Sensors. Anal. Chem. Symp. Ser., T. Seiyama et al. (eds.), Elsevier, Amsterdam, 17 (1983) 125-141

211 H. Moormann, D. Kohl and G. Heiland, Surface Sci., 100 (1980) 302 - 313

212 H. Jacobs, W. Mokwa, D. Kohl et al., Surface Sci., 160 (1985) 217 - 222

Semiconductor Sensors in Physico-Chemical Studies
L.Yu. Kupriyanov (Editor)
© 1996 Elsevier Science B.V. All rights reserved.

Chapter 2

The theory of adsorption-induced response of electrophysical characteristics in semiconductor adsorbents

The methods of quantitative detection

2.1. General principles of selection of semiconductor adsorbents used as a operational sensor elements

Accounting for the designation of semiconductor gas sensors one can conclude that the ultimate importance should be attributed to requirements put forward to such devices in question such as high sensitivity and selectivity in respect to the gas analyzed or the type of active particles studied. Similar requirements involve reproducibility of the signal amplitude during multiple measurements at identical conditions as well as the option to regenerate adsorbent, i.e. the availability of techniques enabling one to bring the measurable characteristics of adsorbent to their initial values. One should also consider the property of reversibility of the signal of the sensor, i.e. the opportunity of the characteristics of adsorbent to be measured to monitor reversible changes in the content of the gas analyzed in ambient volume as a sufficiently important feature. The sensors should also exhibit low inertia of the signal, i.e. small time constant τ necessary to attain the new value of the measured characteristics while changing the content of the gas providing at the same time desired accuracy of the measurements. Specific requirements are also applied to the value of the temperature coefficient of the feature to be monitored, for instance the adsorbent resistivity, since its value determines the necessary accuracy in temperature control to obtain a specific resolution while monitoring concentrations of interest. There are also specific requirements to ensure the level of power consumption, i.e. the value of the operational temperature of the sensor and finally its operational life time without regeneration which is determined as a time over which the parameters of adsorbent providing a required measurement accuracy retain.

The general principles underlying solution of above problems have been examined in Section 1.1. Here it is relevant to note that in each specific case (which can be either reduced to solving an exact analytical problem on detection of extremely small concentrations of given particles when the content of ambient medium is known [1] or manufacturing an industrial sensor for a given gas satisfying requirements on allowable composition and conditions of the ambient atmosphere [2]) the methods to attain the necessary sensitivity, selectivity, reversibility, operation rate, etc. are specific ones, depending on requirements to be met and conditions under which the sensor will operate.

Thus, in physical and chemical studies there are such experimental situations when the reaction volume contains simultaneously both atoms and molecules having a notable impact on electric conductivity of adsorbent, e. g. atom and molecular oxygen. In this case the well known combined probes are used to separate signals from different components [3]. The operation principle of these probes is based on the fact that free atoms and radicals cannot deeply penetrate into a thick level of porous adsorbent due to their heterogeneous recombination, whereas molecules penetrate this level fairly well. Due to this reason the combination of a thin film and a thick rod of oxide would provide a differential signal on a measuring equipment only in case of adsorption of atoms or radicals but not of molecules which almost identically change electric conductivity of both adsorbents [3]. Such situations as those linked with the necessity to detect small amounts of atom oxygen against the background of substantial concentrations of molecular oxygen arise in studies of processes of heterogeneous recombination of atom particles on various surfaces, during the studies of homogeneous and heterogeneous oxidation reactions or, while determining concentration of oxygen atoms in various oxygen-containing media. We should note that similar task can be addressed by a PbO sensor described in [4]. This sensor makes it feasible to detect selectively O-atoms and O_2 molecules owing to their different reaction capability resulting in the donor effect in the change of electric conductivity caused by chemisorption of O-atoms against the background of acceptor effect inherent to the molecular oxygen.

As it has been mentioned in Chapter 1, application of various additives to the surface of adsorbent is effective from the stand-point of obtaining the required selectivity to a specific type of active particles. Thus, doping the surface of ZnO by zinc made it possible to reduce the sensitivity of such sensors to H-atoms and, vice versa, increase to O-atoms. The highest selectivity is obtained at a specific doping degree [5].

It is worth mentioning that in practice one often encounters situations when the benefits of meeting a specific requirement brings about a disadvantage in another area. The optimization problem often arises when it is necessary to match simultaneously several requirements: sensitivity – selectivity, operational temperature – sensitivity, etc. For instance a hydrophilic silica was introduced as a binder into the initial paste [6] to increase the mechanical strength of an adsorbent which led both to an increase in the strength and abrupt increase of "stray" sensitivity to the humidity content in the gas analyzed. Therefore, in subsequent operations the hydrophilic silica was replaced by a hydrophobic one which led to obtaining a reliable CO sensor [7]. The adsorbent obtained was weakly sensitive to humidity alterations and the temperature of surrounding medium.

Generally speaking, (and this coincides with an opinion of Morrison [8]) today there are four most general approaches to solve the problem regarding selectivity of semiconductor sensors. They entail: a) the use of catalysts and promotors, b) the application of the method of temperature control, c) the control of specific surface additives ensuring development of specific adsorption, and, finally, d) the implementation of different filters.

Thus, it has been established that the change in the temperature of adsorbent over substantially wide range results in obtaining broad peaks in electric conductivity profiles plotted against the temperature which is caused by influence of various gases present in the ambient medium. This indicates that adsorbents manifest a maximum sensitivity (from the stand-point of the change in electric conductivity) at various, gas-specific temperatures [9]. However, the high broadening of peaks and resultant weak resolution do not make it possible to conclude that the method of temperature-dependent control is a suitable option to obtain required selectivity.

In several cases application of various additives to the surface of a semiconductor adsorbent, specifically adsorbing or reacting with particles to be detected enables one to improve selectivity. As an example we can mention the use of hygroscopic salts to bind water in humidity sensors, the application of particles of sulfanilic acid to the surface of NiO to detect NO_2 [10]. However, the high operational temperature in majority of semiconductor sensors deprives the method of specific surface additives of its general character.

Fairly good results can be achieved using various filters while identifying different components of gaseous mixtures and selective detection of a specific components. These filters can be either setup in the flux of gaseous particles approaching detector surface or can be used as a covering layers applied to the adsorbent. Such filters can be manufactured from materials active from the stand-point of annihilation-recombination, chemical binding etc. for a specific type of active particles present in mixture. Thus, silver filters are active in removal of O-atoms whereas filters on the basis of CO_3O_4 can be used to remove singlet oxygen [11]. Various zeolites can be used as adsorption agents for numerous gases to distinguish the component of interest [12]. At the same time the filters possessing "transparency" only to detected particles became widely popular. Platinum or palladium films penetrable to hydrogen [13], zirconium penetrable at high temperatures for oxygen [14] and several others provide examples of filters of this kind. The use of filters is the most direct and consistent approach to achieve selective detection but, unfortunately, it can only be applied to a restricted number of gases.

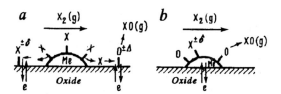

Fig. 2.1. Plausible mechanisms of influence of metal particles deposited on adsorbent on adsorption-caused response of its electrophysical characteristics:
a − chemical sensitization; the dope particles cause the activation and spill-over of adsorbate; *b* − electronic sensitization; the dope particles become donors or acceptors of electrons in dependence on the conditions in gas phase.

The use of catalysts and promotors of various reactions applied as a fine dispersion phase to the surface of semiconductor adsorbent became most popular in providing a required selectivity of sensors with respect to a given gas. As it has been established in experiments (see for instance [8] and the reference list therein), apart from obtaining required selectivity application of such additives results in increase of sensitivity of the sensor with respect to a given gas. However, as of today there is no clarity with regard to understanding the mechanism of effect of catalytic additives on the sensor effect nor in optimization of the choice of catalysts applied.

The spill-over effect and effect related to the change of the Fermi level in semiconductor adsorbent (Fig. 2.1) were considered as two plausible mechanisms of effect of application of catalysts on the value and characteristics of adsorption-caused response of the sensor. In brief, this is manifested in the following. The dissociation adsorption of molecular gases such as oxygen and hydrogen taking place on the surface of metal catalysts (palladium and platinum) is accompanied by the "flowing" of atoms on the surface of substrate (the semiconductor adsorbent) [15, 16]. The subsequent chemisorption of these molecules (charged form of adsorption) or their high chemical activity in respect to other chemisorbed particles or electrically active surface defects results in substantially more high change in electric conductivity of semiconductor if compared to measurements caused by adsorption of respective molecular gases. In detail the role of spillover effect in various heterogeneous processes and reactions as well as its impact on manufacturing of sensors of the "surface type" to detect particles migrating across the surface is considered in Chapter 4.

The effect of applied catalysts on sensitivity and selectivity of semiconductor sensors can be determined through adsorption change of characteristics of the area occupied by spatial charge situated directly under the catalytic additive − semiconductor contact. In particular this can be the contact metal − semiconductor. Indeed, the change in electric subsystem accompanying adsorption on metal microcrystals leads to the change in characteristics of the Shottky barrier which is present at the metal − semiconductor interface which results in the change of concentration of free carriers in SCR participating in current transfer. The detailed semiquantitative model of this phenomenon will be considered in Chapter 5. Here we would like to point out that in several cases it is absolutely unnecessary to overlap SCR created by adjacent metal microcrystals as it was the case in barrier model put forward in [8] to detect the above effect. The point is that besides polycrystal adsorbents characterized by high intercrystalline barriers which limit the transfer of current carriers (see Section 1.10) the sensors whose operational elements are provided by sintered oxide semiconductors are used more often in practice. The structure of such adsorbents is characterized by thin bridges linking neighboring microcrystals, the conductivity of these bridges determines by conductivity of the whole sample [17, 18]. Owing to small sizes of the bridges (~ 10^{-5} cm) not exceeding the shielding length appearance on their surface even occasional microcrystals of metal may substantially affect their conductivity.

It is evident that the multitude of plausible effects of application of catalysts on sensitivity and selectivity of semiconductor sensors cannot be only reduced to above two mechanisms. One should keep in mind the possible influence of contact field spread over substantial area of the adsorbent surface and situated close to metal additives on reaction capacity of adparticles [19] as well as plausible direct catalytic effect of additives accompanied by creation of electrically active products of reaction from non-active reagents.

Nevertheless, we should admit the major conclusion of the study by Morrison [8] that the task of attaining selectivity of gas sensors while using catalysts differs from that to obtain selectivity in catalytic reaction as justified. During gas detection one should make use of selectivity of reactants and the stress is placed on the study of capability to accelerate or slow down reaction of above particles whereas catalysis requires a selectivity with respect to products obtained necessitating the studies and monitoring of all stages of the process. Therefore, the direct use of catalysts of known reactions does not ensure obtaining desired results. However, in the cases when side products and reaction products do not substantially affect the electrophysics of adsorbent the direct use of a catalyst of a certain reaction developing in presence of the particle to be detected may prove to be useful. As an example we can point out the development of a low temperature selective sensor to detect carbon oxide [20] whose operational element is provided by CO oxidation catalyst on the basis of titan doped $\alpha - Fe_2O_3$ containing fine gold fraction.

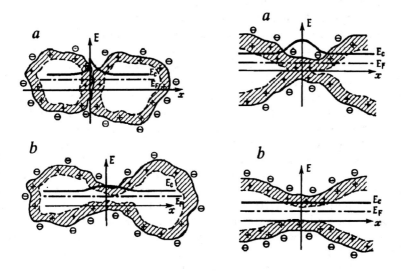

Fig. 2.2. Intercrystalline contacts of a barrier (a) and bridge (b) types.

Fig. 2.3. The bridges of closed (a) and open (b) types.

Initial excitation of detected particles should be classified as one of important techniques of selective detection. This initially deals with molecules non-active (with respect to the change of electrophysical characteristics of adsorbent) in the ground state. The electron or vibration excitation results in bringing them to the active state which enables one to assess the total concentration of the particles of the type studied through detection of concentration of excited particles [21] (see Chapter 5).

In conclusion of this Section let us dwell on another important topic related to possible from our stand-point classification of sensors on crystal type of adsorbent.

Semiconductor adsorbents used as operational elements in sensors can be monocrystals or monocrystal films as well as polycrystals. The latter can be arbitrary divided into two groups differing in properties of contacts between monocrystals. The first group contains those polycrystalline adsorbents in which contacts between crystals can be represented as a double Shottky barrier or, in more general case, isotype heterotransition [22] (Fig. 2.2, *a*). Such contacts are inherent to particulate, pressed or applied polycrystalline adsorbents which were not subject to high temperature treatment as well as to a certain degree to a polycrystals sintered in oxidation medium. The second group involves polycrystalline semiconductors in which contacts between separate crystals are characterized by availability of thin connecting bridges [17, 18, 23] (see Fig. 2.2 *b*). Depending on such parameters as thickness of the bridge, degree of doping of material, or concentration of defects in bridge to be exact, the value of the surface charge localized on SS of various types, the above bridges can be or cannot be barriers hindering the motion of current carriers in a semiconductor. Figure 2.3 illustrates these two situations. It is apparent that the first case characterized by availability of high energy barrier is related to superposition of depletion regions which is caused by small thickness of bridge ($h < 2L_D$) and high value of the surface charge. In the second case there is no such superposition and the barrier to impede motion of carriers is not being formed. In what follows the bridges characterized by the energy barrier with height exceeding kT (the heat energy of carriers) will be referred to as bridges of closed type in consistence with current terminology [24, 25]. The bridges without such barrier will be referred to as bridges of open type.

The above bridge structures usually occur during sintering of several polycrystalline oxides the most branched (lace structure) is formed during sintering in vacuum, inert or reduced gas atmosphere. This is related to the fact that in ionic crystal the motion of both cation and anion are necessary to form and grow thick bridges in ion crystals. Therefore, the total rate of the process is determined by the rate of motion of least mobile ion, usually this is anion as is the case with ZnO, Al_2O_3, NiO [26]. It is obvious that various factors affecting the rate of transfer of above particles change the sintering rate. Thus, the increase of amount of oxygen vacancies taking place during heating of the sample in atmosphere of interest or in vacuum results in development of the process of sintering and in increase of its rate. It is absolutely evident that decrease in free energy occurring due to decrease of the surface area leads to thickening of thin bridges formed at initial stages of sintering. This is illustrated by microphotographs of the structure of polycrystalline zinc oxide taken at various sintering times (Fig. 2.4). It is obvious that with increase in sintering time clearly visible protracted bridges disappear and the unified pattern structure is being formed which is characterized by alternate thickness of branches. Heating the polycrystal sample in oxygen media results in decrease of concentration of oxygen vacancies which causes retardation of the process of sintering and increase in the fraction of thin protracted bridges and bridge-free contacts in the total number of intercrystalline contacts.

Fig. 2.4. Microphotographs of sintered ZnO films with different structures: a – structure consists of microcrystals connecting each other by thin crystal bridges; b – lace structure is characterized by variety of branch thickness. Magnification $2 \cdot 10^4$.

Based on above considerations we can draw several conclusions in respect to domains of applicability of semiconductor sensors made of polycrystalline adsorbents prepared under different conditions. Namely, we can conclude that polycrystalline adsorbents sintered in oxidizing atmosphere are most suitable for application as an operational element to trace low concentrations of reducing or combustible gases (which are active from the stand-point of electrophysical characteristics of adsorbent) in air or in oxygen containing media. In reduction media, in atmosphere of inert gases or in vacuum the application of above adsorbents (especially at elevated temperatures) is not purposeful due to constant drift of the values of their electrophysical characteristics caused by an ongoing oxide reduction.

On the contrary, polycrystalline adsorbent sintered in reducing atmosphere are most suitable (from the stand-point of effect on electrophysics of adsorbent) as operational elements to determine small concentrations of various gaseous admixtures under vacuum conditions and in atmosphere of inert gases. Their use in air especially at high temperatures may be disadvantageous due to plausible development of oxidizing processes.

Thus, the domain of applicability of the first type of polycrystalline adsorbents with intercrystalline threshold contacts is provided by the common gaseous analysis of air media. Accurate vacuum studies and detection of traces of admixtures in reducing and inert media are a typical applicability domain of polycrystalline adsorbents with intercrystalline contacts of the open bridge type.

We have to mention that the third type of adsorbents − monocrystalline oxides − is also feasible. From our stand-point their applicability domain deals with the studies of elementary stages of gas − solid body interaction, the results obtained being useful to manufacture sensors sensitive for specific gases.

Differences in the structure of monocrystalline, threshold or bridge type polycrystalline adsorbents are to be manifested in the shape of adsorption − caused response of electrophysical characteristics [25]. The basic models of adsorption − induced response of monocrystalline and barrier polycrystalline adsorbents have been considered in Chapter 1. Here we describe various theoretical models of adsorption-induced response of polycrystalline adsorbents having intercrystalline contacts of the bridge type and their comparison with experimental results.

2.2. Sintered polycrystalline adsorbents

Above we have considered the possibility of existence of various types of polycrystalline semiconductor adsorbents differing in the character of contacts of specific microcrystals. Let us consider the effect of this difference on adsorption and electrophysical properties of adsorbents in more detail and, which is more important, address the item of the mechanisms of the change of electrophysical properties of semiconductor caused by its interaction with gaseous phase.

Following the widely cited paper by Hatson [18] let us depict the cross-section of a portion of a sintered sample as is shown in Fig. 2.5. The shaded area shows the domain of spatial charge.

As it is shown in the figure separate microcrystals are either linked by thin bridges formed as a result of sintering (contacts 1 and 2), or simply touch each other at the interface (contact 3) [17, 25]. Figure 2.6 displays two possible types of above bridges in a more detailed scale as well as band diagrams for cross-section under provisions of existence of an a priori surface-adjacent regions of depletion of the principal current carriers (electrons for the n-type semiconductor under consideration). It is obvious that the bridge shown in Fig. 2.6, a is an open-type bridge, whereas that shown in Fig. 2.6, b belongs to a bridge of closed type. The energy diagram of the $B' - B'$ section for the bridge of closed type illustrates the effect of formation of plane bands inherent to most thin bridges underlying existence of high energy barrier along the whole thickness of the bridge with respect to transition of conductivity electrons between adjacent microcrystals (see Fig. 1.3, a).

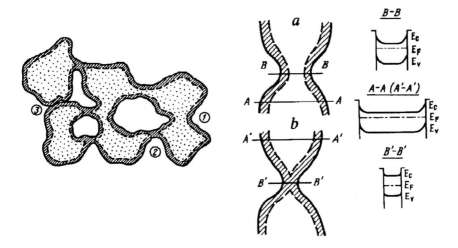

Fig. 2.5. Sintered polycrystalline sample (a portion) [18].

Fig. 2.6. Cross-section of the bridges of open (*a*) and closed (*b*) types with respective schematic diagrams of energy bands.

In the case when the bridges of closed type are dominant they control the resistivity of the whole sample. Almost all the drop of external voltage applied to the sample occurs on these bridges. Once the position of the Fermi level with respect to the conductivity band in the bridge of the closed type is mainly determined by the value of the surface charge the change in electric conductivity of such a material observed under change of external conditions (adsorption, temperature, illumination) is thoroughly dependent on the change in the value of the surface charge. The activation energy of conductivity in such materials is usually compared with the adsorption heat of particles responsible for the value of the surface charge, whereas the conductivity of this type is called the surface – trap – limited conductivity [24, 25, 27].

In case when the material is dominated by the bridges of open type, or, to be more precise there is sufficient amount of continuous paths going through the whole sample which contain the bridges of only this type the energy of activation of conductivity is related to the energy of ionization of the volume dope and the conductivity itself is called bulk – trap – limited conductivity [24, 25, 28]. In this case the conductivity of a specific bridge is proportional to the thickness of a conducting channel $h - 2L_*$, where h is the thickness of the bridge which, according to the estimates made by Bickmans [24], is of the order of 10% of the microcrystal size, L_* being the width of SCR. The effect of adsorption and other external effects resulting in the change of the value of the surface charge influences the width of the depletion layer, too. This is directly linked with the width of conducting channel and, consequently, with the value of resistivity of the bridge.

The contacts of the third type (see Fig. 2.2, *a*) which are interfaces or contact areas separate microcrystals are equivalent (this has been shown in preceding Section) to the double Shottky barrier or, to put it more correctly, to the isotype heterotransitions [22, 29]. As it has been shown in Section 1.10 in detail, the energy of activation of electric conductivity of the material with dominant fraction of contacts of this type is dependent on the heights of intercrystalline barriers. The change in electric conductivity due to effects of various external effects (adsorption in particular) is related to the height of these barriers.

In real polycrystalline materials there are, presumably, all three above types of contacts present. It is assumed that the issue of dominance of a specific type of contacts is related to the technology of manufacturing of a sample and to its subsequent treatment [17, 24, 28, 30]. Moreover, there are various manufacturing techniques enabling one to obtain a sample with dominating type of contacts using the material of another type [30].

We should point out that up to now we have considered only polycrystals characterized by an a priori surface area depleted in principal charge carriers. For instance, chemisorption of acceptor particles which is accompanied by transition-free electrons from conductivity band to adsorption induced SS is described in this case in terms of the theory of depleted layer [31]. This model is applicable fairly well to describe properties of zinc oxide which is oxidized in air and is characterized by the content of surface adjacent layers which is close to the stoichiometric one [30].

Absolutely different situation occurs in case of polycrystalline semiconductors obtained under vacuum conditions. Thus, in paper [30] three types of ZnO samples were studied: ZnO 1 obtained during heating of carbonates containing zinc in air at 600°C; ZnO 2 obtained through vacuum decomposition of carbonate followed by heating at 500°C; ZnO 3 powder of spectrally pure zinc oxide.

Authors of paper [30] studied adsorption and desorption of oxygen, oxygen photoadsorption and photodesorption and finally the rate of isotopic exchange using heavy oxygen isotopes. The results obtained enabled the authors to conclude that the character of development of above processes substantially differs in different samples. Namely, the effect of light resulted in desorption of oxygen from the surface of ZnO 1 and, on the contrary, led to additional adsorption in case of ZnO 2. The isotopic exchange in case of ZnO 2 developed substantially weaker than in case of ZnO 1. Finally, and this is most important from our stand-point, the adsorption-induced behavior was markedly different in ZnO 1 and ZnO 2: adsorption of oxygen on ZnO 2 developed more intensely and was characterized by high irreversibility degree than that on ZnO 1. As for attainable deposition degrees, they were much higher in case of ZnO 1. Such different behavior in ZnO 1 and ZnO 2 can be explain by the fact that ZnO 2 is a partially reduced oxide if compared to ZnO 1, this feature leading to a higher content of superstoichiometric zinc. This conclusion is confirmed by the fact that heating ZnO 1 in hydrogen atmosphere brings its properties close to those of ZnO 2.

High concentration of superstoichiometric zinc in partially reduced zinc oxide (especially in its surface adjacent layers and on the surface of the sample) leads to: a) plausible existence of an a priori enriched surface-adjacent layers, i.e. may result in initial conditions opposite to that postulated in the "depleted layer" model and b) provides the option of direct interaction of adsorbate (in our case oxygen) with surface defects such as interstitial zinc atoms or oxygen vacancies. Experiment situation observed in [32] and described in Section 1.11 provides an example of such an interaction resulting in substantial change in electric conductivity of adsorbent without altering the value of the surface charge. Review by Gopel [33] revealed the important role played by direct interaction of adsorbate with electrically active defects of adsorbent played in effects of adsorption induced change in electric conductivity. This interaction according to Gopel is one of the basic reasons leading to difference between the number of electrons localized on adsorption type SS and the change in concentration of free current carriers which accompanies adsorption.

Thus, in compliance with the dominant viewpoint all sintered polycrystalline semiconductor adsorbents can be classified as belonging to two large groups differing in type of dominant contacts, namely the bridges of open and closed type. The dominance of the contacts of a specific type is provided by the choice of technology to manufacture adsorbent and, which is more important, by sintering technology.

2.3. Electrophysical properties of sintered polycrystalline semiconductors

As of now such semiconductor oxides as ZnO, SnO_2 and TiO_2 are most widely used as operational sensor elements. This is initially explained by the vast amount of experimental data gathered for above compounds and on the other hand by the importance of their being used as catalysts in various reactions. Finally, this can be explained by the fact that they are most suitable from the stand-point of requirements put forward to semiconductors used as operational elements in sensors. In our studies we mostly used zinc oxide adsorbents.

Zinc oxide is a thoroughly studied typical semiconductor of n-type with the width of forbidden band of 3.2 eV, dielectric constant being 10. Centers responsible for the dope electric conductivity in ZnO are provided by interstitial Zn atoms as well as by oxygen vacancies whose total concentration vary within limits $10^{16} - 10^{19}$ cm^{-3}. Electron mobility in monocrystals of ZnO at ambient temperature amounts to 200 cm$^2 \cdot$s^{-1}. The depth of donor levels corresponding to interstitial Zn and oxygen vacancies under the bottom of conductivity band is several hundredth of electron volt [18].

Semiconductor films of ZnO used as operational elements are obtained by oxidation at ~ 500 − 600°C in the jet of purified oxygen of zinc film deposited at vacuum ($P \sim 10^{-7}$ Torr) on substrates made of fused quartz with subsequent sintering at ~ 350°C at high vacuum conditions [34]. As it was concluded in paper [17] the sintered polycrystalline sample obtained in such a manner should not be considered as a set of various separate crystallites touching each other but rather as a monolithic pattern in which microcrystals with diameter of 1 ~ 10 μm are linked with each other by bridges with length and thickness of the order of 0,1 μm (see Fig. 2.4).

Numerous studies of electric conductivity of above films involving alternating and continuous currents [34-37] showed that the major fraction of relatively good conducting material in above systems is separated by thin layers featuring relatively poor conductivity. It has been established that microcrystals or their blocks are responsible for high conductivity, whereas thin vacuum or gaseous caves between microcrystals as well as crystallite interfaces and thin bridges can be considered as domains with poor conductivity. Moreover, it was shown in paper [35] that electric conductivity of ZnO changes during adsorption namely due to the change in conductivity of bridges. However, the availability of such distinctly noticeable bridges (this is confirmed by numerous studies, see for instance [26, 38, 39]) indicates an insufficient sintering stage.

An increase in sintering time makes bridges thicker. The same result, as it has been shown above, can be obtained through various mechanisms affecting the concentration of vacancies. For instance, the addition of dopes with the valence differing from that of the lattice atoms can change concentration of oxygen vacancies and, consequently, sintering rate. Thus, addition of LiO_2 to ZnO leads to increase in sintering rate in the latter whereas introduction of three valence cations of Ga^{3+} and Al^{3+} results in its decrease [26]. The heating of ZnO samples in hydrogen atmosphere brings about the decrease in amount of cation vacancies and increase of anion ones which also leads to a notable rise in sintering rate. As it was shown by electromicroscopic studies [40] with increase in time of vacuum sintering clearly visible bridges disappear in ZnO samples. They are being replaced by a peculiar structure characterized by alternate thickness of its elements (see Fig. 2.4 *b*).

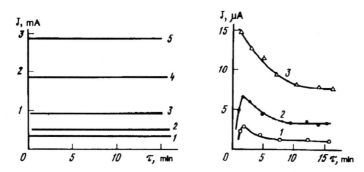

Fig. 2.7. Kinetic curves of the change in current at different values of voltage applied to sintered ZnO film: *1* – 20 V; *2* – 40; *3* – 80; *4* – 160; *5* – 300 V

Fig. 2.8. Kinetic curves of the change in current at various values of voltage applied to pressed ZnO sample: *1* – 100 V; *2* – 300; *3* – 540 V

In order to interpret correctly the results of electrophysical measurements conducted on vacuum – sintered ZnO semiconductor films one should answer the question concerning the origin of contacts between specific crystallites controlling the electric conductivity of the material. This was accomplished in paper [37] using the method of prerelaxation VAC to run a comparative analysis of mechanisms of charge transfer in thin sintered (in vacuum) films and pressed polycrystalline ZnO samples.

To plot these VAC the curves of the change of current in time during application of a step-like voltage to the sample were carried out in a vacuum cell ($P \sim 10^{-6}$ Torr). If a certain voltage drop occurs on a certain barrier this indicates that the current starts flowing through the barrier, the time constant being of the order of the Maxwell relaxation time. However, the potential difference both changes the height of barrier in the direction of current and influences the equilibrium between current carriers in the conductivity band as well as the energy levels localized close to the surface. If SS are not completely occupied the free carriers start migrating from the conductivity band to the surface levels, the height of barrier along the current increases which results in current decrease. As it has been mentioned in Section 1.10 this process proceeds until the new equilibrium gets established between SS and the volume bands which correspond to the value of applied voltage. It is known that the characteristic times of this process contain exponentially high activation factor and may reach $10^2 - 10^7$ s [41]. VAC's were plotted using the maximum values of current in [37]. This means that the sample was as far from the equilibrium as possible. This is this situation that enables one to obtain information regarding barrier statistics in the system. This was shown in [42]. In Figs 2.7 and 2.8 the kinetic curves of the change of current in a thin sintered film and pressed ZnO powder are shown. The samples were kept under vacuum conditions with varying value of applied voltage. As these figures indicate the application of the voltage to the pressed polycrystalline sample results in a sharp increase in current with its subsequent decrease to the stationary value, whereas in case of a sintered film the stationary value is immediately reached. This indicates on various mechanisms of charge transition in samples under studies which is proved by the character of field dependence of electric conductivity (Fig. 2.9). It is obvious that starting with fields of ~ $6 \cdot 10^2$ V/cm the Ohm law is violated for a pressed ZnO sample, the VAC being described by the Shklovsky formula fairly well (see expressions (1.62) and (1.63)).

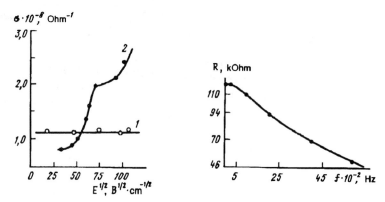

Fig. 2.9. The field dependence of electric conductivity of a thin sintered film (1) and a pressed ZnO sample (2) [37]

Fig. 2.10. The frequency dependence of resistivity of sintered ZnO film [34]

Consequently, the pressed ZnO sample possesses intercrystalline barriers characterized by a wide spread with respect to the height which can be considered as a specific type of intercrystalline contacts. At the same time a rigorous compliance with the Ohm law over the whole interval of applied fields is observed for a ZnO film sintered under vacuum conditions. This result and the fact that the estimation of the average value of the voltage drop per contact is about 0.2 eV $\gg kT \approx 0.025$ eV leads to conclusion that in sintered samples there are conductivity chan-nels free from energy barriers of the height above the thermal energy of the charge carriers. Presumably, the availability of such ohmic pathways to carry the charge occurring during vacuum sintered of polycrystalline samples can be attributed to coalescence of crystallites [43, 44] resulting in removal of barriers, which can be explained by a random crystallo-graphic orientation of facets at crystallite interfaces [45].

The conclusion regarding the fact that constant current conductivity involves not all microcrystals of the sample is proved by results of measurements of electric conductivity in sintered ZnO films in case of alternating current (Fig. 2.10). The availability of barrier-free ohmic pathways is proved by a low value of initial resistivity in sintered sam-ples (~ 1 – 5 kOhm) in addition to exponential dependence of electric conductivity plotted as a function of inverse temperature having activa-tion energy ~ 0.03 – 0.5 eV, which coincides with ionization energy of shallow dope levels. The same value is obtained from measurements of the temperature dependence of the Hall constant [46].

Therefore, the data gathered enables one to conclude that to a cer-tain degree partially reduced sintered polycrystalline films are similar to monocrystals from electrophysics stand-point, but possess much more developed specific surface.

The latter conclusion is reliably confirmed by experimental results [40] in which the studies of effect of the structure on the character of adsorption change in electric conductivity of monocrystal or partially reduced polycrystalline ZnO adsorbents were conducted. The compara-tive studies of the character and the value of response of electric con-ductivity in both types of adsorbents on adsorption of various atoms and molecular particles led the authors to conclusion on identical origin of both the mechanisms of electric conductivity and mechanisms of its adsorption induced change.

Thus, the whole complex of existing experimental data indicates that the major part of polycrystalline contacts in vacuum sintered polycrys-talline oxides are provided by bridges of open type. Moreover, the vac-uum sintering at moderate temperatures ~ 300 – 350°C leads to forma-tion of a unified pattern (see Fig. 2.4, *b*) which cannot be disjoint into specific microcrystals and connecting bridges [37, 40]. The structure of adsorbents obtained presents a complex intertwining of branches of various thickness.

An absolutely different situation occurs in case of polycrystalline ad-sorbent treated at high temperature in air or in other oxygen containing medium. In this case the volt-ampere analysis exhibits sharply non-linear VAC, deviations from the Ohm law being observed at anoma-lously low fields [47]. This indicates an existence of high intracrystal-line barriers in such adsorbents. These barriers can be attributed to crystallite interfaces, as well as to the bridges of closed type characterized by complete intersection of SCR, the concentration of contacts of this type having a dominant character excluding the option of existence of barrier − free pathways going through the whole sample.

2.4. The effect of adsorption on electric conductivity of sintered polycrystalline adsorbents

Having considered the material outlined in preceding sections it is easy to conclude that depending on technology of preparation of polycrystalline adsorbent one can encounter various mechanisms of electric conductivity and, naturally, various mechanisms and peculiarities of adsorption induced change of the value of electric conductivity in adsorbents.

Indeed, in case of partially reduced oxide characterized by substantial degree of non-stoichiometricity and absence of energy barriers in most contacts of microcrystals the electric conductivity is caused by ionization of bulk dopes provided by superstoichiometric metal (interstitial metal atoms or oxygen vacancies). The effect of adsorption-induced in electric conductivity in such adsorbents can be caused either by charging of the surface leading to the change in the width of conducting channel or by the change in concentration of electric active defects responsible for dope electroconductivity or by joint manifestation of above mechanisms.

In case of the use of polycrystalline adsorbent subject to high temperature oxidation and characterized by almost stoichiometric content of the surface its electric conductivity can be linked with penetration of the current carriers through high intercrystalline barriers. The effect of adsorption on electric conductivity of adsorbents of such type is mainly manifested through the change of heights in intercrystalline barriers controlled by the value of the surface charge.

Let us dwell on existing key models describing chemisorption induced response of electric conductivity in semiconductor adsorbent. Let us consider both the stationary values of electric conductivity attained during equilibrium in the adsorbate-adsorbent system and the kinetics of the change of electric conductivity when the content of ambient atmosphere changes. Let us consider the cases of adsorption of acceptor and donor particles separately. In all cases we will pay a special attention to the issue of dependence of the value and character of signal on the structure type of adsorbent, namely on characteristics of the dominant type of contacts in microcrystals.

2.4.1. Adsorption of molecular acceptor particles. Molecular oxygen

One of basic experimental facts is provided by existence of the power law between the value of equilibrium electric conductivity of adsorbent and pressure of oxygen in ambient atmosphere:

$$\sigma \approx P_{O_2}^{-m} \,, \tag{2.1}$$

where m is the numeric factor varying from 0.5 to zero. It should be mentioned that the relationship (2.1) is linked with chemisorption of oxygen, i.e. with its effect on electric conductivity of oxide and not with high temperature equilibrium existing between oxygen of the gas phase and oxygen of oxide lattice described in Section 1.11 resulting in values of m equal to $1/4$ and $1/6$. In numerous studies [25, 48-52] in which sintered films of zinc, titanium and tin oxides were used as adsorbent it was shown that the value of $m \approx 0.5$ was attained at equilibrium chemisorption of molecular oxygen in the range of moderate temperatures $(T < 600°C)$.

There are numerous approaches based on various assumptions explaining the dependence (2.1). They include supposition on dissociative character of chemisorption of O_2 [24, 25, 53] developing through the following schematics:

Scheme 1

$$\frac{1}{2}O_2(g) \underset{\leftarrow}{\rightarrow} O(s) \,, \tag{2.2}$$

$$O(s) + e \underset{\leftarrow}{\rightarrow} O_s^- \,, \tag{2.3}$$

Scheme 2

$$O_2(g) \underset{\leftarrow}{\rightarrow} O_2(s) \,, \tag{2.4}$$

$$O_2(s) + e \underset{\leftarrow}{\rightarrow} O_2^-(s) \,, \tag{2.5}$$

$$O_2^-(s) + e \underset{\leftarrow}{\rightarrow} 2O_s^- \,, \tag{2.6}$$

where $O_2(g)$ is the oxygen in gaseous phase, $O_2(s)$ and $O_2^-(s)$ are neutral and charged molecular forms of chemisorbed oxygen, O_s^- is the atom form. One can easily note that in both cases $\sigma \sim [e] \sim P_{O_2}^{-0.5}$.

The power law (2.1) can also be deduced from assumptions concerning the energy inhomogeneity on the surface of adsorbent [27, 54, 55]. Namely, assuming the availability of exponential distribution with respect to the heat of adsorption, in other words, assuming the validity of the Freindlich isotherm for adsorption of oxygen on oxides, Clifford [54] obtained expression $\sigma \sim P_{O_2}^{-\gamma}$, where $\gamma = kT/E_0$, $0 < \gamma < 1$, the value of E_0 characterizing the width of distribution of adsorption heats.

There are numerous studies in which relationship (2.1) with $m = 0.5$ is explained on the basis that radical O_2^- is the principal adparticle affecting the electric conductivity of adsorbent [56-59]. True, in compliance with EPR and thermodesorption data [28, 60, 61] at temperatures lower than 200°C paramagnetic O_2^- adparticles and a certain amount of O^- are being formed predominantly on the surface of oxides as a results of O_2 adsorption. According to estimates of the authors of study [60] the fraction of oxygen chemisorbed on ZnO as O^- does not exceed 8%. It is assumed that formation of O_2^- and O^- at low temperatures are weakly connected processes. At temperatures higher than 200°C O_2^- desorbs rather than transfers to O^- [61]. Therefore, most likely the scheme 2 (see expressions (2.4) − (2.6)) is inapplicable, the relationship between adsorption of oxygen in O_2^- and O^- forms in opinion of authors [61] being manifested mainly through competition for adsorption centers:

$$O_2(g) + Zn^+ \rightarrow O_2^- \, Zn^{2+}, \qquad (2.7)$$

$$O_2(g) + 2Zn^+ \rightarrow 2\,[O^- \, Zn^{2+}]\;. \qquad (2.8)$$

We should point out that formation of O^- as it is assumed in [61], occurs during adsorption of oxygen molecule both on two centers with subsequent brake in O–O-bond. It was established in paper [60] that formation of O_2^- can develop with participation of free electrons and formation of O^- is observed only on reduced oxides with participation of non-stoichiometric Zn. This is confirmed by conclusions of paper [28] in which it was shown that creation of paramagnetic particles during adsorption of O_2 on partially reduced zinc oxide proceeds even when the intensity of signal with $g = 1.96$ (generated by free electrons) does not change.

In case when the dominant charged form of chemisorbed oxygen is provided by O_2^- and the change in electric conductivity of adsorbent is dependent on formation of O_2^-, the $\sigma \sim P_{O_2}^{-0.5}$ function can be obtained starting with assumption on availability of the following stages of the process:

$$O_2(g) \underset{\leftarrow}{\overset{\rightarrow}{}} O_2(\phi) ,\tag{2.9}$$

$$O_2(\phi) + Me_\bullet^0 \rightarrow (Me^{\delta+}O_2^{\delta-})_s^0 ,\tag{2.10}$$

$$(Me^{\delta+}O_2^{\delta-})_s^0 + e \rightarrow (MeO_2)_s^- ,\tag{2.11}$$

where $O_2(\phi)$ designates weakly bound surface state of oxygen.

Indeed, in this case the formation of neutral surface compound $(Me^{\delta+}O_2^{\delta-})_s^0$ is accompanied by binding of neutral superstoichiometric Me and, therefore, decrease in concentration of donors responsible for dope electric conductivity of adsorbent. In the case when the formed surface state $(Me^{\delta+}O_2^{\delta-})_s^0$ possesses sufficient electron affinity one cannot rule out localization of free electron on it followed by formation of a charged surface complex $(MeO_2)_s^-$. Calculation of such scheme accounting for conditions of electron neutrality $h \cdot [e] + [(MeO_2)_s^-] = h \cdot [Me_\bullet^+]$ and under provision of complete ionization of shallow donors, as well as validity of the Henry isotherm leads to expression

$$\sigma \sim [e] \approx ([Me_\bullet]_0 / k_2)^{1/2} (1 + P_{O_2} / P_C)^{-1/2} ,\tag{2.12}$$

where $[Me_\bullet]_0 \approx [e] \sim \sigma_0$ is the initial concentration of superstoichiometric metal; $k_2(T)$ is the constant of process equilibrium (2.11); $P_C = h / k_1 \Gamma$, h is the thickness of the bridge or in a more general case the ratio of the volume of bridge to the area of its surface, Γ is the Henry constant, $k_1(T)$ is the constant of equilibrium of process (2.10). One can conclude from expression (2.12) that when $P_{O_2} > P_C$

$$\sigma \sim [e] \approx ([Me_\bullet]_0 / k_2)^{1/2} (P_C / P_{O_2})^{1/2} \sim P_{O_2}^{-1/2} ,\tag{2.13}$$

for $P_{O_2} < P_C$

$$\sigma \sim [e] \approx ([Me_\bullet]_0 / k_2)^{1/2} (1 - P_{O_2} / 2P_C) .\tag{2.14}$$

We have to point out that relationship of the type (2.14) was experimentally observed in paper [62] at small pressures in system O_2 using SnO_2, the temperature being $T \sim 180 \div 250°C$.

Despite numerous experimental data regarding the studies of the mechanism of oxygen adsorption on oxides, so far there is no complete clarity in this issue. The point what form of oxygen is responsible for high temperature desorption peak still has to be clarified [28, 61]. Moreover, it is not clear under what conditions and which of all plausible forms of chemisorbed oxygen controls experimentally observed change in electrophysical characteristics of adsorbent. If we add up to this the well known fact dealing with capability of EPR and IR-spectroscopy to determine only a small fraction of all variety of adsorbed oxygen, we can understand the numerous explanations of experimental data linked with effect of oxygen on electric conductivity of oxide adsorbents.

Nevertheless, there is a whole series of experimental results enabling one to propose and substantiate a sufficiently general model consistently describing the effect of oxygen adsorption on electric conductivity of partially reduced oxides observed in experiments. Let us consider these data.

First, let us point out that as it has been mentioned in review [63] the formation of adsorbed radicals of O_2^- is observed only in the course of the process of reoxidation of the surface of initially reduced oxides. During subsequent adsorption of O_2^- oxygen radicals are not formed until oxide gets initially reduced again.

Second, it has been established [64] that notable concentration of O_2^- and main change in electric conductivity occur at different moments of time, namely only when electric conductivity has already stopped changing. In this paper the kinetics of O_2 adsorption of partially reduced TiO_2 as well as assessing the kinetics of increase in adsorbed O_2^- radicals was monitored simultaneously with the change of electric conductivity caused by this adsorption. Only when the electric conductivity stops changing the EPR signal becomes notable and starts growing with time. The EPR signal is originated by O_2^-.

Third, it was shown in paper [65] that there is a motion of donors in partially reduced TiO_2, these donors being provided by oxygen vacancies and/or interstitial titanium atoms influencing the oxygen adsorption. The same effect applied to zinc oxide was mentioned in [17]. The local interaction between titanium enriched TiO_2 surface and oxygen adsorbed is in opinion of authors of paper [65] the cause of observed change in the behavior of adsorption of oxygen if compared to case of almost stoichiometric surface.

The results mentioned together with data outlined in Section 1.11 indicate that adsorption induced change in electric conductivity of sintered and partially reduced oxide is mostly dependent on adsorption related change in concentration of stoichiometric metal atoms which are responsible for dope electric conductivity rather than by charging of the surface of adsorbent due to transformation of radicals of O_2^- and O^-.

In our view the final verification was given to this conclusion in paper [66] in which simultaneous O_2 adsorption on partially reduced ZnO and resultant change in electric conductivity was studied. It was established in this paper that the energies of activation of chemisorption and that of the change of electric conductivity fully coincide. The latter is plausible only in case when localization of free electron on SS is not linked with penetration through the surface energy barrier which is inherent to the model of the surface-adjacent depleted layer.

The above reasoning enables us to formulate the mechanism of effect of oxygen adsorption on electric conductivity of sintered and partially reduced semiconductor oxides.

Namely, the adsorbents of such type are polycrystalline materials with dominant type of intracrystalline contacts in the shape of open bridges enriched in superstoichiometric metal, which is the principal electron donor. Adsorption of oxygen resulting in binding of superstoichiometric metal atoms leads to the change in concentration of free electrons in bridges which results in the change of electric conductivity of the whole adsorbent.

This model was proposed in early studies of Myasnikov and colleagues [48, 59]. It explains a vast amount of experimental data dealing not only with effect of molecular oxygen on electrophysical characteristics of adsorbents of this type but on effects of numerous other molecular, atom or radical particles capable of chemical interaction with highly active from this stand-point superstoichiometric metal atoms as well.

Now we can easily explain the experimentally observed function $s \sim P_{O_2}^{-0.5}$ [59]. Partially reduced oxides are usually highly doped semiconductors. In any case, their surface-adjacent layers can exhibit very high concentrations of excessive metal: $\geq 10^{18}$ cm^{-3} which is a simple electron donor [18, 67] (Fig. 2.11). As is known (see, for instance [68]) the phenomenon of polytrophy of dope is characteristic of highly doped semiconductors. Thus, this phenomenon deals with the fact that even at elevated temperatures the admixture can be ionized not to the full extent. The same comment applicable to oxides was made by Thomas [67]. In the case when similar to dope semiconductor containing not fully ionized admixture of one type the electric conductivity is proportional to the root of dope concentration we have:

$$\sigma \sim [e] \sim \sqrt{[Me_*]} \ , \qquad\qquad\qquad (2.15)$$

where $[Me_*]$ stands for concentration of superstoichiometric metal. Assume that due to low thickness of connecting bridges ($\sim 10^{-5}$ cm) and very high diffusion coefficients of interstitial metal for ZnO $D_{Zn} = 2.7 \cdot 10^{-4} \exp\{-0.55/kT\}$ cm$^2 \cdot$s^{-1} [67] the concentration of superstoichiometric metal is almost constant over the whole volume of the bridge.

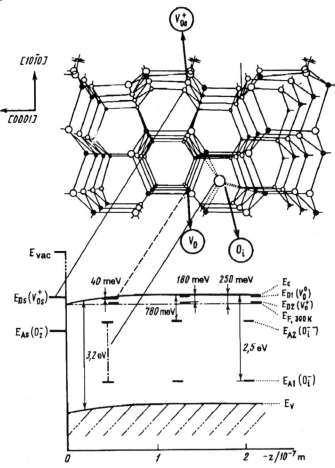

Fig. 2.11. Geometric and electron structure of the surface of ZnO($10\bar{1}0$) [33]: $V_{O_s}^+$ is the surface-adjacent defect; V_O and O_i are the volume defects which are the oxygen vacancies and interstitial oxygen, respectively; E_{DS} and E_{AS}, $E_{D1,2}$ and $E_{A1,2}$ are the surface and volume energy levels of donors and acceptors. The surface level E_{AS} is related to chemisorbed O_2.

Considering chemisorption as chemical interaction, in our case interaction of oxygen with adsorption centers which are modeled by surface-adjacent superstoichiometric metal atoms we can write down

$$O_2(g) \underset{\rightarrow}{\leftarrow} O_2(\phi) \ , \tag{2.16}$$

$$O_2(\phi) + Me_* \underset{\leftarrow}{\rightarrow} (Me^{\delta+}O_2^{\delta-})_s \ , \tag{2.17}$$

On the basis of the law of acting masses we have (under condition of equilibrium)

$$\left[\left(Me^{\delta+}O_2^{\delta-} \right)_s \right] / \left[O_2(\phi) \right] \left[Me_* \right] = K(T) \ . \tag{2.18}$$

Assuming that interaction of acceptor particles (in our case oxygen with superfluous metal) provides the unique reason of the change of concentration of the latter in adsorption process we arrive at the following expression

$$\left[\left(Me^{\delta+}O_2^{\delta-} \right)_s \right] = h \left\{ \left[Me_* \right]_0 - \left[Me_* \right] \right\} \ , \tag{2.19}$$

where $\left[Me_* \right]_0$ is the pre-adsorption concentration of superstoichiometric metal atoms. Under conditions of low surface occupation by adsorbed molecules the Henry adsorption isotherm may become valid, i.e. $\left[O_2(\phi) \right] \approx \Gamma P_{O_2}$. This brings us to situation when merging expressions (2.18) and (2.19) we obtain

$$\left[Me_* \right] = \frac{\left[Me_* \right]_0^{1/2}}{1 + P_{O_2} / P_C} \ , \tag{2.20}$$

substituting the expression obtained into (2.15) yields the dependence of the value of equilibrium electric conductivity plotted as a function of oxygen pressure:

$$\sigma \approx A(T) \frac{\left[Me_* \right]_0^{1/2}}{\sqrt{1 + P_{O_2} / P_C}} \approx \sigma_0 \left(1 + \frac{P_{O_2}}{P_C} \right)^{-1/2} \tag{2.21}$$

It follows from this expression that when $P_{O_2} < P_C = h / K(T)\Gamma$

$$\sigma \approx \sigma_0 \left(1 - \frac{P_{O_2}}{2P_C}\right) , \qquad (2.22)$$

when $P_{O_2} > P_C$

$$\sigma \approx \sigma_0 \left(\frac{P_C}{P_{O_2}}\right)^{1/2} . \qquad (2.23)$$

We should note that often in experimental practice the value characterizing both equilibrium post adsorption value of electric conductivity of
adsorbents and its absolute value during adsorption

$$a \equiv \sigma_0 \, \Delta\sigma / \sigma^2 , \qquad (2.24)$$

is used. Here $\Delta\sigma = \sigma_0 - \sigma$, σ is the equilibrium value of electric conductivity [1, 48, 69]. On the ground of expressions (2.18) and (2.19) one
can easily show that

$$a \equiv \frac{\sigma_0 \, \Delta\sigma}{\sigma^2} = \text{const } P_{O_2} , \qquad (2.24')$$

i.e. the above parameter is a useful characteristics enabling one to assess
the partial pressure of the gas analyzed in the ambient volume.

Finally, the function $\sigma \sim P_{O_2}^{-1/2}$ can be obtained for the case of practically full ionization of superstoichiometric metal atoms or oxygen vacancies as well. This means that the expression is being obtained when
condition $\sigma \sim [Me_\bullet]$ is valid. In this case, assuming the validity of the
process described by expression

$$O_2(\phi) + 2Me_\bullet \rightarrow (Me_\bullet O - OMe_\bullet)_s , \qquad (2.25)$$

i.e. assuming that chemisorbed particle of oxygen binds two interstitial
metal atoms simultaneously, we obtain the following equilibrium condition:

$$\frac{1}{2} \frac{h([Me_\bullet]_0 - [Me_\bullet])}{[O_2(\phi)][Me_\bullet]^2} = K(T) . \qquad (2.26)$$

We can derive from this expression the relationship for the value of
equilibrium electric conductivity of adsorbent as a function of oxygen
pressure:

$$\frac{\sigma_0}{\sigma} \approx \frac{1}{2}\left(1 + \sqrt{1 + 8P_{O_2}[\text{Me.}]_0 / P_C}\right) , \tag{2.27}$$

which yields for the limiting case $P_{O_2} < P_C / 8[\text{Me.}]_0$ that:

$$\sigma \approx \sigma_0 (1 - 2[\text{Me.}]_0 P_{O_2} / P_C) . \tag{2.28}$$

In case of large oxygen pressures ($P_{O_2} > P_C / 8[\text{Me.}]_0$) the relation

$$\sigma \approx \sigma_0 (P_C / 2[\text{Me.}]_0)^{1/2} P_{O_2}^{-1/2} \tag{2.29}$$

is valid.

Thus, over the whole interval of temperatures the function $\sigma \sim P_{O_2}^{-0.5}$ can be dependent on direct interaction of adsorbed oxygen with proper electrically active defects of adsorbent not resulting in notable charging of the surface.

We should note that expressions (2.21) and (2.27) were obtained in application to a specific bridge of the open type characterized by thickness h and initial concentration of superstoichiometric metal $[\text{Me.}]_0$. In real polycrystal with dominant fraction of bridges of this very type there is a substantial spread with respect to the thickness of bridges and to concentration of defects. Therefore, the local electric conductivity of the material in question is a random value of statistical ohmic subgrid formed by barrier-free contacts of microcrystals.

The efficient electric conductivity of such specially-inhomogeneous medium can be calculated in various limiting cases differing by the ratio between the local electric conductivity and its average value using methods developed in papers [70-72]. Namely, if the local electric conductivity $\sigma(r)$ slightly differs from the average one $\langle\sigma\rangle$, i.e.

$$\langle(\sigma - \langle\sigma\rangle)^2\rangle \ll \langle\sigma\rangle^2 , \tag{2.30}$$

then effective electric conductivity can be calculated applying the methods of the perturbation theory [70]:

$$\sigma_{ef} = \langle\sigma\rangle - \frac{1}{3}\frac{\langle(\sigma - \langle\sigma\rangle)^2\rangle}{\langle\sigma\rangle} . \tag{2.31}$$

When the condition $\langle(\sigma - \langle\sigma\rangle)^2\rangle \cong \langle\sigma\rangle^2$ holds the approximation of "apparent medium" is applicable [71]. In case of exponentially high inhomogeneity, i.e. when we have

$$\sigma(r)' = \sigma_0 \exp\{-\xi(r)\} \ , \tag{2.32}$$

where $\xi(r)$ is the distance from the bottom of the conductivity band to the Fermi level expressed in kT units and the inequality

$$\langle(\xi - \langle\xi\rangle)^2\rangle \gg 1 \ , \tag{2.33}$$

is observed the situation can be considered from the stand-point of the percolation theory (see Section 1.10). It is evident that the last limiting case corresponds to the situation considered in Section 1.10 dealing with availability of wide spread in heights of intercrystalline barriers specific for polycrystallites with dominant bridges of closed type in the material or for the case of pressed powder [58, 73].

In the case under consideration — partially reduced oxides — the expression (2.31) can be applied which is valid for small deviations of local σ from average $\langle\sigma\rangle$. This is caused by the fact that in case of constant current conductivity is only provided by barrier-free sections, i.e. the bridges with the lowest values of resistivity. It is these paths characterized by varying thickness of conducting branches and, plausibly, by varying concentration of defects that provide the major input into the electric conductivity of the material. Therefore, the adsorption-induced change (in case of acceptor adsorption) of electric conductivity of these paths provides the change of electric conductivity of the whole adsorbent. By the meaning of selection of low ohmic paths they cannot contain separate sections characterized by either low values of $[Me*]_0$ or low h, because this would have resulted in low values of electric conductivity and would not have aloud them to get included into a low ohmic conducting subgrid. The sections with high (if compared to the average value of concentrations) h or with high value of $[Me*]_0$ provide no practical input into the total resistivity of conducting branch. Thereby, both the value of the initial electric conductivity of adsorbent and the changes induced by adsorption are determined by bridges or, which is more correct, by sections of conducting branches with values h and $[Me*]_0$ close to the average ones in case of sintered adsorbents. Here it is relevant to mention that adsorption reducing the degree of inhomogeneity in adsorbent (which is applicable to acceptor adsorption) results in most rapid decrease in electric conductivity of sections characterized by highest initial values of σ and finally results in applicability of expression (2.31).

Consequently, the adsorption-induced dependence of σ on P_{O_2} has the form

$$\sigma \approx \langle \sigma \rangle \approx \int_{h} \int_{[Me_*]_0} A(T) [Me_*]_0^{1/2} \left(1 + \frac{\Gamma k P_{O_2}}{h}\right)^{-1/2} f(h, [Me_*]_0) dh d[Me_*]_0 , \quad (2.34)$$

where $f(h, [Me_*]_0)$ is the joint density of distribution of the thickness of the branch and initial concentration of superstoichiometric metal. It is absolutely obvious that we do not have access to any reliable information concerning distribution of h or that of $[Me_*]_0$, to say nothing of interrelation of these distributions. Assumption concerning the normal logarithmic distribution of thickness of bridges made in [24] is, unfortunately, groundless. However, taking into account the determining role of sections with average values of h and $[Me_*]_0$ we can write down for post-adsorption value of σ with substantial accuracy

$$\sigma \approx \frac{\langle \sigma_0 \rangle}{\sqrt{1 + \frac{P_{O_2}}{\langle P_C \rangle}}} , \quad (2.35)$$

where $\langle \sigma_0 \rangle$ and $\langle P_C \rangle \equiv \langle h \rangle / \Gamma k(T)$ are the average values of respective characteristics. In case of applicability of inequality $P_{O_2} > \langle P_C \rangle$ we arrive at the following expression

$$\sigma \approx \langle \sigma_0 \rangle \langle P_C \rangle^{1/2} P_{O_2}^{-1/2} , \quad (2.36)$$

which is usually confirmed by experiments.

Note that expressions (2.24') and (2.36) are applicable for equilibrium adsorption of molecular gases. Moreover, this dependence becomes valid during adsorption of oxygen too when the latter is diluted in polar liquids such as H_2O, CH_3OH, C_4H_9OH, C_6H_{12} etc. [59]. The results of these studies related to evaluation of possibility to detect the traces of oxygen in liquid media as well as plausible mechanisms of phenomena will be considered in detail in Chapter 4.

On the basis of consideration of this model one can easily obtain expressions describing kinetics of the change in electric conductivity of sintered partially reduced oxide adsorbent during development of direct and inverse reaction described by expression (2.17). The consumption of superstoichiometric metal atoms which takes place during adsorption of O_2 is described by equation

$$\frac{d[Me_*]}{dt} = -\tilde{k}_1 [Me_*] [O_2(\phi)] , \quad (2.37)$$

where $\tilde{k}_1$ is the constant of direct reaction rate. When the opposite reaction develops, the equation of the balance will take shape

$$\frac{d[\text{Me}_\bullet]}{dt} = \tilde{k}_2\,[(\text{Me}^{\delta+}\text{O}_2^{\delta-})_s]\ , \tag{2.38}$$

where $\tilde{k}_2$ is the constant of inverse reaction rate. Taking into account expression (2.19) the last equation can be rewritten as

$$\frac{d[\text{Me}_\bullet]}{dt} = \tilde{k}_2\,h\,\{[\text{Me}_\bullet]_0 - [\text{Me}_\bullet]\}\ . \tag{2.39}$$

Integrating equations (2.37) and (2.39) under assumption that in case of direct reaction of surface complex formation $(\text{Me}^{\delta+}\text{O}_2^{\delta-})_s$ the reaction of interaction of oxygen with surface metal atoms is the limiting stage rather than formation of physadsorbed oxygen (i.e. assuming that $[\text{O}_2(\phi)] = $ const and it does not change in time) we arrive to the respective expression for kinetics of direct and inverse reactions:

$$[\text{Me}_\bullet] \approx [\text{Me}_\bullet]_0 \exp\{-\tilde{k}_1\,[\text{O}_2(\phi)]\,t\}\ , \tag{2.40}$$

$$\ln\left\{1 - \frac{[\text{Me}_\bullet]}{[\text{Me}_\bullet]_0}\right\} = -\tilde{k}_2 h t + \text{const}\ , \tag{2.41}$$

where const $= \ln\{k\,[\text{O}_2(\phi)]/(h + k\,[\text{O}_2(\phi)])\}$, $k = \tilde{k}_1/\tilde{k}_2$. Assuming, as usual, the applicability of the linear character of adsorption $[\text{O}(\phi)] \approx \approx \Gamma P_0$ we obtain the following expressions for kinetics of the change of electric conductivity for direct and inverse reactions

$$\frac{\sigma_0}{\sigma} = \exp\left\{\frac{1}{2}\tilde{k}_1\,\Gamma P_{\text{O}_2} t\right\}\ , \tag{2.42}$$

$$\ln\left\{1 - \left(\frac{\sigma}{\sigma_0}\right)^2\right\} = -\tilde{k}_2 h t + \text{const}\ . \tag{2.43}$$

We should note that separately expressions (2.42) and (2.43) can describe the kinetics of the whole process (2.17) but only away from the equilibrium position, i.e. when the second process is practically undetectable. The precondition for direct and inverse reaction are provided by not very large times t which means that in case of direct reaction $\sigma(t)$ should not significantly differ from σ_0, whereas in case of inverse reaction – from equilibrium value σ_{eq} provided by expression (2.21).

Taking into account above considerations as well as the fact that $\sigma_{eq} \ll \sigma_0$ we can write down the approximate expression for kinetics of the change in $\sigma(t)$ caused by development of direct reaction

$$\frac{\sigma_0}{\sigma} \approx 1 + \frac{1}{2} \tilde{k}_1 \Gamma P_{O_2} t , \qquad (2.44)$$

In case of inverse reaction we arrive at the following expression

$$\ln \left\{ 1 - \left(\frac{\sigma}{\sigma_0} \right)^2 \right\} \approx -\tilde{k}_2 h t + \text{const} . \qquad (2.45)$$

As it was mentioned in paper [48], expressions (2.44) and (2.45) perfectly describe experimental situation with kinetics of the change of electric conductivity of sintered and partially reduced ZnO film during adsorption and desorption of molecular oxygen. Expression (2.44) describes the kinetics of the change of σ during adsorption of O_2 on ZnO with the surface rich in donors due to photolytic decomposition of ZnO in vacuum fairly well [74].

In case when both direct and inverse reactions have a significant input into the process the general expression for kinetics of the change of electric conductivity caused by adsorption of O_2 has the form

$$\sigma(t) \approx \sigma_0 \left(1 + \frac{P_{O_2}}{P_C} \right)^{-1/2} \sqrt{1 + \frac{P_{O_2}}{P_C} \exp \left\{ -\tilde{k}_2 t \left(1 + \frac{P_{O_2}}{P_C} \right) h \right\}} . \qquad (2.46)$$

When the condition $P_{O_2} > P_C$ is met we obtain the following expression for kinetics $\sigma(t)$

$$\sigma(t) \approx \sigma_0 \left(\frac{P_C}{P_{O_2}} \right)^{1/2} \sqrt{1 + \frac{P_{O_2}}{P_C} \exp \left\{ -\tilde{k}_1 \Gamma P_{O_2} \cdot t \right\}} , \qquad (2.47)$$

It is evident from this expression that the higher the pressure of oxygen in ambient volume and higher the temperature the faster is the rate of the change in adsorbent electric conductivity.

The expression

$$\left. \frac{d\sigma}{dt} \right|_{t \to 0} \approx -\frac{1}{2} \sigma_0 \tilde{k}_1 [O_2(\phi)] , \qquad (2.48)$$

is valid for initial rate of the change of electric conductivity. This relation can be written as

$$\left.\frac{d\sigma}{dt}\right|_{t\to 0} \approx -\frac{1}{2}\sigma_0\tilde{k}_1\,\Gamma P_{O_2}\,. \tag{2.49}$$

in the linear adsorption range, i.e. in case when the Henry isotherm is applicable. The two last expressions obviously indicate on proportionality of the value of initial rate of adsorption induced change in electric conductivity of adsorbent with concentration of the particles occupying its surface or in a more general case with concentration of these particles in ambient volume.

The above result was used as a ground-stone of the well known kinetic method of detection which was initially proposed by Myasnikov [75] more than 30 years ago. Above paper dealt with experimental comparison of the change of relative concentration of CH_3 radicals in gaseous phase using the stationary values of electric conductivity and initial rate of its change. The experiment yielded perfect coincidence of the measured values. Using methyl radicals as example of adsorption it was established that the resolution of this method was better than 10^7 particles per cubic centimeter of the ambient volume [75, 76].

The considered model of the change of electric conductivity of partially reduced surface oxides caused by adsorption of molecular acceptors is based on purely local mechanism of chemisorption [29]. It should be pointed out that the features of the change of electric conductivity of adsorbent established using adsorption of O_2 as an example are applicable to other acceptor molecules, as well. For instance ammonium [77], iodine [78, 79], chlorine were confirmed by various techniques to have the major effect of the influence on electric conductivity of adsorbent by direct interaction with superstoichiometric metal atoms. As it has been shown above the collective effects related to charging of the adsorbent surface and therefore to bending the bottom of the conductivity band resulting in the change in concentration of free charge carriers, are not profound in such systems due to high degree of defects of partially reduced oxides. This is confirmed by results of experiments with atomic beams mentioned in Section 1.12 [80, 81]. We can conclude, therefore, that numerous factors which have a certain inconsistency in explanation of adsorption measurements present in oxides with an a priori surface depletion layer are not important in the case under consideration.

First, this is applicable to recharging of biographic surface state and plausible masking of adsorption caused charging of the surface due to this effect. In our case of partially reduced oxides having high degree of non-stoichiometricity of the surface a fraction of adsorbed oxygen can form charged surface forms of O_2^- and/or O^- (which, naturally, results in possible recharging of biographic surface states). However, despite this feature the major change in electric conductivity, as it has been unambiguously shown by experiments mentioned is caused not by these processes. Small thickness of conducting branches, at the same time, provides for retaining the plane features of bands.

A negligible degree of effect of charging of bridge surface together with existence of plane bands in them is responsible for negligible effects of adsorption on mobility of free charge carriers. For instance, the complex studies of effect of adsorption of both acceptor and donor particles on electric conductivity, concentration and mobility of charge carriers conducted in [46] showed that adsorption induced change in electric conductivity of vacuum sintered ZnO and TiO_2 is mainly caused by the change in concentration of free charge carriers. In particular, the measurements of electric conductivity and the assessment of Hall constant during adsorption of O_2 on ZnO showed that when $\Delta\sigma/\sigma_0 = 0.1$ the value of $\Delta n/n_0$ describing the change in concentration of the charge carriers is about 90% of relative decrease in electric conductivity. We should note that as it can be concluded from the same paper the input caused by the change in mobility of carriers into the change of electric conductivity increases with increase in concentration of chemisorbed particles. This, presumably, is linked with possible origination of high energy barriers in several branches of conducting grid, and, therefore, with increase in "apparent scattering" of current carriers during adsorption of acceptors and with inverse process, i.e. with removal of high barriers during donor adsorption.

Note that besides the kinetics of the change of electric conductivity (2.44) – (2.47) considered, the logarithmic kinetics similar to that of Zeldovich – Roginsky – Elovich is often observed in experiment (see Section 1.10) [30, 65, 82, 83]:

$$\sigma(t) = A - B \ln (t + t') , \qquad (2.50)$$

where A, B, t' are constants. The functions of this kind are usually explained taking into account the availability of inhomogeneity of the surface of adsorbent with respect to the value of activation adsorption energy or accounting for existence of the surface barrier impeding the transition of electrons from the conductivity band to the surface levels of adsorption type.

Indeed, assuming that the surface of adsorbent is characterized by homogeneous distribution in adsorption activation energy we arrive to the Zeldovich – Roginsky – Elovich kinetic isotherm of $N_t \sim \ln \{(t + t')/t'\}$ providing the validity of function (2.50) in case of direct interaction of adparticles with conductivity electrons or electrically active defects. The mechanism based on the model of doped and partially compensated semiconductor where adsorbed acceptors act as a compensating dope was proposed in paper [82]. Note, however, the logarithmic kinetics of the change of electric conductivity is a main peculiarity of the oxidized samples. This is directly confirmed by results obtained by Komuro [65] who showed that dependence of the type (2.50) was observed only in case of practically stoichiometric surface, i.e. in case of existence of surface-adjacent layer depleted in principal carriers. This dependence vanishes when the surface-adjacent layer gets saturated in metal atoms. However, the logarithmic kinetics of oxygen adsorption itself is a peculiarity of oxides with substantial oxidation degree [30]. Only results of the paper by Enikeev [84] who directly examined the effects of oxygen adsorption on the height of surface barrier in ZnO indicate that the logarithmic kinetics of the change in electric conductivity is sooner caused by effects of the surface barrier on kinetics of electron transitions between SS of adsorption origin and volume bands of adsorbent rather than by an a priori inhomogeneity of the surface.

The origin of kinetics (2.50) stemming from existence of a surface-adjacent barrier can be easily understood on the basis of results outlined in Section 1.7. Indeed, at small surface deposition degrees and far from the equilibrium position the kinetic expression (1.54) can be represented as

$$\frac{dn_t}{dt} \approx K_{n_t} N_{ta} \exp\{-\xi(t)\}, \tag{2.51}$$

where, as always, N_{ta} is the surface concentration of chemisorbed acceptors; $n_t(t)$ is the concentration of their charged form at moment t. For the height of the surface barrier under condition $n_t < n_B$, i.e. for substantially high density of biographic SS the deposition degree which is weakly sensitive to adsorption we arrive to expression

$$\xi(t) \approx \left(\frac{n_B + n_t}{\sqrt{2}\,N_D L_D}\right)^2 \approx \xi_0 \left(1 + 2\frac{n_t}{n_B}\right), \tag{2.52}$$

where $\xi_0 = (n_B / \sqrt{2}\,N_D L_D)^2$ whose substitution into (2.51) results in equation of the Zeldovich – Roginsky – Elovich type

$$\frac{dn_t}{dt} \approx A\,e^{-Bn_t}, \tag{2.53}$$

whose solution has the form

$$n_t(t) = \tilde{\alpha} \ln (t + t') - \tilde{\beta} \ . \tag{2.54}$$

Substituting (2.54) into (1.45) ($(\sigma \cong \sigma_0 - q\mu_n(n_t + n_B))$) results in formula (2.50). Therefore, the logarithmic kinetics of the change of electric conductivity during oxygen adsorption is rather the feature of monocrystalline and sintered polycrystalline samples characterized by practically stoichiometric content of the surface, i.e. by existence of the surface-adjacent depletion area with respect to the principal type of free charge carriers. We should point out that in case of sintered polycrystalline samples characterized by logarithmic kinetics of adsorption change of electric conductivity the major type of intercrystalline contacts should be provided by bridges of both closed or open types, still featuring the surface-adjacent depletion domains.

2.4.2. Adsorption of molecular donor particles. Molecular hydrogen. Carbon monoxide

The issue regarding the change in electric conductivity of oxide semiconductor caused by effects of adsorption of donor particles, the molecules of inflammable and reduction gases in particular is not clear so far. There is the whole series of surface and volume processes triggered by adsorption of such gases. The development of these processes results in the change of electric conductivity of adsorbents. Depending on the origin of adsorbent, its preparation history as well as regime and conditions adsorption progress a specific process or reaction would determine the change in electric conductivity and could be chosen as a basic mechanism of sensor performance. As it has been pointed out in review by Clifford [54] the dominant viewpoint now is that detection of reducing and inflammable gases through application of semiconductor oxides is based on one of three possible processes.

1. When reducing gases interact with the oxygen adsorbed in a charged form there is a product created that easily delivers electron into the conductivity zone with fast desorption [85-88]:

$$R(g) + O_s^- \rightarrow B(g) + e \ . \tag{2.55}$$

Here R is the reducing gas; B is the gaseous product. In case of semiconductor of n-type the electric conductivity of adsorbent goes up.

2. The reducing or inflammable gas proper acts as an electron donor. The molecules adsorbed inject electrons into the conductivity band thereby increasing the electric conductivity [85, 86, 89]:

$$R(g) \underset{\leftarrow}{\rightarrow} R_s \underset{\leftarrow}{\rightarrow} R_s^+ + e \ . \tag{2.56}$$

3. Due to participation in oxidation-reduction reactions the reducing or inflammable gases affect the stoichiometricity of oxide and, consequently the concentration of stoichiometric defects which usually control the dope electric conductivity of adsorbent [26, 67, 85, 86, 90]

$$R(g) + MeO_2 \underset{\leftarrow}{\rightarrow} B(g) + (MeO)^{++} + 2e^- \tag{2.57}$$

or

$$R(g) + MeO \underset{\leftarrow}{\rightarrow} F(v) \underset{\leftarrow}{\rightarrow} F^+(v) + e^- \ , \tag{2.58}$$

where $F(v)$ is the volume F-center.

Obviously we are most interested in the use of partially reduced sintered oxides as adsorbents the principal processes affecting electric conductivity being provided by reactions (2.57) – (2.58). Note that now we are referring to the accurate measurements of effects of extremely small concentrations of the gas under study in the known (usually neutral) medium. In this case the major input into the change of electric conductivity in domain of moderate temperatures will be provided by the process (2.56). We should note that effects of the change of electric conductivity of adsorbent caused by reduction processes (2.57) or (2.58) are least attractive from the stand-point of manufacturing of sensors for respective reduction gases which is linked with irreversible character of above processes and resultant irreproducibility of experimental results. As for reaction of reducing gases with chemisorbed oxygen present at the surface of adsorbent (it has been mentioned in [54]), this mechanism seems to be the most probable for chemisorption response of semiconductor sensor of the "gas Taguchi sensor" type [91]. The latter is manufactured during high temperature healing in air. The applicability domain of such sensors is provided by determination of concentration of various gases in air. We should note, however that in several cases mechanism (2.55) can also be responsible for the change in electric conductivity of the partially reduced oxides used in our experiments. This is relevant to the studies of various physical and chemical processes featuring oxygen together with reducing gases.

Apart from the free major types of processes resulting in the change of electric conductivity of oxides during adsorption of donor particles referred to in review [54], reducing and inflammable gases in particular, in our view there is a fourth type responsible for the change in electric conductivity of oxide adsorbent which is observed during adsorption of alcohol and several other water containing compounds. As it will be shown in Section 4.5 in this case oxide plays a two-fold role: it is an active catalyst in dehydration reaction and simultaneously it is a sensitive element of the sensor with respect to atom hydrogen which is formed in course of the reaction. This is the data on concentration of the formed H-atoms that enable one to assess concentration of the gas detected.

Let us dwell in more detail on above mechanisms and their experimental substantiation.

Molecular hydrogen is one of the first particles of donor type whose response to the adsorption-induced change in electrophysics of oxide semiconductors is known. As results of the studies conducted in paper [89] the quantitative features linking chemisorption of hydrogen on partially reduced zinc oxide with its electric conductivity were established. A plausible mechanism of increase in electric conductivity of ZnO in presence of hydrogen linked not to chemisorption of molecular hydrogen which, as experiments indicate develops at low temperatures without change in electric conductivity but rather with the secondary processes, i.e. dissociation of H_2 into atoms with subsequent ionization of chemisorbed hydrogen atom resulting in injection of electron into the conductivity band was proposed as one of mechanisms of increase in electric conductivity of ZnO in presence of hydrogen. It was established that a notable change in electric conductivity of ZnO during adsorption of H_2 results at $T > 100°C$ whereas the chemisorption of hydrogen on ZnO is observed even at very low temperatures ($\sim -78°C$) [92]. On the basis of detailed studies of the processes of adsorption and desorption of hydrogen the following diagram was proposed in paper [89]:

$$H_2(g) \underset{\leftarrow}{\rightarrow} (H_2)_s \ , \tag{2.59}$$

$$(H_2)_s \underset{\leftarrow}{\rightarrow} 2H_s \ , \tag{2.60}$$

$$H_s \underset{\leftarrow}{\rightarrow} H_s^+ + e \ . \tag{2.61}$$

The limiting stage in this reaction is provided by dissociation of H_2. It is assumed that at low temperatures only reaction (2.59) develops which does not result in the change in electric conductivity.

Taking into account the studies of electric conductivity as a function of P_{H_2} as well as the studies of kinetics of electric conductivity accompanying desorption of hydrogen the authors of paper [89] established that apart from processes (2.59) – (2.61) there is a recombination

$$H_s + H_s^+ \underset{\leftarrow}{\rightarrow} (H_2^+)_s \ , \qquad (2.62)$$

resulting in the fact that during equilibrium all elementary stages can be reduced to the consolidated reaction

$$H_2(g) \underset{\leftarrow}{\rightarrow} (H_2^+)_s + e \ . \qquad (2.63)$$

The kinetic equation describing the change in concentration of conductivity electrons and consequently the time dependence of electric conductivity has the following shape

$$\frac{d\sigma}{dt} \sim \frac{d[e]}{dt} = k_1 [H_2]_s - k_2 [H_2^+]_s [e] \ . \qquad (2.64)$$

Here k_1 and k_2 are the kinetic constant of direct and inverse processes (2.62) which are provided by dissociation of H_2 and recombination of the hydrogen ion with electron with subsequent desorption.

The condition of electron neutrality in case under consideration can be written as

$$h[e] = [H_2^+]_s + h [Me_\bullet^+] \ , \qquad (2.65)$$

where as usual $[Me_\bullet]$ is the concentration of superstoichiometric zinc. We will focus our attention on the case of low deposition on the surface and in this case we can assume

$$[Me_\bullet^+] \approx [Me_\bullet^+]_0 = [e]_0 \ ,$$

where "zero" subscript indicates pre-adsorption concentrations. Thus, we arrive at the following equation

$$\frac{d(\sigma / \sigma_0)}{dt} = k_1' - k_2' \left(\frac{\sigma}{\sigma_0} \right)^2 + k_2' \left(\frac{\sigma}{\sigma_0} \right) , \qquad (2.66)$$

where $k_1' = k_1 [H_2]_s / [e]_0$; $k_2' = k_2 [e]_0 h$.

Integrating (2.66) with initial conditions $\sigma (t = 0) = \sigma_0$ we obtain

$$\text{Arth} \left(\frac{2 (\Delta\sigma / \sigma_0) + 1}{\sqrt{1 + 4k'}} \right) = \frac{t}{2} k_2' \sqrt{1 + 4k'} + \text{Arth} \left(\frac{1}{\sqrt{1 + 4k'}} \right) , \qquad (2.67)$$

where $\Delta\sigma = \sigma - \sigma_0$; $k' = k_1' / k_2' = (k_1 / k_2) \left([H_2]_s / h [e]_0^2 \right)$.

We can easily obtain from this expression the relations for kinetics of direct and inverse reaction (2.63):

$$\frac{\Delta\sigma}{\sigma} = k_1[H_2]_s t \ , \tag{2.68}$$

$$\ln\left(1 + \frac{\sigma_0}{\Delta\sigma}\right) = k_2[e]_0 ht + \text{const} \ . \tag{2.69}$$

When the condition of dynamic equilibrium occurs we arrive at expression linking the equilibrium value of electric conductivity with parameters of adsorbate-adsorbent system:

$$\left(2\frac{\Delta\sigma}{\sigma_0} + 1\right)^2 = 4k' + 1 \ . \tag{2.70}$$

In case of applicability of the Henry isotherm, i.e. in range of small H_2 pressures the expression (2.70) acquires the shape

$$\left(2\frac{\Delta\sigma}{\sigma_0} + 1\right)^2 = 4kP_{H_2} + 1 \ . \tag{2.71}$$

The fair agreement of expressions (2.67) and (2.71) with experimental data as well as agreement of independently obtained experimental data concerning kinetics of the change of σ with the data on equilibrium enabled the author of paper [89] to conclude that the proposed mechanism of effect of hydrogen on electric conductivity of semiconductors can be one of active mechanisms. The heat of total reaction (2.63) calculated from the values found was about 4.6 kcal.

We should note that under condition $4k \cdot P_{H_2} > 1$ the expression (2.71) can be written as

$$\sigma \approx \sigma_0\sqrt{k}P_{H_2}^{1/2} \ , \tag{2.72}$$

i.e. it shows the power law which is specific for dependence of the value of chemisorption response of electric conductivity of adsorbent on the pressure of reducing or inflammable gas [54].

The conclusion of the study [89] concerning non-dissociated character of hydrogen adsorption on semiconductors of ZnO type at low temperatures (lower than 70°C) was based on experimental facts indicating that hydrogen atoms which independently arrive at the surface of semiconductor are almost totally ionized with small activation energy (~ 2 kcal), and the change in electric conductivity in this case was observed up to ~ −190°C [93]. Note that during adsorption of molecular hydrogen the activation energy of the change in electric conductivity of adsorbent was about 30 kcal [89].

It should be mentioned, however, that in case under consideration as well as during adsorption of oxygen the principal input into the change of electric conductivity is provided by the change in concentration of the current carriers (in our case conductivity electrons). Yet, in this case this is caused by ionization of the surface states of the donor type formed during adsorption. Similar to previous considerations the charging of the surface occurring during this process has almost no impact of the value of electric conductivity.

We should note that this effect of the hydrogen on electric conductivity is related neither to reduction of oxide nor to the volume dissolution of hydrogen in oxide. The first option was ruled out due to initial heating of adsorbent in hydrogen atmosphere up to establishing the stationary value of electric conductivity at the temperature exceeding the maximum temperature of experiment ($T_{heating}$ ~ 400 ÷ 420°C). The conclusion regarding the absence of additional dissolution of hydrogen under experimental conditions or, at any rate, regarding the absence of notable effect of this process on electric conductivity of adsorbent can be done on the basis of comparing the results obtained in [89] with the well known results obtained for the process of dissolution of hydrogen in zinc oxide [94, 95]. Thus, the heat of consolidated reaction resulting in formation of free electrons amounts to 3.2 eV in case of dissolution of hydrogen [95]. At the same time for the case under consideration the respective value of interaction of H_2 with ZnO is of the order of 0.2 eV. The change in electric conductivity of adsorbent observed during chemisorption coincides with the pre-adsorption values in the order of magnitude [89]. This indicates that the concentration of conductivity electrons change almost by two times. In case when such a change in concentration of carriers is caused by dissolution process the dependence of σ on P_{H_2} is $\sigma \sim P_{H_2}^{1/4}$ in contrast to the function observed with the power index 0.5. Moreover, as it was shown in [95] at T ~ 400°C and P_{H_2} ~ 7 ÷ 8 Torr (i.e., approximately at the same conditions as those described in [89]) the respective change in concentration of conductivity electrons was ~ 10^{15} cm^{-3}, which is 2 − 3 orders of magnitude lower than that observed in paper [89].

Let us dwell now on the issue of non-dissociative form of low temperature chemisorption of H_2 proposed in [89]. As it has been mentioned above absolutely different behavior of electric conductivity of adsorbent observed during adsorption of H-atoms [93] favors such a conclusion. However, approximately at the same time based on IR-spectra of adsorption of hydrogen on zinc oxide Aishens and co-authors showed [96] that hydrogen adsorption resulting in transition of electrons from surface oxygen ions to 4s-levels of regular zinc ions causes formation of two electron bonds – OH and ZnO:

$$ZnO + H_2 \underset{\leftarrow}{\rightarrow} ZnH + OH \ . \tag{2.73}$$

As a prove the authors of paper [96] referred to the fact that adsorption bands occurring during low temperature adsorption are close to the bands of valency oscillations of OH and ZnH in gaseous phase. Moreover, the results of deuterium exchange also indicate on formation of OH and ZnH groups. There is a linear correspondence between intensities in OH and ZnH bands and the ratio of concentrations of OH and ZnH groups is equal to one. The studies of dependence of intensity of OH and ZnH bands for a single ZnO sample plotted against hydrogen adsorbed at 30°C enabled authors of paper [96] to split the curve of the change of intensity into three sections: the first section does not feature bands of OH and ZnH groups despite ongoing hydrogen adsorption; the second section shows a linear increase in intensity of OH and ZnH bands; the third section displays a weak, non-linear rise in intensity of bands. In authors' opinion the he third section is related to adsorption of neutral hydrogen in formation of H^+-ions due to the process $H^0 + + Zn_i^{++} \rightarrow H^+ + Zn_i^+$, i.e. transition of electron from initially neutral H-atom into the interstitial ion of superstoichiometric zinc. According to Aishens and co-authors it is this formed Zn_i^+ that is responsible for the rise in electric conductivity at $T > 60°C$ which is confirmed by decreased background transmission on free carriers.

However, the comparison of the whole series of experimental facts involving IR-spectroscopy of adsorption of molecular and atomic hydrogen as well as the change in electric conductivity of adsorbent is indicative of a more complex phenomenon. For instance, in paper [97] both the spectra of adsorption of adsorbed molecular hydrogen were studied together with those of hydrogen atoms adsorbed from gaseous phase. In case when H_2 are adsorbed in a dissociative manner one would have expected a manifestation of the same bands 3498 and 1708 cm^{-1} or at least one of them inherent to adsorption of H-atoms in the spectrum of ZnO.

In reality one observes the increase in adsorption at 3672 cm^{-1} despite the fact that at 3498 and 1708 cm^{-1} adsorption not only increases, but, on the contrary, decreases as amount of adsorbed H-atoms grows accompanied by increase in adsorbed H-atoms. Together with above change in IR-spectra the increase in electric conductivity and decrease in background transmission were observed. It should be mentioned, however, that the band 3672 cm^{-1} was retained in case of air removal from the ambient volume adjacent to ZnO at room temperature whereas the bands formed during adsorption of H$_2$ completely disappeared during removal of air at the same temperatures according to reports of papers [97] and [96]. Another fact in opinion of the author of paper [97] which does not fit into the concept of low temperature dissociative adsorption of H$_2$ is provided by the fact that above bands 3498 and 1708 cm^{-1} are traced during H$_2$ adsorption on ZnO even in case of very low temperatures (of the order of –150°C). The lower the temperature the lower the pressure of H$_2$ at which above bands appear.

In order to explain all experimental results together with authors of [97] we should assume that under conditions of low temperatures H$_2$ is adsorbed on ZnO in the form which is intermittent between purely molecular and purely atomic one using the schematics given in [96, 98]

$$
\begin{array}{cc}
\boxed{
\begin{array}{c}
\mathrm{H - H} \\
| \quad | \\
-\,\mathrm{O^{2-}} - \mathrm{Zn^+} - \mathrm{O^-}
\end{array}
} - \mathrm{Zn^{2+}} - \quad \text{or} \quad
\begin{array}{c}
\mathrm{H^- - H^+} \\
| \quad | \\
-\,\mathrm{O^{2-}} - \mathrm{Zn^{2+}} - \mathrm{O^{2-}} - \mathrm{Zn^{2+}} -
\end{array}
& (2.74)
\end{array}
$$

The chemical bond in such molecules is relatively loose which, possibly, predetermines the elevated activity of the form of adsorbed hydrogen manifested in IR-spectra, for instance this may be the case in hydration reaction or deuterium exchange [96, 99, 100]. It is the availability of these two closely positioned atoms that causes the probability of the local charge transition between surface ions inside an adsorption complex without participation of the free electrons. The brake up of a weakened bond in admolecules of H$_2$ and the spread of ZnH and OH groups along the surface become feasible with the rise in temperature. In opinion of authors of paper [98] the increase in concentration of free carriers (i.e. the rise in electric conductivity) can be linked with recombination and subsequent desorption of hydrogen atoms adsorbed on Zn-ions resulting in injection of two electrons from each pair into the conductivity band. It should be mentioned that according to the data of paper [89] the activation energy of the change in electric conductivity of ZnO was of the order 30 kcal/mol during adsorption of molecular hydrogen which is pretty close to the value of hydrogen diffusion activation energy (~ 1.12 eV) for zinc oxide [94]. Keeping this in mind one should account for the fact that such high activation energy of the change in electric conductivity during adsorption of molecular hydrogen can be linked with necessary migration of hydrogen adsorbed on Zn-ions in order to form $(O-H)^{2-}$-groups with oxygen ions of the lattice which, according to [67], act as free donors with low activation energy (thus, in case when their concentration is lower than $5 \cdot 10^{16}$ cm^{-3} the energy of ionization being about 0.051 eV). Note, that above brake in H–H-bond accompanied by the rise in electric conductivity of adsorbent can be observed at the room temperature as well due to effect of bombardment of the surface of adsorbent by the flux of slow electrons [101].

The low energy of activation of the change in electric conductivity of zinc oxide observed during adsorption of H-atoms (~ 0.08 eV) [102] can correspond to the ionization energy of $(O-H)^{2-}$-groups formed during direct interaction of H-atoms with O^{2-}-ions of the lattice.

In several cases the adsorption of molecular gases can result in a more complex pattern of the change of electric conductivity in adsorbent. This, for instance, is observed during adsorption of CO on metal oxides. Thus, during adsorption of carbon oxide on TiO$_2$ at $T < 330K$ it acts as acceptor, whereas at more high temperatures it becomes donor [33, 103]. The experiments on thermodesorption indicate a certain amount of CO$_2$ is present in the volume surrounding adsorbent. Such a behavior of CO on TiO$_2$ can be linked with various mechanisms of effects of CO on the process of formation of point defects [33]. Presumably, the formation of the surface defects is also an important stage in the process of the change of electric conductivity of ZnO during adsorption of CO [33, 104]. However, up-to-date there is no clear explanation of the principles of performance of the sensor on CO, i.e. there are no mechanism provided to explain the response of electric conductivity [25]. Windishman and Mark [87] proposed a microscopic model of the phenomenon based on CO oxidation by chemisorbed oxygen present on the surface with subsequent emission of electron into the adsorbent conductivity band. In this paper an attempt was made to explain the following experimental facts: the necessity of presence of oxygen in ambient volume, which means that detection of reducing gas is only possible in non-equilibrium gaseous mixture featuring both the reducing gas to be detected (CO, H$_2$, etc.) and oxygen; the existence of the "temperature window" outside of which there is a notable drop in sensitivity of the sensor, in other words, there is a temperature maximum of response, i.e. the relative change in electric conductivity when the reducing gas is let into the chamber initially grows with the rise of the temperature followed by its drop to zero at substantially high temperatures; the availability of the root dependence of $\sigma \sim P_R^{1/2}$, where P_R is the partial pressure of reducing gas. Note that there is a split in opinions with respect to the latter function. For instance, Morrison [105] referring to the "Figaro catalogue of gaseous sensors" maintains that the square-root function provides a perfect approximation for majority of the known results of effects of reducing gases on electric conductivity of sensors made of TiO_2. Clifford [55] assumes that the conductivity of such sensors is related to the partial pressure of reducing gases through function $\sigma \sim P_R^\beta$, where β satisfies the condition $0 < \beta < 1$ and is proportional to the temperature $\beta = at$. McAleer and co-authors [25] maintains that $0.3 < \beta < 0.8$ for sensors manufactured on SnO_2. In paper by Straisler and Rais [106] various kinetic schemes explaining the above groups dependence with $0.5 < \beta < 1$ were proposed. Nevertheless, all these papers are based on assumption regarding interaction of reducing agent with oxygen adsorbed on the surface of oxide, i.e. oxygen acts as a specific sensitizer providing the change in electric conductivity.

It was assumed in paper [87] that the reducing gas R interacts with oxygen chemisorbed in charged form. Oxygen is created in the course of processes

$$Z_A(\ ) + A(g) \rightarrow Z_A(A) + \Delta H_1 , \tag{2.75}$$

$$Z_A(A) + e \rightarrow Z_A(A^-) + \Delta H_2 , \tag{2.76}$$

where A stands for the sensitizing gas (oxygen); $Z_A(\ )$ is the adsorption center (vacant); $Z_A(A)$ and $Z_A(A^-)$ are respective neutral and charged forms of adsorbed A; ΔH_1 is the adsorption heat of neutral particles; ΔH_2 is the heat of formation of a charged form of adsorption. The condition of creation of $Z_A(A)$ is, naturally, provided by $\Delta H_1 \geq kT_0$, where T_0 is the temperature of the experiment. The electric conductivity of the sensor is changed due to reaction of reducing gas R with $Z_A(A^-)$, namely, reactant is adsorbed in the course of reaction

$$Z_R(\ ) + R(g) \rightarrow Z_R(R) + \Delta H_3 , \tag{2.77}$$

where $Z_R(\ )$ is the adsorption site; $\Delta H_3 \geq kT_0$. Migrating along the surface the reactant reaches $Z_A(A^-)$ and enters into reaction:

$$Z_A(A^-) + Z_R(R) \rightarrow Z_A(B^-) + Z_R(\ ) + \Delta H_4 , \tag{2.78}$$

The chemisorbed reaction product can be characterized by more shallow positioning of the energy level with respect to the bottom of conductivity band if contrasted to $Z_A(A^-)$ which results in emission of electron into the conductivity band in compliance with reaction

$$Z_A(B^-) + \Delta H_5 \rightarrow Z_A(B) + e , \tag{2.79}$$

the condition of development of which is provided by $|\Delta H_5| < |\Delta H_2|$. The final stage of the whole processes is given by desorption of the product:

$$Z_A(B) + \Delta H_6 \to Z_A(\quad) + B(g) , \qquad (2.80)$$

resulting in regeneration of free adsorption centers by sensitizing gas A.

The total detection reaction, thereby, can be written as

$$R(g)+Z_A(A^-)+\Delta H_5+\Delta H_6 \to e+Z_A(\quad)+B(g)+\Delta H_3+\Delta H_4 , \qquad (2.81)$$

We can conclude now that one electron returns to the conductivity band during each act of formation of the vacant site to adsorb sensitizer. Because adsorption centers $Z_R(\quad)$ are not accounted for by (2.81) the energetics of the process does not depend on the manner in which R is closing in A^-, i.e. on the fact which recombination mechanism (either Langmuire-Hinshelwood or Ili-Ridil) takes place.

Thus, the model proposed explains the effect of CO on electric conductivity of several oxides only in case when oxygen is present in ambient volume which was observed in numerous experiments. Accordingly, the fact of existence of relatively narrow temperature interval in which an adsorbent is sensitive to CO becomes clear. This can be linked with the fact that if the operational temperature T_0 is small the reaction products (in case of CO this is CO_2) cannot get desorbed (see expression (2.80)), i.e. regeneration of the centers of oxygen adsorption is not feasible. If T_0 is very high both adsorption of oxygen and reducing gas should be ruled out.

As for the function $\sigma \sim P_R^{1/2}$ it can be explained under further provisions. For instance, it was assumed in [106] that the semiconductor is fully depleted and there are two forms of chemisorbed oxygen on the surface only one of which participates in reaction with reducing gas. At the same time the rate of mutual conversion of these forms is much lower than the rate of surface reducing reaction. The same pattern of the phenomenon was proposed in paper [105] as well for adsorbents characterized by high intercrystalline barriers which provides a substantial change in electric conductivity at small changes of density of the surface charge. In paper [54] the function $\sigma \sim P_R^{\beta}$ was obtained from assumption that reducing gas reacts with physadsorbed oxygen and adsorption of the latter corresponds to the Freindlich isotherm, i.e. the degree of occupation of the surface is already provided by the power function from the partial pressure. Further, in the Shottky limit approximation the height of the intercrystalline barrier is determined as a function of the degree of oxygen coverage on the surface which is obviously dependent on the partial pressure of reducing gas interacting with the former. This yields the dependence of electric conductivity on the pressure of reducing gas.

In all above models an assumption concerning a necessary interaction of reducing gas with adsorbed oxygen proper is a common feature. However, there are other viewpoints as well. Thus, in papers [107, 108] it was assumed that CO interacts with the lattice oxygen and the role of adsorbed oxygen was reduced only to healing of oxygen vacancies formed during this process. Namely, adsorption of CO on ZnO $(10\bar{1}0)$ results in formation of a point defect $(CO_2 \cdot V_O^+)$ (Fig. 2.12) which acts as a weak acceptor [107]. Desorption of CO_2 molecules from these sites results in formation of a strong donor V_O^+ which is a surface oxygen vacancy. Thereby, during joint adsorption of CO and O_2 donor V_O^+ and acceptor O_2^- are being formed. A strong Coulomb interaction between them can explain high catalytic activity of ZnO from the stand-point of oxidation of CO with the maximum lying within the temperature interval $200 - 300°C$. For the rate of formation of the product in the case under consideration we have [107]

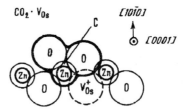

Fig. 2.12. The adsorption complex on the surface of ZnO $(10\bar{1}0)$ formed during interaction with CO [107].

$$\Delta N \, (T, P_{CO}, P_{O_2}) = K(T) \, P_{CO}^{m_1} \, P_{O_2}^{m_2} \, , \tag{2.82}$$

where ΔN is the number of CO_2 molecules formed at a unit surface in unit time; $K(T) = K_0 \exp \{-E/kT\}$ is the reaction rate constant; m_j is the order of reaction with respect to each component. Using the kinetic data it was established for m_1 and m_2 that in the range of high temperatures $(T > 250°C)$ $m_1 = m_2 = 1/2$, whereas for the low temperature range one encounters a more complex behavior in $m_j(T)$. A fraction order of reaction observed at temperature $T > 250°C$ can be explained under assumption that the "healing" of V_O^+ is a limiting stage of the whole process in this temperature domain

$$V_O^+ + e + O_s \rightarrow O_V \ . \tag{2.83}$$

Here O_s is the adsorbed oxygen; O_V is the lattice oxygen. It is assumed that at fast interaction of CO with the lattice oxygen O_V component ($V_O^+ + e$) is being formed:

$$CO + O_V \rightarrow CO_2 + V_O^+ + e \ , \tag{2.84}$$

for which we obtain $m_1 = 1/2$. Due to predominantly dissociative character of adsorption of O_2 at high temperatures ~200°C we have $m_2 = 1/2$. Note that at further rise in temperature the catalytic activity drops. In opinion of authors of [107] it can be explained by decrease in concentration of V_O^+ and O_s.

At low temperatures the reaction

$$V_O^+ + O_2^- \rightarrow O_s + O_V \ , \tag{2.85}$$

is the limiting process. This reaction is controlled by strong Coulomb interaction between the surface donor and acceptor.

Accounting for the fact that ionization degree of both donors V_O^+ and acceptors O_2^- is highly dependent on partial pressures of both components one can understand the complex temperature dependence of m_1 on the ratio of pressures P_{O_2} / P_{CO} observed in experiment [33].

Thus, using CO as an example one can easily see that mechanism of the change of electric conductivity of sensor during adsorption of reducing gases can be a very complex one. This contrasts the fact that a detailed physical and chemical picture of effects of gas adsorbed on electrophysical properties of adsorbent should provide the basis during manufacturing a sensor for a specific gas possessing a predetermined characteristics.

2.4.3. Adsorption of atom and radical particles

As it has been shown in Section 1.11 a comparison of experimental data on effects of adsorption of various particles on electric conductivity of oxide adsorbents results in conclusion that the dominant role in this phenomena is being played by reaction capacity of adsorbed particles. Owing to this conclusion the issue concerning effect of such chemically active particles as free atoms and radicals on electrophysical properties of oxide semiconductors was raised [1, 34, 76, 93]. It was expected that adsorption of above particles on oxide adsorbents would result in substantially high changes of electric conductivity if compared to adsorption of molecular particles. It was verified in experimental studies [93]. The electric conductivity of thin films of metal oxides (ZnO, TiO_2, CdO and others) during adsorption of active particles on them is changed in a most dramatic manner even in the case when concentration of these particles in the ambient space is only of the order of $10^6 -$ 10^8 particles/cm^3. It was established in the studies that above methods are applicable to identify free hydrogen atoms [48, 109 – 111], free oxygen atoms [112, 113], nitrogen [114], atoms of numerous metals [115] as well as simplest free radicals: alkyl [76], amino- and eminogroups [116, 117], hydroxyl [118] and others. All issues regarding experimental studies of behavior of free atoms and radicals in a heterogeneous system applying the method of semiconductor sensors will be considered in detail in Chapter 3. In this Section we will derive relationships linking the value and kinetics of the change in electric conductivity of semiconductor adsorbent with concentration of active particles in ambient volume. As usually, we consider partially reduced sintered oxides of various metals as adsorbents. The experiment shows that a fraction of above particles increases the electric conductivity during their interaction with oxide adsorbents of n-type (atoms of hydrogen and atoms of metals). On the contrary, the fraction of the particles decreases the electric conductivity (atoms N, O, C_2H_5, CH_2, CH_3). Above changes in electric conductivity at elevated temperatures of adsorbents are reversible, whereas at low temperatures these processes cannot be reversed. The heating of adsorbent at T ~ 300 ÷ 350°C results in complete restoration of electric conductivity to its pre-adsorption value. This result correlates with the data obtained in EPR studies of adsorbed atoms and radicals which indicate on a small strength of bond of adsorbed radicals with the surface of adsorbent [119].

The cause of such a high activity of atoms and radicals from the stand-point of the change of electric conductivity of adsorbents is provided, from our viewpoint, with the high chemical activity [93]. Namely, acceptors of electrons such as atoms of oxygen, nitrogen, methyl radicals and others tend to interact with surface defects, the latter being provided by superstoichiometric atoms of the metal proper or admixture atom, i.e. with electron donors. This process (as in case of adsorption of molecular acceptors) results in decrease in electric conductivity of adsorbent, the magnitude of effect being higher.

On the contrary such active particles as hydrogen atoms and metals (electron donors), touching the surface of oxides dopes them increasing the surface concentration of electron donors which results in increase in adsorbent conductivity.

High chemical activity of adsorbed atoms and radicals is manifested not only in their interaction with adsorption centers but in their interaction between each other as well as with active particles and initial molecules incident from gaseous phase [120]. High efficiency of development of these processes results in the fact that adsorption of active particles does not attain its equilibrium value, i.e. the equilibrium – adsorption occupation of the surface is not being attained, yet at substantially high intervals of time the stationary value is being established which is caused by leveling off in rates of direct and inverse processes responsible for occupation of the surface by adsorbed particles and for their removal from the surface. For instance, in case of adsorption of radicals resulting in formation of unstable organometal compounds on the surface of oxides the desorption of radicals [120], their chemical interaction with radicals and molecules from the ambient volume, surface recombination of adsorbed radicals may provide examples of processes leading to removal of adparticles.

Naturally, the process of adsorption of radicals accompanying their interaction with superstoichiometric metal atoms results in decrease in electric conductivity of adsobent. The fact that it is defects of this kind that are the centers of adsorption of radicals possessing substantial value of affinity to electron can be considered as proved today. This is confirmed by numerous experiments on effects of doping of the surface by metal atoms and its effect on the value of adsorption of active particles [1, 121]. This is also proved by the character of kinetics of the change of electric conductivity during radical adsorption [34, 120, 121]. Finally, this is being proved by the fact that adsorption of free radicals results in substantial decrease in intensity of EPR signals caused by metal atoms initially adsorbed on the surface of oxides [122]. Thus, as it was the case with molecular acceptors the decrease in electric conductivity of semiconductor oxide of n-type observed during adsorption of such radicals can be caused by decrease in concentration of superstoichiometric metal which is the donor of conductivity electrons. The inverse process resulting in increase in electric conductivity of adsorbent is linked (apart from usual desorption manifested in case of valence saturated particles as well) with annihilation of radicals caused by the surface recombination and the chemical interaction with molecules and radicals incident from the volume around adsorbent [112].

Taking into account above considerations for the rate of the change in concentration of chemisorbed radicals forming unstable organometal surface compounds we obtain the following expression

$$\frac{dN_S}{dt} = k'N_V (N_{SO} - N_S) - k''N_S - k'''N_S^2 .$$

(2.86)

The first term $k'N_V (N_{SO} - N_S)$ describes the rate of formation of adsorbed complexes; $N_S \equiv [(R^{\delta-} \ldots Me_\bullet^{\delta+})_S]$; N_V is the concentration of radicals in volume or, which is plausibly more correct, their concentration in pre-adsorption state which can be provided by a physical adsorption as well; N_{SO} is the concentration of chemisorption sites, i.e. the concentration of atoms of superstoichiometric metal. Assuming that $N_{SO} \approx h [Me_\bullet]_0$, where h i.e. the thickness of the bridge, $[Me_\bullet]_0$ is the volume pre-adsorption concentration of superstoichiometric metal we stress that almost all interstitial metal atoms available in the bridge can be accessible to the particles adsorbed due to smallness in h and high mobility of these defects. Thereby, $N_{SO} - N_S \approx h [Me_\bullet]_0 - N_S \equiv h [Me_\bullet]$, where $[Me_\bullet]$ is the current value of the volume concentration of donors proportional as it has been mentioned above to the square of electric conductivity of adsorbent.

The second and third terms in equation (2.86) describe kinetics of annihilation of radicals corresponding to above processes developing in the first and second order of magnitude with respect to chemisorbed radicals. Note that the rate constant k'' may be dependent on concentration of free radicals in volume.

Designating $x = N_S / N_{SO}$, $k_1 = k'N_V$, $k_2 = k_1 + k''$, $k_3 = k'''$, let us rewrite equation (2.86) as follows

$$\frac{dx}{dt} = k_1 - k_2x - k_3x^2 ,$$

(2.87)

with solution complying with expression

$$\ln \frac{(x_1 - x)(-x_2)}{x_1 (x - x_2)} = -\sqrt{\lambda} \, t ,$$

(2.88)

where x_1 and x_2 are solutions of the algebraic expression $k_1 - k_2x - k_3x^2$ which represent the points of stable and unstable equilibrium:

$$x_1 = \frac{-k_2 + \sqrt{\lambda}}{2k_3} ; \ x_2 = -\frac{k_2 + \sqrt{\lambda}}{2k_3}, \text{ where } \lambda = k_2^2 + 4k_1k_3 > 0.$$

(2.88')

Simple analysis of the motion of mapping point through the phase trajectory (Fig. 2.13) indicates that x_1 represents the stationary concentration of chemisorbed radicals corresponding to a given external conditions.

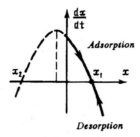

Fig. 2.13. The phase trajectory of the system.

Expression (2.88) can be represented as

$$\operatorname{arth}\left(\frac{k_2 + 2k_3x}{\sqrt{\lambda}}\right) = \frac{\sqrt{\lambda}}{2}t + \operatorname{arth}\left(\frac{k_2}{\sqrt{\lambda}}\right) \quad \text{when } x < x_1 \tag{2.89}$$

or

$$\operatorname{arth}\left(\frac{\sqrt{\lambda}}{k_2 + 2k_3x}\right) = \frac{\sqrt{\lambda}}{2}t + \operatorname{arth}\left(\frac{k_2}{\sqrt{\lambda}}\right) \quad \text{when } x > x_1 \tag{2.90}$$

used during processing the experimental data [34, 112].
Expression

$$N_S(t) = N_{S\infty}\frac{1 - \exp\left\{-\sqrt{\lambda}\,t\right\}}{1 + \gamma \exp\left\{-\sqrt{\lambda}\,t\right\}} \tag{2.91}$$

provides the solution of equations (2.88′) − (2.90) describing the kinetics of formation of chemisorbed radicals during their interaction with
excessive metal atoms when the opposite process of the radical annihilation is being taken into account. Here $N_{S\infty} = x_1 N_{SO}$ is the stationary
concentration of chemisorbed radicals $(R^{\delta-}\ldots\mathrm{Me}_{\bullet}^{\delta+})_S$; $\gamma = x_1/(-x_2) < 1$.
If we take into account the relationship

$$N_S = ([\mathrm{Me}_{\bullet}]_0 - [\mathrm{Me}_{\bullet}])h \tag{2.92}$$

and

$$\sigma = A(T)\sqrt{[\mathrm{Me}_{\bullet}]} \tag{2.93}$$

we can easily arrive to the expression giving the kinetics of decrease in
electric conductivity

$$\sigma(t) = \left[\sigma_0^2 - (\sigma_0^2 - \sigma_\infty^2) \frac{1 - \exp\left\{-\sqrt{\lambda}\, t\right\}}{1 + \gamma \exp\left\{-\sqrt{\lambda}\, t\right\}} \right]^{1/2} \qquad (2.94)$$

where σ_∞ is the stationary value of electric conductivity of adsorbent.

In two limiting cases differing in the values of stationary concentration of chemisorbed radicals and initial electric conductivity of adsorbent the expression (2.94) acquires the following shape:

$$\sigma(t) \approx \sigma_0 - (\sigma_0 - \sigma_\infty) \frac{1 - \exp\left\{-\sqrt{\lambda}\, t\right\}}{1 + \gamma \exp\left\{-\sqrt{\lambda}\, t\right\}} \quad \text{when } \sigma_\infty \lesssim \sigma_0 , \qquad (2.95)$$

$$\sigma(t) \approx \sigma_0\, e^{-\frac{\sqrt{\lambda}}{2} t} \left[1 + \left(\frac{\sigma_\infty}{\sigma_0}\right)^2 e^{\sqrt{\lambda}\, t} \right]^{1/2} \quad \text{when } \sigma_\infty \ll \sigma_0 . \qquad (2.96)$$

Accounting for the fact that at any t in a quality $\gamma \exp(-\sqrt{\lambda}\, t) < 1$ holds the expression (2.95) for small times ($\sqrt{\lambda}\, t < 1$) can be rewritten as

$$\frac{\sigma_0 - \sigma}{\sigma_0 - \sigma_\infty} \approx \frac{\sqrt{\lambda}}{2} t . \qquad (2.97)$$

If we recall that in this case the current values of σ do not significantly differ from σ_0 expression (2.97) at small times t can be used to approximate the experimental kinetics fairly well

$$\frac{\sigma_0 - \sigma}{\sigma_0 - \sigma_\infty} \approx \gamma t . \qquad (2.98)$$

This kinetics was observed during adsorption of atom nitrogen [123]. As for the stationary values of electric conductivity σ_∞ it can be deduced from expressions (2.88) and (2.92), (2.93)

$$1 - \left(\frac{\sigma_\infty}{\sigma_0}\right)^2 = \frac{\sqrt{\lambda} - K_2}{2K_3} , \qquad (2.99)$$

which provides

$$\tilde{a} \equiv \frac{\Delta\sigma \cdot \sigma_0}{\sigma_\infty^2} \cong \frac{K'}{K''} N_V , \qquad (2.100)$$

when we recall that $K_1 = KN_V$, $K_2 = KN_V + K''$. Here $\Delta\sigma = \sigma_0 - \sigma_\infty$. Expression (2.100) is valid either at $K'' > K'''$, i.e. at weak manifestation of the process of recombination of chemisorbed radicals with respect to their annihilation as reaction of the first order, or on the contrary when $K''' > K''$ but in the case of either very small concentrations of radicals in volume: $N_V < (K'')^2 / K' (2K''' - K'') = N_{V2}$ or already at very high concentration: $N_V > 2 (2K''' - K'') / K' = N_{V1}$. Obviously that both these cases guarantee the predominant development of the process of annihilation of adsorbed radicals as reaction of the first order.

In case when $K''' > K''$ and $N_{V1} < N_V < N_{V2}$, i.e. in case of average volume concentrations of radicals and recombination mechanism of their heterogeneous annihilation we arrive at the following expression characterizing the stationary electric conductivity of adsorbent

$$\tilde{a} \equiv \frac{\Delta\sigma \cdot \sigma_0}{\sigma_\infty^2} \approx \sqrt{\frac{K'}{K''}} N_V . \qquad (2.101)$$

We should note that expressions similar to (2.100) and (2.101) obtained in paper [120] have been experimentally confirmed in numerous systems free radical – oxide adsorbent [1, 34, 57, 120].

Annihilation of adsorbed radicals on the surface of semiconductor adsorbent results in increase in electric conductivity of the latter due to making the metal atom available with its subsequent ionization:

$$N_S \Leftrightarrow (R^{\delta-} \ldots Me_\bullet^{\delta+})_s \rightarrow R(\dot{g}) + Me_\bullet \rightarrow R(\dot{g}) + Me_\bullet^+ + e . \qquad (2.102)$$

Assuming that in this case the decomposition of the surface complex $(R^{\delta-} \ldots Me_\bullet^{\delta+})_s$ is the limiting stage rather than establishing electron equilibrium during ionization of the dope the kinetics of the process can be described by the same equation (2.86), although the initial conditions would correspond to the "desorption domain" of the curve on the phase trajectory in Fig. 2.13, i.e. $dN_S/dt < 0$. Usually the process of annihilation of radicals or, to be more exact, the kinetics of the change of electric conductivity of adsorbent caused by it is being studied at sharp annihilation of radicals in the volume surrounding adsorbent. In this case $N_V = 0$ and solution of equation (2.87) is provided by

$$X(t) = X_{max} \frac{K_2}{K_3} \frac{\exp(-K_2 t)}{\left(X_{max} + \dfrac{K_2}{K_3}\right) - X_{max} \exp(-K_2 t)} \; , \qquad (2.103)$$

where X_{max} is the concentration of chemisorbed radicals (in N_{SO} units) at initial moment of development of reversed process. Depending on the method of generation of radicals in experiment the latter may be the moment of turning on the light source, activation of hot filament etc. (see, for instance, [1]). At initial moments of time after activation of the source of generation of radicals the radicals complying with condition $K_2 t < 1$ for concentration of chemisorbed radicals the following expression is applicable

$$N_S(t) \approx N_{S\,max} \frac{1 - K_2 t}{1 + \dfrac{N_{S\,max}}{N_{SO}} K_3 t} \; . \qquad (2.104)$$

Consequently we obtain for kinetics of increase in electric conductivity of adsorbent the following expression

$$\sigma(t) \approx \sigma_\infty + \frac{1}{2} \frac{\sigma_0^2}{\sigma_\infty} \psi \frac{(K_2 + \psi K_3)t}{1 + \psi K_3 t} \; , \quad \psi = (\sigma_0^2 - \sigma_\infty^2) / \sigma_0^2 \; . \qquad (2.105)$$

When the process of recombination of chemisorbed radicals is negligible which is the case when $\sigma_\infty \lesssim \sigma_0$ this expression can be rewritten as

$$\sigma(t) \approx \sigma_\infty \left(1 + \frac{\sigma_0 \, \Delta\sigma}{\sigma_\infty^2} K_2 \, t \right) \; . \qquad (2.106)$$

We should note that we used $N_{S\infty}$ as $N_{S\,max}$ (which is the concentration of chemisorbed radicals at the moment of activation of the source of radicals) in expressions (2.105) and (2.106). This means that we have assumed that the process of chemisorption is already stationary by this moment of time.

In the situation we obtain the following expression for initial rate of increase in electric conductivity of adsorbent ($t < K_2^{-1}$) caused by the process of annihilation of radicals:

$$\frac{d\sigma}{dt} = \frac{\sigma_\infty}{2}\left(\frac{\sigma_0^2}{\sigma_\infty^2} - 1\right)\frac{\left(1 - \frac{\sigma_\infty^2}{\sigma_0^2}\right)K_3 + K_2}{1 + \left(1 - \frac{\sigma_\infty^2}{\sigma_0^2}\right)K_3 t} , \qquad (2.107)$$

which yields

$$\left.\frac{d\sigma}{dt}\right|_{t\to 0} \approx \frac{\sigma_0^2 - \sigma_\infty^2}{2\,\sigma_\infty}\left[K_2 + \left(1 - \frac{\sigma_\infty^2}{\sigma_0^2}\right)K_3\right] . \qquad (2.108)$$

We can deduce from expression (2.108) that in various limiting cases differing in the features of annihilation of chemisorbed radicals (monomolecular or biomolecular reaction with respect to the concentration of the latter) the rate of the change in electric conductivity during turning off the source of generation of radicals is proportional to the concentration of the radicals existing in the ambient volume up to the moment of desactivation. Indeed, because $\sigma_0^2 - \sigma_\infty^2 \sim N_{S\infty} = (\sqrt{\lambda} - K_2)/2K_3$ then in case when $K_2 > K_3(\sigma_0^2 - \sigma_\infty^2)/\sigma_0^2$ we obtain

$$\left.\frac{d\sigma}{dt}\right|_{t\to 0} \sim \frac{\sqrt{\lambda} - K_2}{2K_3}K_2 \sim K_1 \sim N_V . \qquad (2.109)$$

If $K_2 < K_3(\sigma_0^2 - \sigma_\infty^2)/\sigma_0^2$, then

$$\left.\frac{d\sigma}{dt}\right|_{t\to 0} \sim \left(\frac{\sqrt{\lambda} - K_2}{2K_3}\right)^2 K_3 \sim \left(\sqrt{\frac{K_1}{K_3}}\right)^2 K_3 \sim K_1 \sim N_V . \qquad (2.110)$$

We should point out that the initial rate of the change in electric conductivity during adsorption of radicals follows above proportionality as well. Using the expression (2.94) we also obtain

$$\left.\frac{d\sigma}{dt}\right|_{t\to 0} \sim \frac{\sqrt{\lambda} - K_2}{2K_3}\sqrt{\lambda}$$

yielding

$$\left.\frac{d\sigma}{dt}\right|_{t\to 0} \sim K_2\frac{K_1}{K_2} \sim K_1 \sim N_V , \qquad (2.111)$$

when $K_2 > 4K_1K_3$. In the opposite case when $K_2 < 4K_1K_3$

$$\left.\frac{d\sigma}{dt}\right|_{t\to 0} \sim \sqrt{\frac{K_1}{K_3}}\sqrt{K_1K_3} \sim K_1 \sim N_V \ . \tag{2.112}$$

Thus, the rigorous solution of kinetic equation describing the change in electric conductivity of a semiconductor during adsorption of radicals enables one to deduce that information on concentration of radicals in ambient volume can be obtained measuring both the stationary values of electric conductivity attained over a certain period of time after activation of the radical source and from the measurements of initial rates in change of electric conductivity during desactivation or activation of the radical flux incident on the surface of adsorbent, i.e.

$$\left(\frac{a}{a_0}\right) = \left(\frac{v}{v_0}\right)_{on} = \left(\frac{v}{v_0}\right)_{off} = \left(\frac{N_V}{N_{V0}}\right) \equiv \frac{[\dot{R}_V]}{[R_{V0}]} \ , \tag{2.113}$$

where as usual $a = \sigma_0 \Delta\sigma / \sigma^2$; σ is the stationary value of electric conductivity; v_{on} and v_{off} are initial rates in change of electric conductivity $(d\sigma/dt)_{t\to 0}$ during switching on and switching off of radical flux; $(N_V / N_{V0}) = [\dot{R}_V] / [R_{V0}]$ is a relative concentration of radicals. All values in parenthesis are relative magnitudes of respective parameters.

The opposite change in electric conductivity of adsorbent occurs during adsorption of such active particles as atoms of hydrogen and atoms of metals [115, 124,125]. The similar result is obtained during radiolysis of hydrocarbons [126] due to formation and chemisorption of H-atoms. Both the rate of adsorption caused change in electric conductivity and the value of its stationary values are determined in this cases by all the processes accompanying chemisorption [127].

Numerous experiments on adsorption of H-atoms performed using various oxides of both monocrystal and polycrystal type unambiguously proved the existence of charged and non-charged forms of chemisorbed H-atoms on the surface of oxides [89, 109, 102, 110, 125, 128]. Considering the vast experimental data it was initially proposed [127] that the following schematics describes consistently the chemisorption of H-atoms:

$$
\begin{array}{c}
(H_2)_v \\
\quad I \quad\; III \uparrow_{K_3+H_v} \quad II \\
H_v \; \overset{K_1}{\underset{\tilde{K}_1}{\rightleftarrows}} \; H_s \; \overset{K_2}{\underset{\tilde{K}_2}{\rightleftarrows}} \; H_s^+ + e \\
\quad IV \downarrow_{K_4+H_s} \qquad V \downarrow_{K_5+H_v} \\
(H_2)_s \to (H_2)_v \quad (H_2)_v
\end{array}
\qquad (2.114)
$$

where as usual indexes s and v designate adsorbed and free states, respectively. On the grounds that the issue on identification of chemisorbed state of atomic hydrogen was considered in Section 2.4.2 in substantial detail let us dwell on the value and kinetics of the change in electric conductivity of polycrystalline sintered in reduction medium adsorbent caused by adsorption of H-atoms.

In general case the solution describing the development of the process according to the scheme (2.114) is very bulky. Indeed, description of kinetics of the change in electric conductivity during adsorption of H-atoms would need the studies of behavior of dynamic system on the plane of the following type:

$$
\frac{d[H_s]}{dt} = K_1[H_v]\left(1 - \frac{[H_s]}{[H_{s0}]}\right) + \tilde{K}_2[H_s^+][e] - K_2[H_s]\left(1 - \frac{[H_s^+]}{[H_{s0}]}\right) -
$$
$$
- K_3[H_v][H_s] - K_4[H_s^2] , \qquad (2.115)
$$

$$
\frac{d[H_s^+]}{dt} = K_2[H_s]\left(1 - \frac{[H_s^+]}{[H_{s0}]}\right) - \tilde{K}_2[H_s^+][e] - K_5[H_v][H_s^+] , \qquad (2.116)
$$

which might be conducted by standard approach [125] but can be replaced by a straightforward and sufficiently transparent analysis of scheme (2.114) in various limiting cases corresponding to various temperatures T and concentrations of free H-atoms $[H_v]$.

Let us initially consider such ranges in T and $[H_v]$ where we can neglect the inverse process at stage II and side processes III – V resulting in surface recombination of H-atoms, i.e. let us consider a more simple schematic:

$$
H_v \; \overset{I}{\underset{K_1}{\longrightarrow}} \; H_s \; \overset{II}{\underset{K_2}{\longrightarrow}} \; H_s^+ + e . \qquad (2.117)
$$

It is obvious that the rate of accumulation of H_s on the surface is controlled in this case by the following expression

$$\frac{d[H_s]}{dt} = K_1 [H_v] (1 - \theta) - K_2 [H_s] (1 - \theta') , \qquad (2.118)$$

where $\theta = [H_s] / [H_{s0}]$ is the coverage of the surface by H_s-atoms; $[H_{s0}]$ is the maximum possible coverage of the surface by chemisorbed hydrogen, i.e. the number of its adsorption sites; θ' is the coverage of the surface by H-atoms chemisorbed in charged form. Note that in this specific case the major change in electric conductivity of bridges is caused by "doping action" of H-atoms. As it has been mentioned above (O–H)$^{2-}$-groups formed during interaction of H-atoms with oxygen ions of the lattice can act as newly formed donors. Small thickness of bridges h in this case results in the fact that possible effects of charging of the surface do not practically result in the band bending and do not affect the kinetics of the process nor the transfer of current carriers. Assuming that in domain under consideration T and $[H_v]$ $\theta' \ll 1$ one can easily show that solution of equation (2.118) is provided by the following relation

$$\theta = \frac{\dfrac{K_1}{K_2} \dfrac{[H_v]}{[H_{s0}]}}{1 + \dfrac{K_1}{K_2} \dfrac{[H_v]}{[H_{s0}]}} \left[1 - \exp\left\{ -K_2 \left(1 + \frac{K_1}{K_2} \frac{[H_v]}{[H_{s0}]} \right) t \right\} \right] . \qquad (2.119)$$

If we take into account that $[H_s^+]$ is the concentration of chemisorbed in charged form H-atoms, i.e. H-atoms forming double electron bond with the surface we arrive at the following expression

$$[H_s^+] = ([e] - [e]_0)h , \qquad (2.120)$$

where $[e]$ and $[e]_0$ is the current and initial concentration of electrons of conductivity. Thereby, if we account for the kinetics describing the stage II of schematics (2.117) we obtain the following expression

$$\frac{d[e]}{dt} = \gamma \frac{d\sigma}{dt} = K_2 [H_s] h^{-1} , \qquad (2.121)$$

which keeping in mind relation (2.119) yields

$$\frac{d\sigma}{dt} = \frac{\alpha [H_v]}{1 + \beta [H_v]} \left[1 - \exp\left\{ - K_2 (1 + \beta [H_v]) t \right\} \right] , \qquad (2.122)$$

where $\alpha = (h\gamma)^{-1}K_1 \equiv q\bar{v}K_1h^{-1}$; q is the charge of electron; $\bar{v}$ is the mobility of the current carriers; $\beta = K_1 / K_2 [H_{s0}]$. One can easily see from (2.122) that at initial moments of time $t < [K_2(1 + \beta [H_v])]^{-1}$, and when $d\sigma/dt \cong \alpha K_2[H_v]t$ the value of the rate of electric conductivity reach the stationary values

$$\frac{d\sigma}{dt} = \frac{\alpha [H_v]}{1 + \beta [H_v]} \ . \tag{2.123}$$

In the range of substantially high temperatures when condition $[H_v] \ll$ $\ll K_2 [H_{s0}] / K_1$ holds the expression (2.123) acquires a more simple shape:

$$\frac{d\sigma}{dt} \approx \alpha [H_v] \ . \tag{2.124}$$

Indicating that the relative concentration of H-atoms in ambient volume can be expressed through the rate of the change in its electric conductivity as follows:

$$\frac{[H_v]_1}{[H_v]_2} = \frac{(d\sigma / dt)_1}{(d\sigma / dt)_2} \ . \tag{2.125}$$

In the more general case described by expression (2.123) we obtain for the concentration of H-atoms

$$\frac{[H_v]_1}{[H_v]_2} = \frac{\tau_2 - \gamma}{\tau_1 - \gamma} \ , \tag{2.126}$$

where $\gamma = \beta / \alpha$; $d\sigma / dt = 1 / \tau$.

The condition $d\sigma/dt$ = const is retained only in case of substantially short period of time after which $d\sigma/dt$ becomes decreasing which at substantially high temperatures is caused by manifestation of inverse reaction of the stage II and annihilation of H-atoms due to reactions III, IV and inverse reaction of the step I. Above condition results in the fact that for substantially high temperatures one should consider schematics (2.114) instead of (2.117) assuming that step II provides a limiting stage in the change of electric conductivity. This stems from smallness (at elevated temperatures) of both energy of activation of stage I and the stationary concentrations of H_s-atoms which is attained quick enough.

160

The relationship between the stationary concentrations of free H-atoms in the volume and chemisorbed H_s-atoms would be determined by what reaction (III, IV or inverse one I) is a dominant one under given conditions of experiment.

In case when the inverse reaction I dominates this relationship will be of the form

$$\frac{\theta}{1-\theta} = \frac{k_1}{\overset{\leftarrow}{k_1}} ,$$

(2.127)

indicating that at high concentrations of $[H_v]$ the degree of occupation of the surface by H_s-atoms becomes independent on concentration of H-atoms in volume.

If reaction III is a dominant one then

$$\frac{\theta}{1-\theta} = \frac{k_1}{k_3}[H_v] ,$$

(2.128)

and finally for reaction IV when the latter is the dominant process one obtains the following expression

$$\frac{\theta^2}{1-\theta} = \frac{k_1}{k_4}[H_v] ,$$

(2.129)

which in case $\theta \ll 1$ acquires the form

$$\theta \approx \left(\frac{k_1}{k_4}\right)^{1/2} \sqrt{[H_v]} .$$

(2.130)

Thereby, occupation of the surface of adsorbent by H_s-atoms can either be independent of the change in $[H_v]$ or change proportionally with $[H_v]^{1/2}$ or $[H_v]$. The first case should be anticipated for high concentrations in $[H_v]$ and large degrees of occupation of the surface. The second case takes place for substantially large degrees of occupation of the surface and small $[H_v]$ and is mainly controlled by mobility of H_s-atoms. The third case is encountered at small value of the binding energy of H_s-atoms with the surface of adsorbent.

For the rate in the change of concentration of conductivity electrons we have the following equation

$$\frac{d[e]}{dt} = k_2[H_s] - \bar{k}_2[H_s^+][e] \ . \tag{2.131}$$

Taking into account above consideration let us assume that the stationary value $[H_s]$ is attained very quickly and is controlled by one of expressions (2.127) – (2.130). Condition of electric neutrality (2.120) yields

$$h \cdot [e] = [H_s^+] + [Me_s^+]h \equiv [H_s^+][e]_0 h \ ,$$

where $[Me_s^+]$ as usually designates the concentration of superstoichiometric metal atoms the equation (2.131) can be rewritten as

$$\frac{d[e]}{dt} = k_2[H_s] - \bar{k}_2([e] - [e]_0)[e]h \ , \tag{2.132}$$

with solution provided by the following expression

$$\text{arth}\frac{2\dfrac{\Delta\sigma}{\sigma_0} + 1}{\sqrt{1 + 4K}} = \frac{1}{2}K't\sqrt{1 + 4K} + \text{arth}\frac{1}{\sqrt{1 + 4K}} \ , \tag{2.133}$$

where $K = \dfrac{K_2}{\bar{K}_2 h}\dfrac{[H_s]}{[e]_0^2}$; $K' = \bar{K}_2[e]_0 h$. This expression describes the kinetic of the change in electric conductivity of adsorbent. Using the technique of limiting transitions one can easily show that kinetics of direct reaction of the II step is described by expression

$$\frac{\Delta\sigma}{\sigma_0} = K_1't \ , \tag{2.134}$$

where $K_1' = K_2[H_s] / [e]_0$, and kinetics of inverse reaction is given by

$$\ln\left(1 + \frac{\sigma_0}{\Delta\sigma}\right) = K_2' + \text{const} \ , \tag{2.135}$$

where $K_2' = h\bar{K}_2[e]_0$; $\text{const} = \ln(1 + \sigma_0/\Delta\sigma_{\max})$.

For the stationary values of electric conductivity equation (2.132) leads to the following expression

$$\frac{\Delta\sigma \cdot \sigma}{\sigma^2} = K \ , \tag{2.136}$$

where $K = K_2(\bar{v}q)^2[H_s] / \overline{K}_2 \, \sigma_0^2 \, h$. Note that owing to expressions (2.127 – 2.130) the dependence of parameter $\Delta\sigma \cdot \sigma / \sigma_0^2$ to be experimentally measured as a function of concentration $[H_v]$ can be either linear or a root one. We would like to mention in advance (details will be given in Chapter 3) that at not too high concentrations of H-atoms a root function is being obtained

$$\frac{\Delta\sigma \cdot \sigma}{\sigma^2} = \text{const} \sqrt{[H_v]} \tag{2.137}$$

The relative change in concentration of hydrogen atoms in the volume of reaction vessel can be expressed through the stationary value of electric conductivity of adsorbent as follows:

$$\frac{[H_v]_1}{[H_v]_2} = \left\{ \frac{(\Delta\sigma \cdot \sigma)_1}{(\Delta\sigma \cdot \sigma)_2} \right\}^2 . \tag{2.138}$$

These expressions describe the kinetics and the value of response of electric conductivity in oxide adsorbent during adsorption of atomic hydrogen. In general case they are applicable for adsorption of any radical particles of the donor type. However, in each specific case any of the processes of schematics (2.114) controlled by specificity of each system [129] may be a dominant process controlling both the kinetics and the stationary value of electric conductivity.

For instance, in case of adsorption of metal atoms [46, 115, 130-132] it was established in experiment that kinetics of the change in electric conductivity of ZnO due to effects of metal vapors can be described by expression

$$\Delta\left(\frac{\sigma}{\sigma_0}\right) = \mu \, e^{-\alpha t} \ , \tag{2.139}$$

up to the stationary value well enough, where $\Delta(\sigma/\sigma_0) = (\sigma_\infty/\sigma_0) - (\sigma/\sigma_0)$.

This function becomes obvious if, as it was the case of adsorption of hydrogen atoms if we assume that the change in σ is dependent on the rate of transition of adsorbed metal atoms, whose concentration $[Me_s]$ is assumed to be constant to the charged form Me_s^+ :

$$\frac{d\sigma}{dt} = K_1[\text{Me}_s] \left(1 - \frac{[\text{Me}_s^+]}{[\text{Me}_s^+]_\infty} \right) = K_1[\text{Me}_s] \left(1 - \frac{\Delta\sigma}{\Delta\sigma_\infty} \right) . \tag{2.140}$$

Integration of equation (2.140) results in expression (2.139) with the following parameters: $\alpha = K_1[\text{Me}_s]/\Delta\sigma_\infty$, $\mu = \Delta\sigma_\infty/\sigma_0$.

In Chapter 3 we will provide experimental verification of expression obtained in this Section linking the concentration of active particles in ambient volume with the change in electric conductivity of adsorbent under stationary and kinetic conditions as well as experimental prove of validity assumptions made while deriving above expressions.

Thus, we have considered in detail various theoretical models of effect of adsorption of molecular, atom and radical particles on electric conductivity of semiconductor adsorbents of various crystalline types. Special attention has been paid to sintered and partially reduced oxide adsorbents characterized by the bridge type of intercrystalline contacts with the dominant content of bridges of open type because of wide domain of application of this very type of adsorbents as sensitive elements used in our physical and chemical studies.

The adsorption of particles of various type results in the change in electric conductivity of such bridges mainly due to local chemical interaction of adsorbed particles with electrically active defects which are electron donors and resulting, thereby, in decrease of their concentration or, on the contrary, in increase due to creation of new defects of this type. In both cases as it has been shown above there are substantially straightforward and easily verified relationships linking both the initial rates in the change of electric conductivity and the stationary values reflecting concentration of adsorbed particles in ambient volume.

The regular availability of linear dependence between above values enabled us to substantiate two methods of detection the stationary and the kinetic one. The very names of these methods indicate that the first one makes it feasible to use the stationary values of adsorption-induced changes in electric conductivity of adsorbent to judge concentration of the particles detected. The second method enables one to obtain the same information using the measurements of initial kinetics of the change of electric conductivity.

In our studies we mainly used the kinetic method of detection which was initially proposed in paper [75]. The advantages of this method if compared to the method of measurements of the stationary electric conductivity deal (apart from fast availability of the data) with the fact that in several cases it enables one to obtain information concerning concentration of detective particles even during development of certain surface reactions in which these particles directly participate. The most simple example is provided by the surface recombination whose study using the stationary method of detection is not possible. However, the major advantage of kinetic method is provided by the option of very long use of adsorbent without regeneration of its surface. This is caused by the fact that at such detection method the occupation of the surface of active element of the sensor by adsorbate necessary to measure the current value of $(d\sigma/dt)_{t\to0}$ is extremely small if compared to the stationary occupation corresponding to the volume concentration of detected particles. The importance of this item is stressed by introduction of such a notion as "critical capacity of adsorbent" which determines the number of possible changes within which the initial proportionality between the value of this signal of the sensor and concentration of detected particles is maintained. It is absolutely obvious that the number of measurements which is necessary to conduct without regeneration of the surface of adsorbent during the use of kinetic method is much higher than that allowed by the stationary method.

These and others issues linked both with experimental substantiation of the detection methods themselves as well as with the experimental verification of above relationships between the values of the signals of sensor with concentrations of particles detected will be considered in detail in the next chapter.

References

1   I.A. Myasnikov, Mendeleev ZhVKhO, 20 (1975) 19 - 32
2   A.I. Buturlin, T.A. Gabuzyan, N.A. Golovanov et al., Zarubezh. Electron. Tekhnika, 10 (1983) 3 - 29
3   I.A. Myasnikov and G.V. Malinova, DAN SSSR, 159 (1964) 894 - 903
4   G.V. Malinova and I.A. Myasnikov, Zhurn. Fiz. Khimii, 49 (1975) 1326 - 1369
5   A.I. Lifshits, E.E. Gutman and I.A. Myasnikov, Zhurn. Fiz. Khimii, 52 (1978) 2953 - 2960
6   Jap. Pat. Gaz. 75-23317, Aug. 1975
7   M. Nitta and M. Haradome, IEEE Trans. Electron Devices, 26 (1979) 247 - 249
8   S.R. Morrison, Sensors and Actuat., 12 (1987) 425 - 438
9   R. Larause, N.D. Bui and C. Pijolat, Chemical Sensors: Anal. Chem. Symp. Ser, T. Seiyama et al. (ed.), Elsevier, Amsterdam, 17 (1983) 47 - 62
10 S.R. Morrison, Sensors and Actuat., 2 (1982) 329 - 343
11 L. Elias and E.A. Ogryzlo, Canad. J. Chem., 37 (1959) 1680 - 1688

12 G.J. Advani and A.G. Jordan, J. Electron. Mater., 9 (1980) 29 - 34

13 L.A. Harris, J. Electrochem. Soc., 127 (1980) 2657 - 2664

14 H. Okamoto, H. Obayashi and T. Kudo, Solid State Ionics, 1 (1980) 319 - 331

15 N.E. Lobashina, N.N. Savvin and I.A. Myasnikov, DAN SSSR, 268 (1983) 1434 - 1439

16 V.M. Zaloznykh, I.A. Myasnikov and N.N. Savvin, DAN SSSR, 280 (1985) 920 - 924

17 I.A. Myasnikov I.N. Pospelova and T.A. Koretskaya, DAN SSSR, 179 (1968) 645 - 651

18 A.R. Hytson, Semiconductors, N.B. Khennei (ed.), Inostrannaya Literatura Publ., 1962

19 I.V. Miloserdov and I.A. Myasnikov, DAN SSSR, 224 (1975) 1352 - 1358; DAN SSSR, 227 (1976) 387 - 391

20 T. Kobayashi, M. Haruta, H. Sano et al., Sensors and Actuat., 13 (1988) 339 - 352

21 I.A. Myasnikov, E.I. Grigoriev and V.I. Tsivenko, Uspekhi Khimii, 55 (1986) 161 - 205

22 V.Ya. Sukharev, Zhurn. Fiz. Khimii, 62 (1988) 2429 - 2437

23 V.Ya. Sukharev and I.A. Myasnikov, Zhurn. Fiz. Khimii, 60 (1986) 2385 - 2392

24 N.M. Beekmans, J. Chem. Soc. Faraday Trans. I., 69 (1973) 1 - 13

25 J.F. McAleer, P.T. Moseley, J.O. Norris et al., J. Chem. Soc. Faraday Trans. I., 83 (1987) 1323 - 1331

26 N. Henney, Chemistry of Solids, Mir Publ., Moscow, 1971 (translation into Russian)

27 S. Baidyarov and P. Mark, Surface Sci, 30 (1972) 53 - 58

28 J.O. Cope and I.D. Campbell, J. Chem. Soc. Faraday Trans. I, 69 (1973)

29 V.F. Kiselev and O.V. Krylov, Electron Phenomena in Adsorption and Catalysis on Semiconductors and Dielectrics, Nauka Publ., Moscow, 1979

30 T.I. Barry and F.S. Stone, Proc. Roy. Soc. A, 255 (1960) 124 - 129

31 S.P. Morrison, Chemical Physics of the Surface of Solids. Mir Publ., Moscow, 1982 (translation into Russian)

32 W. Gopel, G. Rocker and R. Feierabend, Phys. Rev. B, 28 (1983) 3447 - 3462

33 W. Gopel, Prog. Surface Sci., 20 (1985) 9 - 46

34 I.A. Myasnikov, Chemisorption of Free Radicals on Oxide Semiconductor Adsorbents and Single Acts of Some Radical Processes, PhD (Doctoral) Thesis, Moscow, 1965

35 E.J. Hahn, J. Appl. Phys., 22 (1951) 855 - 857

166

36   E.V. Ferway, Semiconductor Materials, Inostrannaya Literatura Publ., Moscow, 1963 (translation into Russian)
37   V.Ya. Sukharev, N.E. Lobashina, N.N. Savvin et al., Zhurn. Fiz. Khimii, 57 (1983) 405 - 502
38   Ya.E. Geguzin, Physick of Sintering, Nauka Publ., Moscow, 1967
39   V.I. Tsivenko and L.T. Nazarov, DAN SSSR, 78 (1951) 749 - 755
40   N.E. Lobashina, N.N. Savvin, I.A. Myasnikov et al., Zhurn. Fiz. Khimii, 56 (1982) 1719 -1723
41   M.K. Sheinkman and A.Ya. Shik, Fiz. i Tekhn. Poluprovodnikov, 10 (1976) 209 - 233
42   A.Ya. Vinnikov, A.M. Meshkov and B.N. Savushkin, FTT, 22 (1980) 2989 - 2995; FTT, 24 (1982) 1352 - 1363; FTT, 272 (1985) 1929 - 1937
43   D.W. Pashley, M.T. Stowell, M.N. Jacobs et al., Philos. Mag., 10 (1964) 127 - 135
44   J.W. Mathews, Physics of Thin Films, Mir Publ., Moscow, 1970 (translation into Russian)
45   G. Matare, Electronics of Defects in Semiconductors, Mir Publ., Moscow, 1974 (translation into Russian)
46   B.S. Agyan, I.A. Myasnikov and V.I. Tsivenko, Zhurn. Fiz. Khimii, 47 (1973) 980 - 984, 1292 - 1301, 294 - 2908
47   V.V. Chistyakov, V.Ya. Sukharev and I.A. Myasnikov, FTT, 29 (1987) 2305 - 2312
48   I.A. Myasnikov, Zhurn. Fiz. Khimii, 31 (1957) 1721 - 1728
49   E.E. Gutman, I.A. Myasnikov, A.G. Dovtyan et al., Zhurn. Fiz. Khimii, 50 (1976) 1721 - 1728
50   P. Amigues and S.J. Teichner, Discuss. Faraday. Soc., 41 (1966) 362 - 375
51   K. Ihokura, New Matter. and New Proces., 1 (1981) 43 - 55
52   A. Jones, T.A. Jones, B. Mann et al., Sensors and Actuat., 5 (1984) 75 - 83
53   F.A. Kroger, Nonstoichiometric Oxides, O. Toft (Ed.), Acad. Press, Sorensen, 1981
54   P.K. Cliffird, Chemical Sensors: Anal. Chem. Symp. Ser., T. Seigama et al (eds.), Elsevier, Amsterdam, 17 (1983) 135 - 136
55   P.K. Cliffird and D.J. Tuma, Sensors and Actuat., 3 (1983) 233 - 242; 255 - 261
56   I.A. Myasnikov, Izv. AN SSSR. Ser. Fiz., 21 (1957) 192 - 205
57   E.E. Gutman, Zhurn. Fiz. Khimii, 58 (1984) 801 - 808
58   V.Ya. Sukharev and I.A. Myasnikov, Zhurn. Fiz. Khimii, 61 (1987) 577 - 592
59   I.A. Myasnikov, Zhurn. Fiz. Khimii, 55 (1981) 1278 - 1282; 183 - 191; 2053 - 2058; 2059 - 2068
60   K. Tanaka and G. Blyholder, J. Phys. Chem., 76 (1972) 3184 - 3187

61 K.N. Spiridonov and O.V. Krylov, Problemy Kinetiki i Kataliza, Nauka Publ., Moscow, 16 (1975) 7 - 11

62 J. Gutierrez, F. Cebollada, C. Elvira et al., Mater. Chem. and Phys., 18 (1987) 265 - 277

63 V.B. Kazanski, Problemy Kinetiki i Kataliza, Nauka Publ., Moscow, 15 (1973) 77 - 84

64 V.B. Kazanski, V.V. Nikisha and B.N. Shalimov, DAN SSSR, 188 (1969) 112 - 119

65 M. Komuro, Bull. Chem. Soc. Jap., 48 (1975) 756 - 761

66 L.M. Zavyalova, I.A. Myasnikov and S.A. Zavyalov, Khim. Fizika, 6 (1982) 473 - 478

67 D. Tomas, Semiconductors, N.B. Henney (ed.), Foreign Literature Publ., 1962 (translation into Russian)

68 V.I. Fistul, Highly Doped Semiconductors, Nauka Publ., Moscow, 1967

69 I.A. Myasnikov, Electron Phenomena in Adsorption and Catalysis on Semiconductors, F.F. Volkenshtein (ed.), Mir Publ., Moscow, 1969

70 L.D. Landau and E.M. Lifshits, Electrodynamics of Solids, Nauka Publ., Moscow, 1982

71 S. Kirkpatrick, Rev. Mod. Phys., 45 (1983) 574 - 596

72 B.I. Shklovski and A.L Efros, Electron Properties of Doped Semiconductors, Nauka Publ., Moscow, 1979

73 V.Ya. Sukharev and V.V. Chistyakov, FTT, 31 (1989) 264 - 271

74 E. Arijs, F. Cardon and V. Maenhout-Van der Vorst, J. Solid State Chem., 6 (1973) 326 - 333

75 I.A. Myasnikov, Proceedings of the 2nd National Conference on Radiation Chemistry, USSR Ac. of Sci. Publ., Moscow, 1962

76 I.A. Myasnikov and E.V. Bolshun, DAN SSSR, 135 (1972) 1164 - 1172

77 V.I. Tsivenko and I.A. Myasnikov, Zhurn. Fiz. Khimii, 41 (1967) 3 - 8

78 B.S. Agayan, I.A. Myasnikov and V.I. Tsivenko, Zhurn. Fiz. Khimii, 54 (1980) 2704 - 2707

79 S. Laroch and T. Tarkevich, Surface Sci., 20 (1970) 192 - 203

80 I.A. Myasnikov, E.U. Gutman, S.A. Zavyalov et al., The 5th Joint USSR-Japan Seminar on Catalysis, TAshkent, 1979

81 N.V. Ryltsev, E.U. Gutman and I.A. Myasnikov, Zhurn. Fiz. Khimii, 57 (1983) 381 - 386

82 H.E. Matthews and E.E. Kohnke, J. Phys. Chem. Solids, 29 (1968) 653 - 661

83 E.Kh. Enikeev, L.Ya. Margolis and S.Z. Roginski, DAN SSSR, 129 (1959) 372 - 380

84   E.Kh. Enikeev, Development in Kinetics and Catalysis, 10 (1960) Nauka Publ., Moscow, 88 - 113

85   T. Seiyama and S. Kagawa, Anal. Chem, 38 (1966) 1069 - 1081

86   G. Heiland, Sensors and Actuat., 2 (1982) 343 - 352

87   A. Windishmann and P. Mark, J. Electrochem. Soc., 126 (1979) 627 - 632

88   Y. Morooka and A. Ozaki, J. Catal., 5 (1966) 116 - 120

89   I.A. Myasnikov, Zhurn. Fiz. Khimii, 32 (1958) 841 - 857

90   E.M. Logothetis, K. Park, A.H. Meitzer et al., Appl. Phys. Lett, 26 (1975) 209 - 222

91   N. Taguchi, Inc. US Pat. # 3676820, July 11, 1972

92   H. Taylor and S. Ling, J. Chem. Soc., 69 (1947) 1306 - 1315

93   I.A. Myasnikov, DAN SSSR, 120 (1958) 1298 - 1306

94   E. Mollwo, Ztschr. Phys., 138 (1954) 478 - 489

95   D.G. Thomas and J.J. Lander, J. Chem. Phys., 25 (1956) 1136 - 1141

96   R.P. Eischens, W.A. Pliskin and J.D. Low, J. Catalys., 1 (1962) 180 - 183

97   N.N. Savvin, Mechanism of Adsorption of Atoms, Radicals and Some Simple Molecules on Metal Oxides According to the Data on Electroconductivity and IR-spectroscopy, PhD (Chemistry) Thesis, Moscow, 1980

98   V.V. Platonov, I.E. Tretyakov and V.N. Filimonov, Uspekhi Fotoniki, 2 (1971) 99, Leningrad University Publ., Leningrad

99   E.E. Gutman, N.N. Savvin, I.A. Myasnikov et al., Application of Optical Spectroscopy in Adsorption and Catalysis, Nauka Publ., Alma-Ata, 1980

100 E.E. Gutman, A.I. Lifshits, I.A. Myasnikov et al., Zhurn. Fiz. Khimii, 55 (1981) 995 - 998

101 A.M. Panesh and I.A. Myasnikov, Zhurn. Fiz. Khimii, 39 (1965) 2326 - 2328

102 I.N. Pospelova and I.A. Myasnikov, Kinetika i Kataliz, 10 (1969) 1097 - 1100

103 G. Rocker, PhD Thes. Montana State University, Bozeman, 1958

104 V.I. Yakerson, L.I. Lafer and A.M. Rubinshtein, Problemy Kinetiki i Kataliza, Nauka Publ., Moscow, 16 (1975) 49 - 68

105 S.R. Morrison, Sensors and Actuat., 11 (1987) 283 - 292

106 S. Strassler and A. Reis, Sensors and Actuat., 4 (1983) 465 - 467

107 P. Esser, R. Feierabend and W. Gopel, Ber. Bunsenges. Ges. Phys. Chem., 85 (1981) 447 - 459

108 M.A. Babaeva, D.S. Bystrova and A.A. Tsyganenko, Uspekhi Fotoniki, 9 (1987) 69 - 106

109 G. Heiland, Ztschr. Phys., 148 (1957) 15 - 28

110 I.N. Pospelova and I.A. Myasnikov, DAN SSSR, 167 (1966) 625 - 633

111 J. Villermaux, J. Lede and D. Chery, J. Chem. Phys., 65 (1968) 1963 - 1970

112 G.V. Malinova and I.A. Myasnikov, Kinetika i Kataliz, 10 (1969) 328 - 335

113 H. Nahr, H. Hoinkes and H. Wilsch, J. Chem. Phys., 54 (1971) 3022 - 3028

114 V.I. Tsivenko and I.A. Myasnikov, Zhurn. Fiz. Khimii, 43 (1969) 2664 - 2667

115 I.A. Myasnikov and E.V. Bolshun, Kinetika i Kataliz, 8 (1967) 182 - 189

116 V.I. Tsivenko and I.A. Myasnikov, Problemy Kinetiki i Kataliza, Nauka Publ., Moscow, 14 (1970) 198 - 201

117 V.I. Tsivenko and I.A. Myasnikov, Zhurn. Fiz. Khimii, 45 (1971) 1844 - 1957

118 I.N. Pospelova and I.A. Myasnikov, Zhurn. Fiz. Khimii, 46 (1972) 1016 - 1026

119 V.B. Kazanski, Problemy Kinetiki i Kataliza, Nauka Publ., Moscow, 12 (1968) 36 - 73

120 I.A. Myasnikov, E.V. Bolshun and E.E. Gutman, Kinetika i Kataliz, 4 (1963) 867 - 872

121 I.A. Myasnikov, Vestn. AN SSSR, 8 (1973) 40 - 56

122 B.S. Agayan and I.A. Myasnikov, Zhurn. Fiz. Khimii, 51 (1977) 2965 - 2971

123 V.N. Tsivenko, Adsorption of Nitrogen Atoms and Nitrogen-Hydrogen Radicals on Oxide Adsorbents by Electroconductivity Technique, PhD (Chemistry) Thesis, Moscow, 1973

124 I.N. Pospelova and I.A. Myasnikov, DAN SSSR, 170 (1966) 1372 - 1379

125 I.N. Pospelova and I.A. Myasnikov, Zhurn. Fiz. Khimii, 41 (1967) 1990 - 1997

126 I.A. Myasnikov, DAN SSSR, 155 (1964) 1407 - 1412

127 I.A. Myasnikov and I.N. Pospelova, Zhurn. Fiz. Khimii, 41 (1967) 567 - 572

128 N.V. Ryltsev, E.E. Gutman and I.A. Myasnikov, Zhurn. Fiz. Khimii, 54 (1980) 2516 - 2520

129 A.A. Andronov, A.A. Vitt and S.E. Khaikin, Theory of Oscillations, Nauka Publ., Moscow, 1981

130 D.G. Tabadze, I.A. Myasnikov and M.A. Dembrovski, Zhurn. Fiz. Khimii, 44 (1970) 821 - 827

131 D.G. Tabadze, E.V. Bolshun and I.A. Myasnikov, Zhurn. Fiz. Khimii, 44 (1970) 1804 - 1809

132 I.A. Myasnikov, E.V. Bolshun and V.S. Raida, Zhurn. Fiz. Khimii, 48 (1973) 2349 - 2355

Semiconductor Sensors in Physico-Chemical Studies
L.Yu. Kupriyanov (Editor)

170

Chapter 3

Experimental studies of the effect of adsorption of active particles on the conductivity of semiconductor sensors

3.1 Production of sensitive elements of sensors. Application of sensors to detect active particles

The majority of heterogeneous chemical and physical-chemical processes lead to formation of the intermediate particles – free atoms and radicals as well as electron- and oscillation-excited molecules. These particles are formed on the surface of solids. Their lifetime in the adsorbed state τ_a is determined by the properties of the environment, adsorbed layer, and temperature. In many cases τ_a of different particles essentially affects the rate and selectivity of heterogeneous and heterogeneous-homogeneous physical and chemical processes. Therefore, it is highly informative to detect active particles deposited on surface, determine their properties and their concentration on the surface of different catalysts and adsorbents.

However, making an even small step to qualitative assessment of availability of active particles on the surface under regular thermodynamic conditions is difficult. This is especially difficult if we are faced with the problem of quantitative evaluation of particles' origin and role in specific heterogeneous processes.

Up to date, several experimental techniques have been developed which are capable of detecting some of these particles under ordinary thermodynamic conditions. One can use these methods to keep track of transformations of the particles. For instance, it is relevant to mention here the method of electron paramagnetic resonance (EPR) with sensitivity of about 10^{14} particles per cm^3 [1]. However, the above sensitivity is not sufficient to study physical and chemical processes developing in gaseous and liquid media (especially at the interface with solids). Moreover, this approach is not suitable if one is faced with detection of particles possessing the highest chemical activity, namely, free radicals and atoms. As for the detection of excited molecular or atom particles deposited on the surface of a solid body, this task still remains exceptionally difficult because their relaxation time through physical and chemical mechanisms [2] under such conditions are extremely short, sufficient, however, for interaction with the solid body itself and with the layer of molecules, atoms, and radicals adsorbed on it.

Besides EPR, in many cases one can make use of spectroscopic, electronic, and kinetic methods. These techniques require tedious procedures on obtaining representative samples. Additionally, applying the methods mentioned one comes across numerous experimental problems, particularly if the experiment should be performed in situ.

A detailed description of analytical techniques is given in a number of original articles and books [3]. We will focus our interest on comparison of capacities of the mentioned physical and chemical methods with those of semiconductor detectors (SCD) or semiconductor sensors (SCS). These detectors are growing popular in experimental studies. They are unique from the stand-point of their application in various branches of chemistry, physics, and biology. They are capable of solving numerous engineering, environmental and other problems.

The vast capacity of SCS-based methods are related to simplicity of manufacturing and application of sensitive elements as well as with their unique response to adsorption of chemically active particles of different nature and structure.

Numerous studies of broadband oxide semiconductors made in the form of a relatively thin monocrystal or a thin film sintered on a dielectric substrate revealed that if oxides ZnO, TiO_2, SnO_2, CdO, or similar ones are used, it is possible to determine with sufficient accuracy trace concentrations of such active molecules, atoms and radicals as O_2, Cl_2, Br_2, J_2, H_2, H, N, O, OH, Cl, $\dot{O}H$, $\dot{C}l$, $\dot{C}H_3$, $\dot{C}_2H_5$, $\dot{C}_3H_7$, $\dot{N}H_2$, $\dot{N}H$ as well as atoms of many metals (Na, Ag, Zn, Cd, Pb, Fe, Pd, Pt, etc.) without the need of prior activating adsorbates.

Among the oxides shown above, zinc oxide made in the form of a thin monocrystal [4] or thin sintered film is known to have widespread use for detection of active particles in different phases. As shown by experimental data, it is purposeful to use relatively pure fused quartz, ceramics, or glass ceramic as a substrate. Glass quickly deteriorates the semiconductor film by substantially altering its initial electrical characteristics, particularly when the glass substrate with the applied oxide film is heated. This fact is evidently related to the diffusion of foreign "poisoning" impurities from the glass into the film.

Depending on the task to be solved, sensitive elements of sensors are made in the form of 2 to 4 mm plates of different shapes (round, square, rectangular). As a rule, they have a small size (linear dimensions may vary from less than 1 to 10 mm). Such plates are either polished or, in order to provide better adhesive properties, made rough. In compliance with the shape and the application area of the sensitive element, miniature contacts are applied using special platinum paste followed by welding of thin platinum pins. If necessary, the pins may be soldered in vacuum through the walls of experimental cells.

Zinc oxide paste made from fine suspension of preparation thoroughly purified and ground in an agate mortar is applied on the prepared substrate. Applied semitransparent layer is dried out and then gets gradually heated in vacuum conditions (presence of oil or mercury vapors should be ruled out!) up to 350°C along with constant monitoring of the rise in its conductivity. When the conductivity of the sample reaches plateau the thermostat maintaining the temperature at 350°C is switched off. On completion of this process oxide films should be "flashed" by pure hydrogen (filtered through palladium filter) which must be then evacuated until a constant film conductivity is reached. The zinc oxide sensitive element prepared in this way is ready for use.

Disadvantage of the described technology of obtaining zinc oxide semiconductor film deals with its poor capacity to withstand fast heating and cooling. It is quite susceptible to vibrations caused by rapid gas inflow or pumping out, etc. Such films exhibit peeling from the substrate or develop cracks with time becoming unsuitable for further applications.

Obtaining of oxide film through sedimentation or by electrophoretic depositing from water, alcohol, or benzene solution also makes it possible to produce a fairly thin film with sufficient sensitivity to adsorption of above particles. However, poor mechanical features of the films obtained are their major drawbacks.

The most reliable and easy to control method of oxide films application is provided by either sputtering of an oxide layer (e. g., by a laser beam), or evaporation of a thin metal zinc film (its resistance is usually 5-10 Ohm) from tantalum vessel onto a thoroughly prepared clean quartz substrate in a vacuum of $\approx 10^{-7}$-10^{-8} Torr (containing no oil and mercury vapors!), followed by oxidation in a flow of pure oxygen or air for 4 to 5 hours at ≈ 500°C. On completion of above process, the semitransparent oxide film gets "flashed" and warmed up in hydrogen at 350°C, then the hydrogen is evacuated. The resistance of the film thus obtained is of the order 10 to 20 kOhm in vacuum. The semiconductor oxide film prepared in such manner is fairly strong. It can be used in a laboratory or industrial installation under condition that it is not exposed to effects of detrimental gases (or even their traces). These gases comprise vapors of mineral acids, phosphides, arsenides, various sulphide compounds. Should the need arise to store prepared films for a long time, the best would be to use pure atmosphere of hydrogen, nitrogen, and inert gases for these purposes. These storage recommendations are given because the film resistance tends to slowly but irreversibly increase in the presence of even small amounts of oxygen at room temperature due to oxidation of superstoichiometric zinc atoms on the film surface which are the impurity concentration centres of zinc oxide conductivity. However, it should be pointed out that only a minor part of them is not susceptible to reduction after heating of the film in nitrogen. Besides, in the presence of oxygen the film "ages" faster, i. e. its sensitivity to the adsorption of various active particles irreversibly deteriorates. It is evidently related to irreversible structural variations caused by oxidation processes which lead to film cracking and sharp increase in resistance. The latter makes the sensor unoperational for it is impossible to restore its properties in this case.

Of all existing methods to monitor electrical properties while using semiconductor sensors, only two [5] have become widely implemented both in experimental practice and in industrial conditions. These are kinetic method, i.e. measurement of various electrical parameters under kinetic conditions, and stationary (equilibrium) method based on the measurement of steady-state parameters (conductivity, work function, Hall's electromotive force, etc.).

The first the methods mentioned is based on the monitoring of initial changes in electric parameters (mainly the dope conductivity σ) of semiconductor film of sensor caused by adsorption of active particles. In the limiting case this value may be estimated as

$$\left(\frac{\Delta\sigma}{\Delta t}\right)_{t\to 0} = \left(\frac{d\sigma}{dt}\right)_{t=0}.$$

Both calculations and experimental data indicate that at a reasonably small volume concentration of active particles and small degree of particle deposition on the film surface the following relationship [6] is rather strictly obeyed (see Fig. 3.1.):

$$[A]_\Gamma = \alpha\left(\frac{d\sigma}{dt}\right)_{t=0} = v, \tag{3.1}$$

where $[A]_\Gamma$ is the active particle concentration in gaseous (liquid) phase close to the interface of semiconductor film; α – is the proportionality factor. The right part of the equation (3.1) is denoted by v for the brevity sake.

Kinetic detection provides an opportunity to assess the concentration of active particles in cell's volume through monitoring the electric parameters caused by active particles adsorption on the surface of semi-conductor at a notably small coverage θ. Experimental results suggest that when sensitivity of a semiconductor film lies within 10^{-6} to $10^{-7}\%$ of monolayer, i.e. 10^{-7} to 10^{-6} particles per cm^2 [7] the sensor may be used without regeneration for a long period of time to obtain stable and re-producible signals ϑ, even in the absence of processes accompanying ad-sorption and resulting in partial "removal" of adsorbed particles from the surface (desorption, recombination, emission, etc., or excitation re-laxation, if we are dealing with adsorption of excited particles).

However, with aging under fixed experimental conditions the same value measured by the sensor will diverge from its maximum value as the semiconductor film gets covered up, if the rate of inverse processes reducing σ of the film is small or there are no such processes at all. If the admissible deviation is of the order of several per cent, the number of measurements of a parameter of interest may amount to several doz-ens or even hundreds without regeneration of the film applying, for in-stance, long heating at high temperature. The interval of admissible de-viation in conductivity $\Delta\sigma_e = \sigma_0 - \sigma_t$ of the sensor during measurement of a physical or chemical parameter (e. g., minor concentration of active particles) is called "film capacity" [8]. According to experiment, it equals several per cent of the initial (prior to measurements) film con-ductivity, i. e.

$$\Delta\sigma_e \,/\, \sigma_0 =\approx 2 \div 3\% . \qquad (3.2)$$

The above value is not accidental; it can be explained by the fact that the linear dependence between the values ϑ and concentration of the active particles in the space adjacent to the film (see equation (3.1)) holds with a few per cent accuracy in this very range. The dependence deduced in the film thickness range $\Delta\sigma$ shown above enables one to de-termine accurately and in a simple manner (through direct measure-ment) and calculate such values as the energy of molecular binding, specific heat of evaporation of metals, features of emission of various particles from adsorbent and catalyst surfaces, as well as many other values (see Chapter 4 for details) on the basis of the data obtained, us-ing simple and inexpensive miniature semiconductor sensors and ordi-nary electrical equipment. The most accurate and reliable results can be obtained when the sensor is made of two adjoin films with similar pa-rameters with surface of one film covered by a thin layer of evaporated quartz or other material non-sensitive to active particles and preventing them from reaching the film's surface. These films are connected accord-ing to an appropriate circuit diagram based on the principle of subtrac-tion of signals. One of them is generated both by adsorption of active particles on the open film, and – although to a lesser degree – by various interference mechanisms (e. g., thermal or other directional sources, etc.). Another signal is generated by interferences only. In this case the sensitivity of a sensor designed in this way significantly increases and its readings get more stable even at varying outer conditions. It should be noted that when performing on-line gas monitoring using the differential method described above, two identical (within 10% range) open-surface semiconductor films installed in different vessels are often used. To arrange for comparison, one of the vessels is filled by pure carrier gas, whereas the other is filled by the same gas with impurities to be monitored. In this case 10^{-6} vol%, for instance, oxygen in hydrogen or nitrogen, or other gases, can be reliably measured. There is also a number of other methods usually making use of a comparison of sensor signals depending on properties of components to be analyzed. For example, oxide semiconductor sensor may easily distinguish between adsorption signals of electrically active atom and molecular gases in mixture. The difference detected is based not only on the value of signals (atom particles are chemically more active) but rather on the known property of atoms and radicals to recombine on the surface of solid bodies, e. g. in pores of the sorbent itself. For example, a thin zinc oxide semiconductor film matched with a small porous rod (pin) also made of zinc oxide can be used for these purposes. Both sensitive elements of the sensor respond differently to the radicals and molecules of the impurities being analyzed in the flow of the carrier gas. The thin film changes its conductivity when both atoms and molecules (e.g., oxygen) are present in the carrier gas. On the other hand, the pin only reacts to molecules capable of diffusing into the porous oxide to any depth thereby changing drastically its conductivity [9] (which is not the case with atoms and radicals). Figures 3.1 and 3.2 show signals of semiconductor sensitive elements when above components are present in gas. While implementing the same principle one can make use of an oxide film coated by other porous materials simulating its natural porosity. Finally, the response of semiconductor films due to monatomic particles (radicals) and chemically active molecules can be calibrated through different location of sensors put in the carrier gas flow. The higher the pressure of the carrier gas and the lower its velocity, the smaller the distance between sensors ensuring reliable distinguishing the signal of the first film caused by overall impurity adsorption (atoms and molecules) from that caused by adsorption of impurity component reaching the second sensor. On changing the flow rate of the carrier gas, one can vary the distance between the films. When monitoring an impurity component not too chemically active to the sensor, an automatic valve may be used for alternate supply of pure gas and gas of interest to the sensor. The higher concentration of active impurities in the gas being analyzed, the higher should be the switching rate. For example, if one is faced with monitoring traces of oxygen in nitrogen, it is important to slightly activate the gas being analyzed by small portions of hydrogen or some other reducing gas which removes oxygen from the film surface during the switching, thus bringing it faster to initial condition. Other applicable methods will also be discussed.

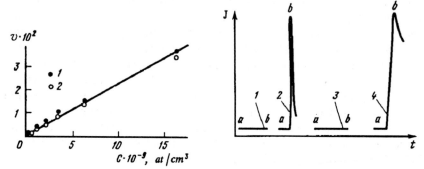

Fig. 3.1. Initial change in ZnO film conductivity at 310°C (*1*) and at 380°C (*2*) as a function of concentration of hydrogen atoms in gaseous phase.

Fig. 3.2. The profile of electric conductivity of complementary sensor caused by adsorption of molecules H_2 (*1*) and O_2 (*3*); radicals CH_3 (*2*); atoms H (*4*). The temperature during the experiment is 300°C. *a* – depicts the moment when particles enter the vessel; at moment *b* – the particles disappear (the source is switched off).

It should be stressed that if there is a need to use semiconductor sensors in physical or chemical experiments at minor gas pressure (10^{-3}-10^{-5} Torr or lower), one has to develop novel detection techniques. If there are free atoms and radicals in the system under these conditions, one should make use of both chemical reactions and heterogeneous recombination of active particles on the walls of experimental cells as well as recombination developing in special porous walls coated by various metal films (Pd, Pt, Pb, Sb, Ti, etc.) in order to improve their activity. The coating can also comprise Pd, Pt, etc. to accelerate recombination rate, or Pb, Sb, etc. absorbing active impurities. These membranes will only let penetrate molecular gases, filtering active components. In this case one can make use of the principle of electric subtraction or other circuit designs. The same method can be applied to analyze mixtures of radicals and atoms having different chemical properties, for instance, H-atoms and alkyl radicals. It should be noted that if it is necessary to divide mixtures of some molecular gases, polymer membranes can be successfully used.

We have already pointed out that most of ordinary and complex molecules are inactive as to the influence of their adsorption on electrical properties of semiconductors. In some cases (if one discards the influence of dipole components) this is caused by the absence of chemical affinity between the component being analyzed and the semiconductor film, or by high activation energy of chemical interaction with the semiconductor resulting in generation of electrical signal. In order to decrease substantially the activation energy of such an interaction, one can resort to preliminary activation of reagents or semiconductor, namely, if we are dealing with detection of gases consisting of diatom homonuclear molecules (O_2, H_2, N_2, etc.) they can be activated through initial dissociation into single atoms using a source placed close to the sensor. This results in interaction of chemically active atoms (O, N, H, etc.) with detector. In case of more complex molecules like saturated and unsaturated hydrocarbons, etc. the detector responds to effect of sufficiently reactive simple radicals (CH_3, C_2H_5, C_3H_7, etc.). In order to improve selectivity of the sensor one can use various techniques which we have discussed above.

Preliminary activation may be performed not only by means of dissociation of the components being analyzed, but also by electronic and vibrational excitation, either in the gaseous phase, or even better, directly on the film of semiconductor sensor. It should be also noted that this method is applicable to dissociation in the adsorbed layer. Excitation of the molecules in adsorbed layer (we are referring to physically adsorbed particles) can be performed optically, by an electron (ion) beam, or by an electronically excited atom beam, by Hg, for example [10, 11].

To dissociate molecules in an adsorbed layer of oxide, a spillover (photospillover) phenomenon can be used with prior activation of the surface of zinc oxide by particles (clusters) of Pt, Pd, Ni, etc. In the course of adsorption of molecular gases (especially H_2, O_2) or more complex molecules these particles emit (generate) active particles on the surface of substrate [12], which are capable, as we have already noted, to affect considerably the impurity conductivity even at minor concentrations. Thus, the semiconductor oxide activated by cluster particles of transition metals plays a double role of both activator and analyzer (sensor). The latter conclusion is proved by a large number of papers discussed in detail in review [13]. The papers cited maintain that the particles formed during the process of activation are fairly active as to their influence on the electrical properties of sensors made of semiconductor oxides in the form of thin sintered films.

Instead of making use of direct absorption of the excitation energy by the molecule being analyzed it is possible (as is the case of oxygen molecules) to achieve the preliminary activation by initial absorption on a light-excited dye layer [14] or on molecular dispersed oxide V_2O_5 [15] applied to a developed glass, quartz, or other substrate surface. Under appropriate conditions (gas pressure, temperature, light intensity, degree of surface development, surface accessibility to light, etc.) such systems exhibit generation of chemically active singlet oxygen due to a triplet-triplet energy exchange between the excited dye layer and oxygen. This oxygen gets emitted from the surface of dye into the gas phase contacting with sensor. In this case the electric conductivity "response" of the sensor to effect of molecules $^1O_2$ (singlet oxygen) is even higher if compared to adsorption of non-excited (triplet) oxygen. This method may be used to detect other molecules like benzene, naphthalene, anthracene, etc.

It is of great interest to consider sensors with semiconductor films activated by dyes or other agents capable of transmitting electron excitation not to the semiconductor but to molecules of adsorbed layer under illumination. Under excitation these molecules interact with semiconductor substrate with lower activation energy leading to generation of conductivity. In absence of such an arrangement the signal would not exist or would be very weak. In this particular case we are dealing with semiconductor photosensors activating and detecting weakly active molecular particles. It should be noted that the principle of electric subtraction of signals collected from complementary semiconductor films is applicable to this case, too.

3.2. Donor particles

3.2.1. Hydrogen atoms

It was first discovered in the study [16] that in contrast to molecular hydrogen adsorption of atomic hydrogen at liquid nitrogen temperature (77 K) notably increases conductivity of semiconductor oxide films, the effect being specifically profound in case of zinc oxide applied to quartz and other substrates. Independently of above study a similar effect has been discovered for monocrystals of zinc oxide [17]. An idea of using semiconductors to detect atom hydrogen as well as other atoms and radicals in various media and to measure their concentrations has been put forward in paper [18]. Study [19] experimentally confirmed suitability of semiconductor oxides as sensors of molecular oxygen. Although effect of oxygen on electrical properties of semiconductors ranks below that of free atoms and simple radicals, but it is profound enough to achieve objectives mentioned. In order to move forward with application of sensors to detect active particles in various media, it became essential to proceed with scrupulous investigation of the impact of adsorption of active particles on electrical properties of oxides both at kinetic and stationary conditions.

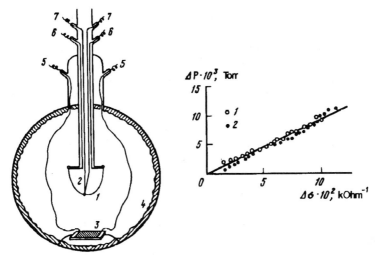

Fig. 3.3. Reaction vessel to investigate adsorption of H atoms on ZnO and conductivity. 1 – Filament; 2 – Pt/PtRh = Thermocouple; 3 – Plate to measure conductivity of ZnO films; 4 – ZnO film (adsorbent); 5 – Contacts; 6,7 – Filament and thermocouple terminals

Fig. 3.4. ZnO film conductivity as a function of amount of adsorbed hydrogen atoms. 1 – Film temperature -196°C; filament temperature 1,000°C; 2 – Film temperature -196°C; filament temperature 1,100°C

On the one hand, the choice of a proper oxide is based on necessity to obtain reliable signal, i.e. to ensure high sensitivity of conductivity σ and the release work $\Delta\varphi$ of electrons of the semiconductor on adsorption of active particles. On the other hand, it is important to ensure the chemical strength of oxide, i.e. to arrange for the absence of chemical interaction between oxide and the agent to be studied. The latter is important for providing the desired reproducibility of the signal, durability of the sensor within operational temperature range, etc. Zinc, titanium, and cadmium oxides and others turned out to be such compounds for atom hydrogen, meanwhile tungsten, molybdenum and some other oxides [7] were found to be not as good. The sensors for these purposes were produced in accordance with the description given in previous Section.

It was necessary to confirm in the scheduled experiments that there was an unambiguous reproducible relationship between the concentration of, for instance, atom hydrogen present in a medium (gaseous, liquid or solid) adjacent to the oxide film surface, and signal ϑ (when kinetic method was applied), or the steady state change in conductivity $\Delta\sigma$ (in case of stationary technique). To achieve this objective we made use of several physical and chemical methods to measure adsorption of active particles. The results obtained were then compared to variation of electric properties of semiconductor adsorbents due to adsorption of these particles. Thus, using classic volumetric method and, subsequently, the method of atom beam it became feasible to compare the values of adsorption of H-atoms on films of zinc oxide with the change in conductivity of the films caused by adsorption of H-particles applying the Langmure effect at low temperatures.

The experiment was carried out in a reaction cell shown in Fig. 3.3 with inner walls covered by a zinc oxide film having thickness 10 μm [20]. The surface area of the measuring film on the quartz plate was about $1/445$ of the total film area on the wall of the vessel. The results of direct experimental measurements obtained when the adsorbent temperature was -196°C and temperature of pyrolysis filament (emitter of H-atoms) 1000°C and 1100°C, are shown on Fig. 3.4. One can see a satisfactory linear dependence between parameters $\Delta\sigma$ (the change in film conductivity) and ΔP_{H_2} (reduction of hydrogen pressure due to adsorption of H-atoms), i.e. relations

$$\Delta\sigma = k'\Delta P_{H_2} = ka \,, \tag{3.3}$$

where a is the value of adsorption of H-atoms; k and k' are the proportionality factors. Figure 3.5 shows the results of similar experiments conducted at temperatures -78°C and -33°C. On further heating the walls we failed to notice the reduction of hydrogen pressure which can be explained by growing rate of surface recombination and reduction of adhesion coefficient of H-atoms as the temperature increases.

Based on the experimental data obtained one can easily evaluate the degree of ionization of adsorbed H-atoms ϑ. This value can be expressed as

$$\vartheta = \frac{\left[H_{hs}^+\right]}{\left[H_{hs}\right]+\left[H_{hs}^+\right]} \,, \tag{3.4}$$

or, taking into account that $\left[H_{hs}^+\right] = \Delta[e]$,

$$\vartheta = \Delta[e] \, / \, a \, , \tag{3.5}$$

where $\Delta[e]$ is the incremental number of conductivity electrons on the film and vessel walls; a — is the total number of adsorbed hydrogen atoms on the film when charged (H_{hs}^+), or neutral (H_{hs}).

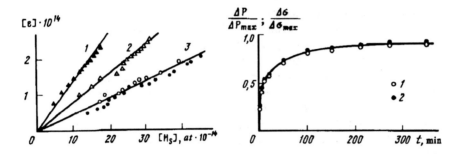

Fig. 3.5. Relationship between the number of adsorbed H-atoms on the film and the number of electrons caused an increase in its conductivity. The calculation has been performed on the basis of experiments with the following film temperatures: 1 — -33°C; 2 — -78°C; 3 — -196°C. Filament temperature was fixed at 1,100°C.

Fig. 3.6. H-atom chemosorbtion kinetics ($\Delta P/\Delta P_{max}$) (1) and ZnO film conductivity $\Delta\sigma/\Delta\sigma_{max}$ (2). All quantities are reduced to their maximum values. Experiment temperature = -196°C; filament temperature = 1,400°C; [H_V] in filament adjacent area = 2×10^{13} cm^{-3}.

Referring to results obtained in study [16], we can assume that the conductivity of crystals in sintered ZnO film increases due to increase in number of conductivity electrons in the surface layer. Taking the sample mobility of electrons μ as 10 cm^2·s^{-1}·V^{-1}, the temperature dependence being $\mu \approx T^{-1/2}$ (the data borrowed from [21]) one estimate the value $\Delta[e]$ from the following expression: $\Delta\sigma = q\mu\,\Delta[e]$, where $\Delta\sigma$ is the conductivity increment, in Ohm^{-1}·cm^{-1}, q is the electron charge, in Coulumbs, μ is the electron mobility in the crystal expressed in cm^2·s^{-1}·V^{-1}. The Table below lists the results for ϑ

Temperature, °C	-196	-78	-33
ϑ, %	5	8	13

We emphasize that these are rough calculations which depend on validity of approach and assumptions made on the values of mobility of current carriers in zinc oxide. To demonstrate convincingly that there is a strong proportionality between adsorption of hydrogen atoms on ZnO film a and the variation in its conductivity $\Delta\sigma$ up to the limiting concentration we have plotted the critical curves of adsorption a and variation of values $\Delta\sigma$ caused by adsorption. These curves are shown in Fig. 3.6 for the range 0 to 400 min, the temperature of the hot filament emitting H-atoms being 1400°C, and at the temperature of walls with ZnO film amounting to -196°C.

On the basis of the experimental data given here one can maintain that kinetics of chemosorbtion and kinetics of conductivity variation caused by the chemosorbtion are described by the same function.

Figure 3.7 shows a linear relationship between the stationary values of $\sigma\,\Delta\sigma$ and v/v_0 [6] in log-log coordinates. It proves the wide applicability of ZnO-sensors for systems which require monitoring the concentration of hydrogen atoms either under kinetic or stationary conditions up to its limiting value on the surface of semiconductor films. The latter is mainly dependent on the charge of the particles, their temperature, and the concentration of active particles in the volume of a cell. Below will be discussed the applicability of semiconductor sensors to examine heterogeneous and homogeneous-heterogeneous processes developing in presence of H-atoms in the volume or on the surface of solid bodies as well as in various gas media.

We were the first to apply the method of molecular (atom) beams to prove the existence of a rigorous quantitative relationship between the number of hydrogen atoms adsorbed on a ZnO film prepared using the method described above and adsorption-caused increase in its conductivity. In these experiments we compared the numbers of single collisions of H-atoms in the beam incident on a ZnO target with increase in conductivity of a ZnO film [22]. Figure 3.8 schematically depicts the beam method used in this study. Profile of the beam of H-atoms is given in Fig. 3.9. It was obtained through horizontal shift of a ZnO-sensor. Using a titanium atomizer meant for applying layers of titanium onto inside walls, it was possible to reduce the "drift" of hydrogen and oxygen atoms scattered by inner parts of the device almost to zero. Owing to the property of the unit shown above, it became feasible to study the interaction between the active particles and the target surface provided the collisions with the surface being analyzed and with the surface of the film were only single ones. The intensity of the beam of active particles was controlled by adjusting the temperature of pyrolysis filament. During the experiment the pressure of molecular hydrogen in the unit was 6.3×10^{-5} Torr, the temperature of pyrolysis filament was 1550°C, the temperature of reflecting target (e. g., in case of H and O) varied from 23 to 350°C, the sensor being kept at the temperature of 23°C. The intensity of incident beam of H-atoms was controlled within 10^8 to

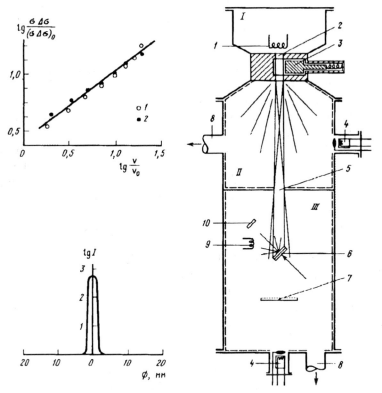

Fig. 3.7. Relationship between the stationary values of conductivity $\sigma\Delta\sigma$ and the values of initial conductivity variation rates $(d\sigma/dt)_{t\approx0} = v$ for various temperatures of hot filament (i. e., various concentrations of H-atoms in the vessel volume). *1* – 300°C; *2* – 380°C.

Fig. 3.8. Experimental set-up to examine interaction of atom particles with the surface of a solid body by means of atom beam reflection. *I* – Chamber with atom particles source installed; *II, III* – Intermediate and main chambers; *1* – Pyrolysis filament; *2* – Collimation channel; *3* – Beam chopper; *4* – Titanium atomizer; *5* – Collimation slot; *6* – Target; *7* – Deflector; *8* – To vacuum pump pipe; *9* – Filament; *10* – ZnO semiconductor sensor

Fig. 3.9. Intensity profile *I* of H-atoms beam incident on the target. Beam pressure in the main chamber is 6.3×10^{-5} Torr; temperature of H-atoms generator pyrolysis filament is 1550°C.

10^{12} atoms/cm^2·s range. Removal of air was provided by oil and mercury free pumps. First of all, the experimental set-up enabled us to confirm the unambiguous correlation between the intensity of the beams incident onto the active particles film-target (H- and O-atoms) and the rate of variation of conductivity under conditions of rare deposition of adsorbed particles on the film's surface. Additionally, comparing intensities of incident and reflected beams it became possible to determine the scattering coefficient ω. Finally, the rates of the change in electric conductivity of targets brought about the conclusion on the ionization coefficient ϑ of atom hydrogen and oxygen on the ZnO film.

Using obtained values of ω and ϑ [23] we could calculate the absolute values of flow intensity of free H- and O-atoms based on the initial rates of conductivity variation in ZnO-sensor caused by the incident atom flows. To prove the validity of approach proposed let us give here several numeric examples [22]. The flow of O-atoms amounting to 7.89×10^9 at·cm^{-2}·s^{-1} obtained in this experiment and calculated on the basis of ZnO target conductivity data fairly correlates with the calculation of the same flow performed on the basis of equilibrium concentrations of atoms close to the surface of the hot filament [24] taking into consideration the geometry of the unit [25]. In the latter case we obtained 7.91×10^9 at·cm^{-2}·s^{-1}. Similar data was obtained for atom hydrogen and for atom metals which will be discussed below.

One can conclude that thin ZnO films may serve as absolute sensors for H- and O-atoms as well as for other atoms and radicals affecting their conductivity (CH_3', OH, N, and others) provided that values ω and ϑ are determined by above technique.

Based on the data of hydrogen and oxygen atoms scattering by the surface of ZnO film obtained by the method of atom beams we calculated the degree of ionization ϑ of chemosorbed particles on the film using the variation of its conductivity (see Table 3.1) [23].

TABLE 3.1.

Particle	Ionization Degree	
	23°C	100/150°C
H	0.26	0.45
O	0.45	1.00

It follows from the above that

$$\vartheta_H = \left[H_{hs}^+\right] / \left(\left[H_{hs}^+\right] + \left[H_{hs}\right]\right) = \Delta n_e / I_A(1 - \omega), \qquad (3.6)$$

where H_{hs}^+ and H_{hs} are chemosorbed hydrogen atoms, charged or uncharged; Δn_e is the variation of the number of electrons in the conductivity area of the semiconductor film due to chemosorbtion of the atoms; I_A is the flow of particles. We should mention that close results for hydrogen had been obtained in study [20] by the volumetric method. Thus, the volumetric approach revealed that $\omega_H = 0.1$ for ZnO at the temperature of -33°C which does not contradict to the data obtained by the atom beam method.

The comparison of experimental data on determination of concentrations of H- and O-atoms in the gas volume or on the surface using semiconductor ZnO films given in this Section with results of calculations based on the formulas derived in Chapter 2 we can conclude that even extremely small concentrations of active particles (10^5 to 10^7 m^{-3}) can be assessed quantitatively by sensors both in the gaseous phase and on the surface of solid bodies (see below).

3.2.2. Metal atoms

Hydrogen atoms are the simplest donor particles. Particles known as hydrogen-like donors which are atoms of many metals stay next to them. Due to the general ideas concerning the structure of their outer electron shell in the medium having a fairly high dielectric constant $\varepsilon \cong 10$, valence electrons of these atoms are similar to that of atom hydrogen. In contrast to vacuum they have extended orbitals with radius proportional to ε^2 of the medium, in other words, their radius is ε times as much as the radius of the same atoms in vacuum [26]. Similar to H-atoms, in this case the energy of binding with nucleus screened by the remaining electron charge, drops by ε^2 (i.e., hundreds) times. Even at room temperature of the solid adsorbent this effect leads to a partial ionization of these hydrogen-like particles placed onto a solid matrix with a sufficient dielectric constant (about 10 for ZnO). Hydrogen-like metal atoms [26] placed on the surface of a solid body have these properties as well. Orbitals of these superficial atoms are asymmetrically extended inside the surface layers of the solid body.

If the above comparison of the properties of metal atoms with those of hydrogen deposited on the surface of a solid body (semiconductor) is correct, the effect of their adsorption on electric properties of semiconductor oxide films will be similar to features accompanying adsorption of hydrogen atoms. The atoms of hydrogen are very mobile and, in contrast to metal atoms, are capable of surface recombination resulting in formation of saturated molecules with strong covalent bond.

Metal atoms applied to the surface of a solid body under mild conditions (low target temperature and flow intensity) e.g., by means of an atom gun [25] (the method of molecular beams), provided that diffusion into crystal is minimal, do not obey the Warvey valence rule [27] as in most cases they form interstitial atoms on the surface of the target. At low ionization energy these atoms can bring the valence electron into the semiconductor conductivity band thus increasing concentration of free carriers.

Above reasoning can be confirmed by a number of experimental results which showed that although with some peculiarities irrelevant to the properties of semiconductor sensors the correlation between the amount of the atoms in the flux incident on the target, or their surface concentration, and the variation (increase, if we are dealing with semiconductor of n-type) of the target conductivity takes place [28]. Based on the relations cited in Chapter 1, one can estimate concentrations (i. e., flow intensities) of these particles in vacuum or in gaseous medium if these values are quite small, using the values of conductivity variation of the semiconductor film.

Here it is relevant to mention results of some experiments. They were carried out specifically to substantiate the applicability of semiconductor sensors to solve numerous problems dealing with metal atoms and clusters (both in vacuum and on the surface) in cases when the use of other techniques does not yield sound results.

To conduct experiments correctly with metals it is strongly recommended to reject the use of oil and mercury vacuum pumps because there is no guarantee that aerosols of oil or mercury get past most sophisticated traps at cooling down to 77 K and lower. Under experimental conditions the surface of objects to be analyzed and the inner walls of the vessel get gradually covered with oil film.

To solve above problems one can resort to use discharge, cryogenic, adsorption, turbomolecular or other types of vacuum pumps. To our opinion, the most appropriate methods to conduct such studies are low intensity atom metal beams with rigorous thermal control. For example, to analyze the adsorption of silver atoms on oxide films one should ensure that the atom flow completely covers the target made of the material of interest to be analyzed. In such experimental arrangement thin platinum contacts are used for conductivity monitoring. Fig. 3.10 shows a soldered experimental cell made of pure fused quartz which can be heated higher than 1000°C for cleaning purposes. The cell was used to investigate silver and palladium. These metals relatively easily evaporate from a platinum evaporating belt whose temperature is strictly controlled by means of a thermocouple. The metals being analyzed in form of foil with dimensions 3 per 3 mm were attached to the platinum belt in such manner that the evaporator centre was situated exactly opposite the centre of a 3 by 3 mm ZnO film. The intensity of the metal flow incident on ZnO film was calculated based on the evaporator temperature,

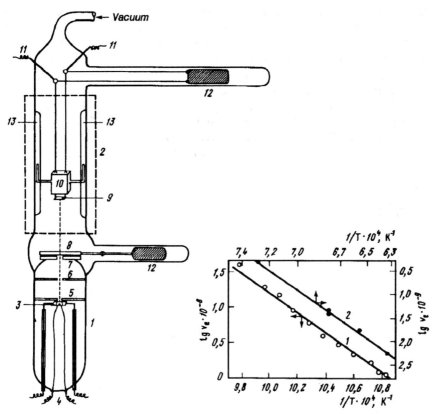

Fig. 3.10. Reaction cell. *1* – Atom gun; *2* – Thermostate; *3* – Metal evaporator; *4* – Pt/Pt-Rh thermocouple; *5/7* – Collimation holes (diameter 3 mm); *8* – Shutter; *9* – ZnO semiconductor sensor; *10* – Mobile quartz weight; *11* – Platinum contacts terminals; *12* – Vitrificated iron bars controlled by a magnet; *13* – Quartz guides

Fig. 3.11. Dependence of $v_e = (d\sigma/dt)_{t \to 0}$ on the evaporator temperature plotted in the Arrenius coordinates for evaporation of silver (*1*) and palladium (*2*).

geometry of the vessel and respective evaporation heats in vacuum. Thus, the bombardment velocity of the target I_Z, and consequently the adsorption velocity (adhesion coefficient s for these metals is close to 1) was calculated according to the formula $I_Z = I_A S$, where I_A is the beam intensity, S is the target area. The increase in the number of electrons (current carriers) $\Delta[e]$ in ZnO film was calculated accounting for the exposure time of the target to the beam using well-known relationships:

$$\Delta\sigma_0 = \Delta[e]\mu \cdot 1,6 \times 10^{-19}, \quad \Delta\sigma = \frac{S}{l}\Delta\sigma_0, \tag{3.7}$$

188

where $\Delta\sigma_0$ and $\Delta\sigma$ are respective variations of total and specific film conductivity; l and S are the length and the section of the film; $\mu = 10$ cm^2/s·V is the mobility of current carriers for ZnO. The results obtained in paper [29] brought authors to conclusion that the mobility of current carriers does not depend on the degree of surface occupied by metal atoms if it remains sufficiently small. Based on the data listed in Table 3.2, profiles of functions $\lg v_e$ vs. $1/T$ are plotted in Fig. 3.11. These graphs were used to calculate evaporation heats of silver and palladium (64 and 95 kcal/g-at respectively) which is coincident (within a few percent) with the values given in the Nesmeyanov's Table [30].

TABLE 3.2.

Metal	Film temperature, °C	Evaporator temperature, °C	Intensity of incident beam, $I_{A_2}\cdot10^{10}$, at·cm⁻²·s⁻¹	Film bombardment velocity, $I_2\cdot10^{-9}$, at·s⁻⁹	Velocity of concentration variation in the film, $v_e\cdot10^{-9}$ electron·s⁻¹	$\alpha = \dfrac{I_z}{v_e}$ Intensity evaluation	Checkup: isotope method	Determination based on the Hall electromotive force
		650	0.28	0.39	0.11	3.5		
		670	0.49	0.68	0.23	3.0	2.7	
Ag	20	710	2.49	3.48	0.88	3.9	4.4	
		730	4.93	6.88	1.89	3.6	3.4	
				0.65	0.174			3.8
		1110	8.72	12.2	2.42	5.8		
		1200	81.5	114.0	23.2	5.0		
Pd	150	1200	81.5	144.0	26.4	4.2		
		1250	224.2	314.0	62.0	5.0		
		1300	605.0	846.0	188.1	4.4		
	180			0.65	0.515			1.3

Similar to the case of H-atoms the results obtained fully confirm the validity of expression $v = \vartheta I_z$, where ϑ is the degree of ionization depending on adsorbate, adsorbent, and the temperature. This means that ZnO films (it is also correct both for a CdO layer, and for other chemically stable semiconductor oxides) may be used as very sensitive miniature sensors to determine intensity of atom flow for detected noble metals Ag and Pd (see Table 3.2). If the sensitivity of the measuring equipment is brought up to 10^{-16}A one can measure atom flows equal to 10^3 at/cm^2·s. It should be noted that the degree of surface occupation was only 10^{-4} to 10^{-2}% during these experiments. There was no conductivity relaxation detected at this surface occupation degree after shutting down the atom flow even at high temperatures of ZnO films (around 100 to 180°C). Conductivity relaxes to its initial value at considerably higher occupations under the same conditions. This phenomenon will be detailed in the next Chapter.

We feel it is interesting and important to dwell on other experiments which characterize the relationship between the number of atoms of different metals and the variation of conductivity in a film of a semiconductor sensor as well as the dependence of the signal value on the degree of surface occupation.

To resolve the problem applying methods of collimated atom beams, equilibrium vapour as well as radioactive isotopes, the Hall effect and measurement of conductivity in thin layers of semiconductor-adsorbents using adsorption of atoms of silver and sodium as an example the relationship between the number of Ag-atoms adsorbed on a film of zinc oxide and the increase in concentration of current carriers in the film caused by a partial ionization of atoms in adsorbed layer were examined.

The experiments with atom beams of radioactive silver Ag^{110} have been carried out in a cell (see Fig. 3.12) with air pumped out using adsorption and heteroion pumps [31] to the residual pressure of 10^{-8} Torr. The cell can operate with the collimated beam of labelled silver atoms which was 70 mm long, 2 mm in diameter and had the angular divergence of 1 deg. The target area was 0.1 cm^2. Either one of the sections of the mobile quartz plate or the film sensor have been exposed to atom beam as a target in these experiments. In the first case it was meant to accumulate a certain amount of radioactive silver on the target sufficient to conduct a quantitative analysis, in the second case the problem in question dealt with determination of variations of the film-sensor (detector) conductivity after being exposed over a long period of time to the atom beam with the same intensity as in plate section. The measurements of film-sensor conductivity have been performed before and after the deposition of silver on one of the plate sections by an atom beam of a given intensity. The second measurement was meant to make sure that the intensity of the silver atom beam was constant throughout the experiment.

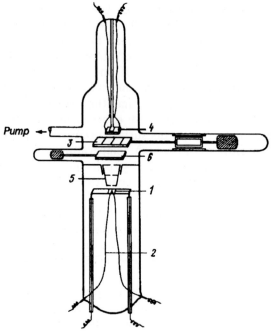

Fig. 3.12. Reaction cell. *1* – the evaporator; *2* – thermocouple; *3* – beam splitter; *4* – semiconductor sensor (ZnO); *5* – collimating holes; *6* – atom beam shutter

After each series of experiments with beams of various intensity the section plate would be removed from the cell and disassembled, with radioactive silver washed out by nitric acid. Radioactivity of the solutions obtained was measured by a multichannel spectrometric scintillation γ-counter with sensitivity of up to 10^{-9} G, i. e. around 10^{13} of Ag^{110} atoms which, according to calculations, is 10^5 times lower than sensitivity of ZnO sensor: 10^{-14} G or 10^8 of Ag^{110} atoms respectively [28]. This difference in sensitivity lead to great inconveniences when exposing of targets was used in above methods. Only a few seconds were sufficient to expose the sensor compared to several hours of exposure of the scintillation counter in order to let it accumulate the overall radioactivity. It is quite evident that due to insufficient stability during a long period of exposure time an error piled up.

Table 3.3 shows the data obtained while measuring intensities of a Ag^{110} beam by isotope and semiconductor sensor method using a ZnO oxide film as a sensitive element. The data reveals that in this case there is also a satisfactory linear dependence between the number of silver atoms incident onto the surface of a ZnO film and variation in its conductivity, i. e. the variation in the number of conductivity electrons of the film produced by Ag-atoms incident onto the film. We have estimated this value based on the results of monitoring the Hall characteristics during adsorption [29] and through assessment of the scattering coefficient of silver atoms ω by ZnO film. The coefficient turned out to be close to 1 provided that air is pumped out by an oil-free pump. The calculation was performed applying the well-known formula (see expressions (3.6) and (3.7)). It should be noted that the degree of surface occupation of a ZnO film in these experiments was no more than 10^{10} cm^{-2}, i. e. 10^{-5} of a monolayer, or 10^{-3} %. With the surface population of higher than 10^{10} cm^{-2} it is necessary to account for relaxation of the film conductivity due to aggregation of the surface Ag-atoms for clusters of Ag-atoms affect conductivity of the semiconductor film in a different manner (see below for details).

TABLE 3.3.

Evaporation temperature, °C	Ag^{110} atoms flow intensity, isotope method, $I_A \cdot 10^{-9}$ atoms s^{-1}	Current carrier concentration variation rate in the film, semiconductor sensor method, $v_e \cdot 10^{-9}$, electrons s^{-1}	$\alpha = \dfrac{I_A}{v_e}$
670	10.70	4.6	2.33
680	14.67	5.58	2.63
690	22.05	7.44	2.96

Note: I_A is the intensity of the atom beam incident onto the film, at/s^{-1}; v_e is the current carrier concentration variation rate in the film.

Besides the experimental data mentioned above, the kinetic dependencies of oxide adsorption of various metals are also of great interest. These dependencies have been evaluated on the basis of the variation of sensitive element (film of zinc oxide) conductivity using the sensor method. The deduced dependencies and their experimental verification proved that for small occupation of the film surface by metal atoms the Boltzman statistics can be used to perform calculations concerning conductivity electrons of semiconductors, disregarding the surface charge effect as well as the effect of aggregation of adsorbed atoms in theoretical description of adsorption and ionization of adsorbed metal atoms. Considering the equilibrium vapour method, the study [32] shows that sodium, zinc, indium, cadmium, and lead vapours as well as the vapours of some other metals when adsorbed on a ZnO film from gaseous medium drastically increase its conductivity while this is not the case for mercury vapours which can probably be explained by a greater mobility of Hg-atoms on the surface. The minimum pressure of metal vapours at which one can detect their effect on the film conductivity was about $10^{-7} - 10^{-10}$ Torr. Kinetics of ZnO film conductivity variation under the effect of the adsorption of atoms of various metals up to the plateau value may be described by the following equation [33]:

$$\Delta\left(\frac{\sigma}{\sigma_0}\right) = A\,e^{-\alpha t},$$

(3.8)

where $\Delta\left(\dfrac{\alpha}{\sigma_0}\right) = \left(\dfrac{\sigma}{\sigma_0}\right)_\infty - \left(\dfrac{\sigma}{\sigma_0}\right)$; σ_0 and σ are ZnO film conductivity

before and during the process of atom adsorption; ∞ designates the stationary final value at a given temperature; α and A are the constants whose values depend on the temperature. The following formula is the result of differentiation of equation (3.2):

$$\left(\frac{d\sigma}{dt}\right) = \tilde{A}\,e^{-\alpha t},$$

(3.9)

where $\tilde{A} = A\alpha\sigma_0$ is the coefficient. Formula (3.9) expresses temporary dependence of the rate of variation of σ in film under the effect of metal vapour adsorption (see expression (2.14)).

Equations (3.8) and (3.9) given above describe fairly well the adsorption of atoms metals investigated as well as that of hydrogen atoms. They can be derived on the basis of the general approach to formal chemical kinetics for charged particles (electron semiconductor conductivity included) featuring Boltzman statistics in case when one can neglect the charge of semiconductor surface due to adsorbed particles. Verification of equations (3.8) and (3.9) proves the validity of assumptions made while deriving them which in turn attests that we have a solid theoretical ground to proceed with manufacturing semiconductor sensors of metal vapours to monitor quantitatively numerous physical and chemical processes.

In conclusion it should be emphasized that the principle objective of this section was to prove the physical-chemical suitability of semiconductor oxide sensors for quantitative measurements of extremely small concentrations of atoms of many metals both in vacuum and deposited on the surface of semiconductor adsorbents. Given above there are the tables showing sodium vapour pressures measured by the semiconductor sensor approach (Table 3.4), and evaporation heats of various metals (Table 3.5).

TABLE 3.4.

Sodium temperature, °C	$v_e \cdot 10^{-9}$, electron·s	$I_z \cdot 10^{-10}$, at·cm^{-2}·s^{-1}	P, Torr·10^{10}	
			Calculation	Table data
20	1.25	1.25	0.3	0.13
30	4.20	4.20	1.0	0.40
37	7.00	7.00	1.7	1.00
42	10.00	10.00	2.4	2.00
45	12.50	12.50	3.0	2.50

Note. Z is the number of atom collisions per 1 cm^2 of the surface in 1 s.

TABLE 3.5.

Element	Experimental temperature range, °C	Experimental vapour pressure range, Torr	H, kcal/Mol	
			Semiconductor sensor method	Literature
Sodium	20–120	10^{-10}–10^{-6}	20.0	23.4
Zinc	100–160	10^{-9}–10^{-6}	23.8–30.0	27.0–30.0
Cadmium	105–180	10^{-7}–10^{-4}	21.0–22.0	23.9
Lead	350–420	10^{-8}–10^{-6}	43.0–44.0	43.0
Silver	690–860	10^{-8}–10^{-6}	59.0–60.0	60.7
Lead	2290–2650	10^{-8}–10^{-6}	17.0	17.6

3.3. Acceptor particles

3.3.1. Atoms and molecules of simple gases

This class of particles consists of those free atoms, radicals, and molecules which are capable of capturing conductivity electrons of oxide semiconductor in the adsorbed layer. In case of n-semiconductor (ZnO, etc.) the resultant dope conductivity gets decreased whereas for p-type semiconductors it features and increase. This group includes a wide variety of particles which may quite differently influence physical properties of an oxide semiconductor. The group includes numerous free radicals of various hydrocarbons as well as hydroxyl, imine, amine and many other radicals. It follows from experiments that nitrogen atoms (in contrast to the complete inactivity of molecular nitrogen), atoms and molecules of oxygen, chlorine and other active gases have strong acceptor properties in the adsorbed layer of the oxide semiconductors. Study [34] revealed that electronically and vibrationally excited particles O_2, N_2, H_2, as well as excited atoms of noble gases, have quite different effect on conductivity of these adsorbents depending on the temperature and type of semiconductor.

In this part we dwell on the properties of the simplest radicals and atoms in the adsorbed layer of oxide semiconductors as well as analyse the quantitative relationships between concentrations of these particles both in gaseous and liquid phases and on oxide surfaces (mostly for ZnO), and effect of former parameters on electrophysical parameters. Note that describing these properties we pursue only one principal objective, i. e. to prove the existence of a reliable physical and physical-chemical basis for a further development and application of semiconductor sensors in systems and processes which involve active particles emerging on the surface either as short-lived intermediate formations, or are emitted as free particles from the surface into the environment (heterogeno-homogeneous processes).

As with donor particles, in order to resolve the posed problem it is initially necessary to prove experimentally a rigorous validity relationships derived in above domain of parameters (pressure and temperature) based on substantially wide number of experimental results. It is known that when preparing such experiments it is recommended to avoid various "reefs" which may provoke an experimentalist to take wrong assumptions for real, or to hide from a theorist, for example, simple functions in relationships analysed. Pioneering experiments conducted with acceptors involved such active particles as molecular oxygen which on the one hand possesses strong acceptor properties, and on the other is a fairly widespread and chemically sufficiently stable element,

unlike free atoms and radicals tending to recombine as well as interact
in other manner both in the volume of the system being analyzed (gas,
liquid, solid body) and on its walls. In order to study electrophysics of
"oxygen-semiconductor" system one can use both kinetic and equilibrium detection technique thereby trying to draw a possibly rigorous
comparison of results obtained which is not always possible concerning
the adsorption of valence-unsaturated particles like atoms and radicals.
As we mentioned in part 2, the study [19] revealed quantitative relationships between concentration of molecular particles (molecular oxygen) in gaseous phase and that of conductivity electrons in sintered
semiconductor film, although the latter did not exceed the degeneration
threshold for electron gas. Experimental results were analysed applying
the method of formal kinetics to both molecules and conductivity electrons. In other words, properties of all the particles of the system being
analyzed satisfied requirements of the Boltzman statistics. Besides, the
above conclusions were additionally restricted by degree of depositing of
adsorbed particles on the semiconductor film. Thus, for the equilibrium
case an equation was derived linking concentration of molecular particles with that of conductivity electrons, and was shown that simple
expression (3.10) adequately describes the experimental data within the
temperature range 250 to 300°C and at oxygen pressure 10^{-3} to 1 Torr.

$$\left(\frac{\sigma}{\sigma_0}\right)^2 = \frac{B}{P_{O_2}} \text{ with } P_{O_2} > 0, \qquad (3.10)$$

where σ_0 and σ are the equilibrium values of the impurity conductivity
at the beginning of the experiment and at different values of equilibrium
P_{O_2}.

Figure 3.13 can serve as a proof of the above. It shows the deduced
dependence to be linearized in log-log coordinates. Angular coefficients
of the straight lines obtained for different temperatures correspond to
requirements of the equation and equal to $-1/2$. It confirms applicability of the deduced expression (3.10). Experiment shows that the dependence (3.10) is valid at low temperatures as well though the time
necessary to reach equilibrium is fairly long in this case.

Based on results obtained we built [19] a sensitive oxygen analyzer
which we have widely used in our further studies (see next chapter for
details).

It should be noted however that the method related to the measurements of equilibrium or stationary oxygen concentrations by means of
sensors is not always applicable and convenient even in this particular
case. Sometimes too much time is required to conduct these measure-

ments. Besides, the method always implies considerable coverage on the semiconductor film surface or the pin, even when measuring low concentrations of gases which inevitably leads to some residual phenomena related to partial irreversible oxidation of the sensor surface. It is clear that this process does not makes it difficult to reproduce sensor readings for consecutive identical experiments.

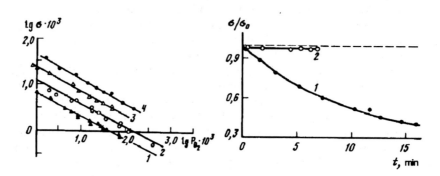

Fig. 3.13. Conductivity of the film of a sensor plotted vs. oxygen pressure in bilogarithmic coordinates. 1 – 250°C; 2 – 260°C; 3 – 467°C; 4 – 620°C

Fig. 3.14. Comparison of the change in relative conductivity of a ZnO film caused by chemisorbtion of O-atoms (1) and O_2 molecules (2) at 383°C. 1 – $P_{O_2} = 4 \cdot 10^{-4}$ Torr ($N_0 = 10^{12}$ particles/cm³); 2 – $N_0 = 10^9$ at/cm³

In relation to the results obtained from experiments on detection of molecular oxygen, there appeared an issue of surface physical-chemical and electrophysical properties of non-molecular acceptor particles, i.e. more simple but equally aggressive free atoms of various substances. First of all, these are atoms of oxygen, chlorine, etc. Experiments have shown that atoms of elementary substances like oxygen or nitrogen which have acceptor properties similar to donor particles (for instance, atoms of hydrogen) are far more active regarding the adsorption effect on oxide conductivity, if compared with molecular particles. This fact is related to their greater chemical activity, i.e. to minor activation energy of adsorption (chemisorbtion) on oxides compared to molecular particles whose adsorption often develops through surface dissociation which requires considerable activation energy. Oxygen and nitrogen provide the most vivid examples of such acceptor particles whereas dissociation of chlorine on atoms of the surface of an oxide semiconductor occurs with much lower activation energy, the difference in effect of its molecular and atom particles on semiconductor electrophysics being masked at high temperatures. One can distinguish the influence of molecular and atom particles on electrophysical properties of semiconductors only using a combined semiconductor sensor [9] provided that both types of above mentioned particles are electrically active. The principle of combined semiconductor sensor is based on a thin sintered semiconductor film and a sintered ball or a pin made of the same semiconductor material and connected into a parallel electrical circuit. It is noteworthy that molecular and atom oxygen, chlorine, and many other pairs, for example, excited and non-excited electrically active particles like triplet, singlet, and vibrationally excited molecular oxygen, excited molecular nitrogen, hydrogen, etc. should be grouped with the above particles pairs.

It was first demonstrated in the studies [35, 36] that the effect of adsorption of O-atoms on conductivity of zinc oxide (or other oxides) films is many dozen times greater than that of molecular oxygen in the range of moderate and low temperatures, all other factors being equal. But it should be noted that even at room temperature adsorption of oxygen atoms on oxides is accompanied by a progressive irreversible oxidation of interstitial zinc atoms which are the centres of adsorption of O-atoms. The more the temperature and the surface occupation degree, the faster the process proceeds. It is not virtually detected at low temperatures. Figure 3.14 displays the variation of relative conductivity of a zinc oxide film during chemisorbtion of atom and molecular oxygen in low pressure domain ($P_{O_2} = 4 \cdot 10^{-4}$ Torr). According to the calculation, concentration of O_2 in this experiment was 10^{12} particles/cm^3, and that of atoms was 10^9 at/cm^3. Nevertheless the variation rate and the stationary value of variation of oxide film conductivity when adsorbing O-atoms even when their number is critically low on the surface of oxide is considerably higher than the similar feature of the same semiconductor due to adsorption of molecular oxygen, with a far greater surface occupation degree. This indicates that O-atoms have a greater chemical activity as well as means that there is no evident surface atom dissociation with chemisorbtion of O_2 molecules, at least in the moderate temperature range.

On the one hand, experiments involving nitrogen atoms on the background of molecular nitrogen and semiconductor sensors are much simpler to conduct owing to the inertia of nitrogen molecules but on the other they grow more complicated due to a greater bond strength between nitrogen atoms. It hinders the use of the method of thermal dissociation of molecular nitrogen thus necessitating the use of powerful high-temperature generators. Implementation of the latter is not desirable because of a high probability of emission of foreign particles and inclusions from the heater of the generator into the gas volume together with N-atoms. Regarding the use of discharge units in this and other similar cases, it becomes even more complicated because it may lead to a formation of foreign particles of various nature which are fairly difficult to get rid of during the experiment. Moreover, in contrast to capability of monitoring of concentration or flow rate of atom particles during pyrolysis of molecular gases [24], it is virtually impossible to produce necessary atom particles in discharges of various types without using other measuring devices. It makes the experiment far more complex and in general devaluates all the convenience and the ease of use of sensors.

It was first shown in study [37] that adsorption of N-atoms on films of zinc oxide reduces its conductivity to a certain stationary value which depends, as with oxygen atoms, both on the stationary concentration of particles in the volume adjacent to the sensor's film and on the temperature.

Stationary concentration of adsorbed acceptor particles of O- and N-atoms on a film of zinc oxide is attained for the most part due to the competition between the chemisorbtion of particles and their interaction, i. e. mutual recombination on the adsorbent surface, and with free atoms "attacking" the adsorbed layer of the adsorbent from outside.

Over large periods of time, the conductivity of oxide film grows linearly with time during chemisorbtion of atoms of oxygen and nitrogen at ultra low concentrations on the zinc oxide film over wide time interval at moderate and elevated temperatures, i. e. the following simple kinetic equation is valid:

$$\sigma_t \ / \ \sigma_0 = 1 - kt \ \text{ or } \ \Delta\sigma \ / \ \sigma_0 = kt \ , \qquad (3.11)$$

where σ_0 and σ are the conductivity of ZnO film before and during the adsorption; k is the kinetic constant. The above linear dependence is likely to be caused by a low degree of occupation of adsorbent surface θ by atoms of oxygen or nitrogen during experiment. In this case the volume concentration of active particles was only 10^7–10^9 at/cm^3. Kinetic mechanisms for oxygen and nitrogen atoms become more complicated at higher concentrations of active particles due to the increase of their velocity of interaction with each other on the surface of semiconductor.

Figure 3.15 shows the validity of above simplest equation for adsorption of O-atoms provided that there are different concentrations of interstitial zinc atoms on the zinc oxide surface. In case of oxygen atoms the experiment has been carried out in absence of molecular oxygen so that effect of its adsorption on change in conductivity was ruled out. O-atoms were produced by means of pyrolysis of carbon dioxide. From this figure we notice that zinc atoms (superstoichiometric) applied onto the surface of the zinc oxide film are the active centres of adsorption of

O-atoms. Similar situation has also been shown for adsorption of nitrogen atoms [38]. In this case the equation (3.11) holds over wider time interval and for higher volume concentrations of active particles.

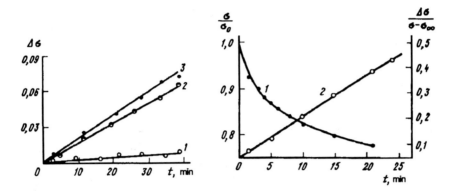

Fig. 3.15. Kinetics of conductivity of ZnO film at 20°C during adsorbtion of oxygen atoms: *1* – Prior to treatment by zinc vapour; *2, 3* – Two consecutive experiments after adsorption of zinc atoms. O-atoms are obtained by pyrolysis of CO_2 at $P = 0,9$ Torr, $T_{filament} = 1160°C$.

Fig. 3.16. Kinetics of conductivity of ZnO film at room temperature during adsorbtion of N-atoms.

It should be also noted that the kinetic equation derived from the notions of formal kinetics is applicable over a wide range of concentrations of active particles at temperatures above 200°C in case of adsorbtion of O- and N-atoms on ZnO films. It may be expressed as follows:

$$\Delta\sigma / \sigma_t = kt , \qquad (3.12)$$

where t is the time; $\Delta\sigma = \sigma_0 - \sigma_t$ (see Chapter 2, expression (2.44)).

However, applicability of this expression breaks down at the range of moderate (<200°C) and low temperatures. The following equation (see Chapter 2, expression (2.98)) was empirically found in the study [39]:

$$\Delta\sigma_0 / \Delta\sigma = kt , \qquad (3.13)$$

where $\Delta\sigma_0 = \sigma_0 - \sigma_\infty$; σ_∞ is the minimum value of conductivity under the conditions of the present experiment which means that the semiconductor film includes dope conductivity centres which are for some reason not involved in the chemisorbtion of nitrogen or oxygen atoms.

Figure 3.16 shows an adequate applicability of the above equation for nitrogen.

Our comments on adsorption of oxygen and nitrogen atoms lead to conclusion that practically under all conditions the initial rate of variation of conductivity of zinc oxide film due to adsorption of acceptor particles discussed in this section is proportional to the concentration of particles in the space adjacent to the surface of oxide film. This is similar to the case of donor particles. This means that the following equation is applicable:

$$\left(\frac{d\sigma}{dt}\right)_{t\cong0} = k[A],\qquad\qquad(3.14)$$

where $[A]$ is the atom particles concentration in the gaseous phase. This means that the number of adsorbed particles on the ZnO oxide film at reasonably low surface occupation degrees is always proportional to the number of free current carriers which constitutes the difference between their initial (before the adsorption) and the final values. The above holds true for adsorption of both acceptor and donor particles. The conclusion obtained is consistent with results of studies [28, 40] aimed at determination of above relationship using the method of measurement of the Hall electromotive force during the adsorption of atoms of hydrogen, oxygen, and iodine, as well as atoms of many metals, which attests a small and insignificant effect of minor variations of Hall mobility of conductivity electrons on the said proportionality when adsorbing active particles on a zinc oxide film at low degrees of surface occupation.

All the aforesaid on the adsorption of acceptor atom and molecular particles lends support to the general conclusion of suitability of the designed sintered semiconductor oxide films (mainly ZnO) as highly sensitive semiconductor sensors meant to quantitatively detect extremely low concentrations of atom and molecular acceptor and donor particles in a concentration range of $10^7–10^9$ atoms per cubic centimetre in the volume adjacent to a semiconductor sensor.

3.3.2. Free radicals

Electric effects detected in semiconductor oxide films during chemisorbtion of atom particles have been also thoroughly studied for chemisorbtion of various free radicals: CH_2, CH_3, C_2H_5, $C_6H_5OH_2$, OH, NH, NH_2, etc. [41]. It was discovered that all of these particles have an acceptor nature in relation to the electrons of dope conductivity in oxide semiconductors their adsorption, as a rule, being "reversible" at elevated temperatures. It is clear that we deal with reversibility of electron state of the oxide film after it has been heated to more than 250–300°C in vacuum or in a oxide-inert medium rather than reversibility of radical adsorption. The further state of chemisorbed radicals after such treatment is determined by disintegration of surface radicals complexes on chemisorbtion centres (most often these are oxide superstoichiometric metal atoms) and by emission of valence-saturated chemical hydrocarbon and other compounds into the gas volume.

Excrements show that all the alkyl, hydroxyl, and amine radicals which we have studied considerably reduce the conductivity and increase the work function of oxide semiconductors like ZnO, TiO$_2$, CdO, WO$_2$, MoO$_3$, etc. during chemisorbtion. It should be noted that the revealed effects are rather profound especially if we are dealing with the effect of chemisorbtion of active particles on conductivity of a thin (less than 1 μm) sintered polycrystal semiconductor films. Thus, conductivity of such films in the presence of free CH$_3$-radicals with the concentration of even 10^9 cm^{-3} and less may change from initial value by dozens or hundreds percent depending on experimental conditions.

Usually, the decrease in conductivity during chemisorbtion of alkyl radicals on semiconductor oxides of n-type at elevated temperature has a reversible nature. However, the effect value under the same conditions depends on the chemical nature of adsorbent. For example, the following adsorbent activity row can be deduced if the oxides being studied are arranged in a chemisorbtion-induced conductivity descent order. In case of, say, CH$_2$-radicals, the other experimental conditions being the same, we obtain:

ZnO > TiO$_2$ > CdO > WO$_3$ > MoO$_3$.

It can be easily noticed that the left-hand side of the row features oxides whose metals easily form metal-organic compounds with free radicals.

Additionally, it was deduced from experiments that the change in conductivity of a certain oxide (e.g., ZnO) caused by chemisorbtion of various alkyl radicals (the other experimental conditions being the same) is substantially dependent on the chemical nature of free radicals. The adsorbates can be put in the following activity row provided that the simplest alkyl radicals analyzed are ordered according to their effect on the conductivity of films made of the oxide selected:

CH$_2$ > CH$_3$ > C$_2$H$_5$ > C$_3$H$_7$.

It is important to note that the above row correlates well with the reduction in chemical activity of the radicals listed [42] but contradicts to, for instance, electronic affinity between some of them [43]. There is no doubt that the results reported provide the evidence in favour of dominating significance of chemical activity of free radicals on their adsorption (at low degrees of occupation of semiconductor surface), and what is particularly important, on its effect on the conductivity of sintered oxide films, but not on the accompanying charging of adsorbent surface. In order to determine the nature of radical adsorption centres experiments on doping of zinc oxide films by various metals (Na, Zn, Ti, Fe, etc.) have been carried out. Atom beam method and the method of low pressure vapour equilibrium (Na, Zn) have been used to dope the films.

Figures 3.17 and 3.18 show the results of experiments performed to reveal the effect of CH_2- and CH_3-radicals on the conductivity of zinc oxide film doped by zinc [41] and titanium [44] atoms. It is evident that the more the value σ_0, i.e. the more the number of impurity atoms in the sample, the more the value of stationary variation of conductivity, i.e. the higher degree of occupation of the film surface by chemisorbed materials. As the number of adsorption centres which (as results of experiments suggest) are impurity metal atoms increases, the rate of adsorption kinetics of the doped film conductivity increases as well. In accordance with our hypotheses on the alkyl radical nature of chemisorbtion centres we can assume that the more active their interaction with the atoms of given metals like Zn, Ti, Na, Pb, etc., the more they change the conductivity of corresponding oxide at a given temperature. The latter was confirmed by experiments of Panet with metal mirrors [42].

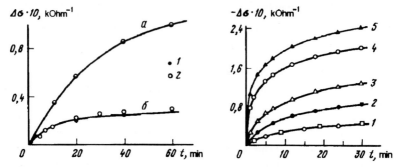

Fig. 3.17. Kinetics of conductivity of a zinc oxide film in the process of adsorption of CH_2-biradicals at 100°C. The radicals have been produced by means of photolysis ketene vapour at pressure $P = 0,5$ Torr. a – after adsorption of Zn-atoms; b – prior (1) and after adsorption of (2) Zn-atoms.

Fig. 3.18. Kinetics of conductivity of ZnO film during adsorption of methyl radicals $\dot{C}H_3$ at room temperature depending on the degree of preliminary alloying of the surface by titanium atoms. 1 – Blank experiment with a clean (Ti-atom free) film (O – before doping; □ – after heating of alloyed film at 350°C, i. e. after the film has been regenerated); 2-5 – Experiments with doped films. Doping degree increases in the following row: 2<3<4<5.

The experiment verifies the assumption made (see activity rows above). It follows from our reasoning that chemisorbtion of alkyl radicals may be schematically represented as follows:

$$R_v \xrightarrow{\ 1\ } (R-Me)_s \xrightarrow{\ 2\ } \left(R^- \text{——} Me^+ \right)_s ,$$

where v and s are the gas and solid phases; Me and Me^+ are the defect (interstitial) metal atoms and ions; R is the radical. The above diagram describes satisfactorily the adsorption of radicals at low occupation degrees when the reaction of surface recombination as well as the reaction of interaction of adsorbed radicals with free radicals and molecules of initial substance can be ignored. On the basis of the above diagram, initial velocity of chemisorbtion of radicals, i. e. the rate of variation in conductivity of oxide semiconductor film can be displayed as follows (the same assumption is valid for the adsorption of other acceptor particles, e. g. N- and O-atoms):

$$-\frac{d\sigma}{dt} \sim \frac{d[\overline{R}]_s}{dt} = k[\dot{R}]_v [Me]_s , \qquad (3.15)$$

where k is the kinetic constant dependent on the temperature. The following equation can be easily inferred from this equation, as shown in Chapter 2 (see (2.44)):

$$\frac{d\sigma}{dt} = -k'\sigma , \qquad (3.16)$$

or, in finite differentials:

$$\frac{\Delta\sigma}{\sigma} = kt . \qquad (3.17)$$

Applicability of the above equation for description of the effect of radical adsorption on conductivity of ZnO film is displayed by the example of CH_3-radical (see Fig. 3.19). The figure shows a fairly linear dependence of the experimental curve *1* plotted in the coordinates $\Delta\sigma/\sigma - t$ over a wide range of time and conductivity variation [41].

It should be noticed that a similar equation describes the effect on conductivity of ZnO film of such acceptor particles as atoms of nitrogen and oxygen, (see (3.12)).

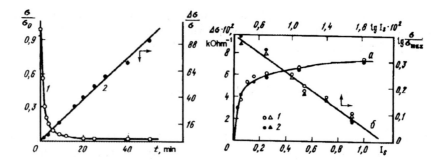

Fig. 3.19. Kinetics of conductivity of ZnO film in the course of adsorption of free CH_3 radicals at 185°C. $1 - \sigma/\sigma_0 = f(t)$; $2 -$ Verification of equation (3.17).

Fig. 3.20. Conductivity of ZnO film σ as a function of light intensity I_s (in arbitrary units) under conditions of acetone photolysis (CH_3-radicals generator) at 300°C; $P_{acetone} = 3,8$ Torr. $a - (\Delta\sigma - I_s)$. $b - \lg(\sigma/\sigma_{max}) - \lg I_s$; $1 -$ Increase in I_s. $2 -$ Decrease in I_s.

The study [39] shows that similar equation is valid for adsorption of NH- and NH_2-radicals, too. There are a lot of experimental data lending support to the validity of the proposed two-phase scheme of free radical chemisorbtion on semiconductor oxides. It is worth noting that the stationary concentration of free radicals during the experiments conducted was around 10^8 to 10^9 particles per 1 cm^3 of gas volume, i.e. the number of particle incident on 1 cm^2 of adsorbent surface was only 10^{12} per second. Regarding the number of collisions of molecules of initial substance, it was around 10^{20} for experiments with acetone photolysis or pyrolysis provided that acetone vapour pressure was 0,1 to 0,01 Torr. Thus, adsorbed radicals easily interact at moderate temperatures not only with each other but also with molecules which reduces the stationary concentration of adsorbed radicals to an even greater extent. As we know now [45] this concentration is established due to the competition between the adsorption of radicals and their interaction with each other as well as with molecules of initial substance in the adsorbed layer (ketones, hydrazines, etc.).

The experiments carried out under conditions when a film with chemisorbed radicals was immersed into liquid acetone for a while with a further return to the gas volume (in situ). This brought about an immediate restoration of initial conductivity. These cycles may be performed as many times as is desired. We are having here a situation identical to the case of heating an oxide film with radicals previously chemisorbed on it immediately in gaseous phase ($P_{acetone} \approx 0,1$ Torr) in order to "clean" the adsorbed layer surface. However, this process takes much longer than when experimenting with liquids.

Taking into consideration the above, it is easy to assume that the stationary value of coverage of chemisorbed radicals on the surface of a film under these conditions is insignificant and is subject to change in direct proportion to the stationary concentration of free radicals in the vessel volume, i.e. may be expressed in this particular case by the following universal relationship:

$$\frac{d[\dot{R}]_s}{dt} = -\alpha \left(\frac{d\sigma}{dt}\right)_{t \approx 0} = \beta [\dot{R}]_v (1 - \theta). \tag{3.18}$$

Provided that the value θ is small enough, there is a simple linear relation between the initial rate of variation in conductivity of oxide film (e.g., ZnO) and concentration of free radicals in the space adjacent to the film surface. When adsorbing the simplest alkyl radicals as well as imine- and amine-radicals, the stationary value $\left(\frac{d\sigma}{dt}\right)_{t \approx 0}$, always falls far short of 1 due to their greater chemical activity with respect to each other and to the molecules of adsorbed layer. Consequently, it is quite acceptable and justified to use a sufficiently strong proportionality between the values of conductivity signals (i.e., the value of θ) and concentration of active particles in the gas volume being analyzed, to measure various physical and chemical parameters in gas and liquid systems as well as on the interface which will be covered in the next chapter.

Our studies [46] of interaction of hydroxyl radicals with the surface of oxide semiconductors show that our reasoning on other radicals is also applicable to these particles as their chemical activity is sufficiently high. With radicals possessing low chemical activity the situation changes drastically becoming close to the adsorption of valence-saturated molecules.

In conclusion to this part it seems noteworthy that in contrast to the effect of adsorption of molecular particles on electrophysical properties of oxide semiconductors, the major peculiarity of this effect for such chemically active particles as the simplest free radicals or atoms of simple gases (H_2, O_2, N_2, Cl_2, etc.) is that they are considerably more chemically active concerning the impurity centres [47]. The latter are responsible for dope conductivity of oxide semiconductors. As for the influence of electric fields on their adsorption due to adsorption-induced surface charge distribution, they are of minor importance which is proved by results of the experiments on assessing field effect on adsorption of molecular and atom particles [48]. These experiments have verified that chemical activity of adsorbed particles plays a significant role in changing dope conductivity of a semiconductor.

To summarize the description of specificity of adsorption of active radicals on the surface of doped oxide semiconductors, we can conclude that we have a substantial experimental basis to draft the following diagram initially proposed in the study [41] and considered in detail in Chapter 2:

$$\dot{R}_v \xrightarrow{\ \ I\ \ } \dot{R}_s \xrightarrow{\ \ II\ \ } \dot{R}_s^- \ ,$$

$$
\begin{array}{cc}
{}_M\!\diagdown\!\!\bigwedge\!\!\diagup_M & {}_M\!\diagdown\!\!\bigwedge\!\!\diagup_M \\
\downarrow_M & \downarrow_M \\
M'\quad M''\quad M''' & M\quad M'\quad M'''(+e_s)
\end{array}
$$

where $\dot{R}_v$ and $\dot{R}_s$ are free radicals of the gas phase chemisorbed on the surface; M are the molecules of the initial substance in the medium volume and in the adsorbed layer; e_s is the quasi-free electron of the solid body; M', M'' and M''' are the molecules of different substances formed as a result of interaction between molecules M and chemisorbed particles.

Due to the competition of the phases shown in the diagram, there appears a stationary concentration of adsorbed particles $\left[A_s^-\right]$ and $[A_s]$ which are the charged and uncharged forms of chemisorbtion and electrons of dope conductivity in the semiconductor. The value of stationary concentration of charged chemisorbed radicals and, consequently, the value of conductivity electrons in the semiconductor film is dependent of the temperature, stationary concentration of free radicals in the volume adjacent to the semiconductor, and concentration of molecules of initial substance.

To detect radicals by means of semiconductor films there may be used both kinetic and stationary methods. The latter is based on relationship between the stationary conductivity of semiconductor film (e. g., ZnO, CdO, etc.) and stationary concentrations of active particles being detected.

The above relationships as we have already said in part 4.3 (see Chapter 2) have been deduced in a series of studies. For instance, the study [49] was the first to reveal that the following equation linking these values holds fairly rigorously (see Chapter 2, expressions (2.99)–(2.101)):

$$\sigma^2 / \Delta\sigma = \text{const}/\left[\dot{R}\right](1 - \theta), \tag{3.19}$$

where $\Delta\sigma = \sigma_0 - \sigma$; σ_0 is the semiconductor conductivity in absence of radicals on the surface; σ is the stationary value of semiconductor conductivity in presence of radicals.

The above formula may be converted into the following approximate expression provided that $\sigma_0 \gg \sigma$, and $\theta \ll 1$:

$$\sigma = \text{const}/\sqrt{[\dot{R}]}. \tag{3.20}$$

The validity of above expression has been proved in this study by the example of chemisorbtion of methyl radicals. They have been produced by means of photolysis of acetone vapour. It should be noted that the deduced expression is identical to that describing adsorption of molecular oxygen on semiconductor, i. e. $\sigma = \text{const}/\sqrt{P_{O_2}}$ [19]. For example, when performing experiments on photolysis of acetone, the change in stationary concentration of radicals as function of absorbed light intensity I_s is consistent with the law $[A]_r \sim \sqrt{I_s}$. Substituting this expression in the previous formula we derive the relationship between I_s and conductivity of semiconductor film σ, i. e.:

$$\sigma = \text{const}/\sqrt[4]{I_s}. \tag{3.21}$$

This formula may be experimentally verified as shown in Fig. 3.20. It displays, in logarithmic coordinates, a good straight line with plotted experimental data for coordinates $\Delta\sigma - I_s$. The tangent of the angle of the constructed straight line is about 0.26. It means that the deduced formula is valid. It evidently follows from the above relationships which link together the change in conductivity of oxide film and concentration of acceptor particles (molecular oxygen, free radicals) in the space adjacent to the film both under the equilibrium conditions (oxygen) and in the stationary phase (radicals), that the expression below has a universal character and valid with adsorption of all acceptor particles:

$$\sigma = \text{const}/\sqrt{[A]}, \tag{3.22}$$

where $[A]$ is the concentration of acceptor particles. Due to the fact that this expression has been derived according to the applicability of the mass law both for charged and uncharged surface particles as well as for conductivity electrons under the conditions of applicability of the Boltzman statistics, low degrees of surface occupation and validity of

208

the Henry law, the above conclusion of versatility of the derived expression is well founded under the said limitations.

Regarding the charging of a semiconductor film surface in the act of forming a charged adsorption, in this particular case it is just an accompanying effect which does not impose any limits on the whole chemisorbtion process.

Calculation yields that in these experiments the stationary concentration of radicals $[A]_r$ for the maximum intensity of light of a mercury lamp SVDSh-500 was no more than 10^{10} radicals/cm^3, the stationary conductivity of a ZnO film at 300°C changing by 300 to 400% compared to its initial pre-adsorption value. Thus, in accordance with the above formula (3.21), concentrations of radicals equal to 10^6–10^7 radicals/cm^3 may be detected by means of semiconductor sensors which is 7 magnitudes lower than existing methods would allow.

Similar results have been derived in generating free radicals through pyrolysis of acetone on a platinum filament [50]. Adsorption of more complex radicals such as C_2H_5, C_3H_7, CH_2C_6, etc. has been studied using the same methods. The above relationship asserts satisfactorily in these cases, too. This provides the evidence for versatility of the found relationship (3.22) which can be successfully applied in the methods involving the use of sensors.

3.4. Semiconductor sensors in condensed media

Up to now we have considered the relationship between the concentration of active particles in systems like gas (vapour) – solid body (semiconductor) and variation of conductivity of a semiconductor. In connection to these systems we mentioned numerous relationships which may be used for quantitative assessment of the content of gaseous media on the basis of data provided by semiconductor sensors when analyzing various active components.

Problems of applicability of semiconductor sensors in gaseous and solid media containing active particles of molecular oxygen, chlorine, bromine, etc. as well as free atoms and radicals might be of even higher importance. Detection of the above particles and particularly atoms and radicals in condensed media is a very complicated problem. Under regular thermodynamic conditions it gets reduced to monitoring ultra low concentrations of these particles. Nevertheless, study [51] shows that it is quite possible to detect low concentrations of, for instance, dissolved oxygen with sufficient accuracy applying semiconductor sensors (thin sintered ZnO films) at room and lower temperatures, even if the concentrations is around 10^{-5} to 10^{-6} vol-%. This analysis may be performed both in a water medium and in organic solvents which (being polar liquids with dielectric permittivity not lesser than 5 to 6) are chemically inert to metal oxides.

Detailed studies of oxygen adsorption on ZnO films in water and organic (both proton and aproton) solvents with various values of dielectric permittivity have been made in studies [51 – 54]. However it should be noted that the analysis of the dependence of detector (sensor) signal on oxygen concentration in vapour phase brought to equilibrium with the liquid (alcohol, water, etc.) revealed that the method of vapour-liquid analysis aimed at determination of oxygen traces in gases (inert gases, nitrogen, air, hydrocarbons, etc.) is also feasible [52] and may be performed even if a semiconductor film is placed in the atmosphere of saturated vapour of a chosen polar liquid at room temperature. A layer of liquid is being condensed on the surface of a semiconductor sensor under these conditions which is sufficient to reveal the effect similar to that of a semiconductor film immersed into a liquid polar medium. It should be noted, however, that in contrast to a sensor immersed into liquid, diffusion restrictions imposed on adsorption of oxygen by a sensor with a thin liquid film applied onto its surface are obviously less rigorous or even negligent. Therefore, in this case the response time of the sensor is mainly controlled by kinetics of physical-chemical processes dependent on the dielectric constant and chemical nature of the medium (solvent), as it is shown in the study [53].

The advantage of the vapour-liquid method of oxygen detection in gases in contrast to liquid technique (e.g., for dissolved oxygen) is that there is no stirring of liquid required to reduce the diffusion retardation leading to significant simplification of design of semiconductor sensor. Moreover, the concentration of oxygen emitted by various polar liquids in some chemical, photochemical, or biological processes developing in liquid media (e.g., photoemission of oxygen by algae under anaerobic conditions) can be determined using the vapour-liquid method. It is worth pointing out that the advantage of this method compared to the liquid method is that the coefficient of oxygen distribution between liquid and gaseous phases (equilibrium concentration ratio) is always less than 1. For instance, it is equal to 0.03 for water and 0.15 for alcohol. Due to that the oxygen produced by one or another process in liquid medium is mainly concentrated (at equilibrium conditions) in the gaseous phase which facilitates its detection for example by means of a "dry" sensor or using a vapour-phase technique.

Touching on drawbacks of vapour-liquid method, we should note that among those there is a lack of control of the thickness of condensed liquid layer on detector surface which may result (for fairly thick liquid layers) in diffusion retardation of oxygen chemisorbtion. The diffusion retardation if any may adversely affect reproducibility of sensor's readings and its sensitivity.

The above potential advantages of application of semiconductor sensors in studies of liquid media put forward a necessity to establish relationships linking electrophysical parameters of a semiconductor film and concentration of dissolved oxygen, or its concentration in saturated solvent vapour. According to the analysis of various experimental data regarding the effect of oxygen adsorption on conductivity of ZnO film in saturated vapour or immersed in an appropriate solvent we can assume [53] that chemisorbtion of oxygen occurs on oxide metal semiconductors (for the most part on ZnO) immersed in polar liquids of various nature, resulting in creation of a charged form of chemisorbtion (similar to the "dry" interface [54]) in the form of surface polar oxygen-containing complexes. Under the effect of polar medium these complexes start interacting with molecules of solvent already at room or even lower temperatures.

Apparently we can assume that chemical activity of surface oxygen complexes in relation to solvent considerably increases in highly polar media due to creation of contact or solvate-divided ion pairs $\left| Zn_s^* \cdots O_2 \right|$, where Zn^* is the dope atom.

Ion-radicals O_2 generated on the surface interact fairly actively with molecules of various solvents. Peroxydes, hydroperoxydes, and other compounds may be produced as a result of such reaction. Due to low concentration of oxygen dissolved in the course of the experiment, small area of the film surface as well as low degree of occupation by charged molecules of chemisorbed oxygen ($\sim 10^{-4} - 10^{-3}$ [55]) the accumulation rate of above products is low enough in this particular case.

It is known that lifetime of solvated peroxyde ion-radicals vary in different solvents. Thus, it amounts to hours in pyridine and dimethylsulphoxide but falls far short of a minute in proton solvents like water or alcohols [56, 57].

It should be noted that dissociation of surface complexes of oxygen in polar solvents on semireduced ZnO films is presumably justified from the thermodynamic point of view as oxygen adsorption heat on ZnO and electron work function are [58] 1 and approximately 5 eV respectively while the energies of affinity of oxygen molecules to electron, to solvation of superoxide ion and surface unit charge zinc dope ions are 0.87, 3.5, and higher than 3 eV, respectively [43].

In order to find the relationship between the stationary concentration of current carriers in semiconductor film and concentration of dissolved oxygen in polar liquid, it is essential to examine the expression for rate of chemisorbtion of dissolved oxygen molecules on ZnO film and its chemical desorbtion from the surface under effect of solvent molecules. In accordance with above reasoning, the velocities of these processes can be expressed for low occupation of the film surface by adsorbed oxygen as follows:

$$\frac{da_{O_2}}{dt} = K_{ads}\left[O_{2\,(ads)}\right]\left[Zn_s^{\bullet}\right]\exp\left[-\left(q\varphi_0 + U + \Delta q\varphi\right)/kT\right], \qquad (3.23)$$

$$\frac{da_{O_2}}{dt} = K_{ds}\left[Zn_s^{\bullet+}......O_2^{-}\right]_{(solv)}[nM], \qquad (3.24)$$

where a_{O_2} is the amount of adsorbed oxygen in charged form; K_{ads} and K_{ds} are the constants of rates of chemical reactions of oxygen adsorption and desorbtion from ZnO film; $q\varphi_0$ and $\Delta q\varphi$ are electron work function from ZnO before oxygen gets adsorbed and its variation caused by dipole moment of adsorbed complexes being formed; U is the adsorption activation energy of non-electrostatic nature; $[nM]$ is the concentration of solvent molecules. Apparently we can write down the following expression for the stationary system:

$$\frac{\left[Zn_s^{\bullet+}...O_2^{-}\right]_{(solv)}}{\left[Zn_s^{\bullet}\right]} = \frac{K_{ads}\left[O_{2\,(ads)}\right]}{K_{ds}}\exp\left[-\left(q\varphi_0 + U + \Delta q\varphi/kT\right)\right], \qquad (3.25)$$

where the surface concentration of oxygen complex may be equated to the difference in the number of zinc dope electrons before and after oxygen adsorption provided that they serve as the only centres of adsorption of oxygen in charged form. Assuming that electron equilibrium is established at any moment when processes (3.23) and (3.24) develop in the system in question, we derive [59]:

$$\left[e_s\right]^2 = \left[Zn_s^{\bullet}\right]\frac{2\left(2\pi mkT\right)^{3/2}}{h^3}\exp\left(\Delta E/kT\right), \qquad (3.26)$$

where $[e_s]$ is the concentration of conductivity electrons in the film; ΔE is the energy gap between the bottom of conductivity band and the dope levels $Zn^{\bullet}$ in the forbidden gap.

Based on equations (3.25) and (3.26) for the linear area of adsorption of dissolved oxygen on a thin semiconductor film (Γ being the Henry coefficient), we can derive the following equation:

$$\left(\frac{\sigma_0}{\sigma}\right)^2 - 1 = \frac{K_{ads}[O]_p}{K_{ds}[nM]}\exp\left[-\left(q\varphi_0 + U + \Delta q\varphi\right)/kT\right], \qquad (3.27)$$

TABLE 3.6.

Water

$$K_{H_2O} = \left(\frac{B}{C_{O_2}}\right)_{cp} = 1{,}10 \cdot 10^3$$

T = 22°C
$\sigma_0 = 6.61 \cdot 10^{-10}$ kOhm^{-1}, $\varepsilon = 81$

$C_{O_2} \cdot 10^5$, vol.-%	$\frac{\Delta\sigma}{\sigma_0}$, %	B	$\frac{B}{C_{O_2}} \cdot 10^{-3}$
4.0	2.00	0.4	1.00
16	7.70	0.159	1.00
60	3.26	0.758	1.26
154	69.70	1.881	1.22
339	120.28	3,852	1.14
711	197.36	7.843	1.10
2952	505.17	35.623	1.25
5700	613.20	49.872	0.88

Methyl alcohol

$$K_{CH_3OH} = \left(\frac{B}{C_{O_2}}\right)_{cp} = 3{,}43 \cdot 10^3$$

T = 22°C
$\sigma_0 = 7.46 \cdot 10^{-10}$ kOhm^{-1}, $\varepsilon = 38$

$C_{O_2} \cdot 10^5$, vol.-%	$\frac{\Delta\sigma}{\sigma_0}$, %	B	$\frac{B}{C_{O_2}} \cdot 10^{-3}$
18.3	28.7	0.656	3.60
42.7	61.5	1.607	3.76
92.0	96.9	2.879	3.13
213.0	189.5	7.378	3.46
457.0	282.8	13.653	3.00
945.0	499.3	34.910	3.46
1921.0	715.7	65.54	3.41
3935.0	1109.0	145.2	3.68
8735.0	1543.7	270.0	3.10

Isobuthyl alcohol

$$K_{CH_9OH} = \left(\frac{B}{C_{O_2}}\right)_{cp} = 15{,}6 \cdot 10^3$$

T = 22°C
$\sigma_0 = 6.65 \cdot 10^{-10}$ kOhm^{-1}, $\varepsilon = 18$

$C_{O_2} \cdot 10^5$, vol.-%	$\frac{\Delta\sigma}{\sigma_0}$, %	B	$\frac{B}{C_{O_2}} \cdot 10^{-3}$
3.6	21.7	0.480	12.8
11.8	60.8	1.585	14.0
22.6	92.8	2.718	12.1
45.2	185.4	7.145	15.8
90.4	302.3	15.185	16.8
180	488.0	33.570	18.6
722.0	1180.0	138.0	19.1

Vacuum

$$K_{Vac} = \left(\frac{B}{C_{O_2}}\right)_{cp} = 4{,}57 \cdot 10^5$$

T = 22°C
$\sigma_0 = 0.506 \cdot 10^{-10}$ kOhm^{-1}, $\varepsilon = 1$

$C_{O_2} \cdot 10^5$, vol.-%	$\frac{\Delta\sigma}{\sigma_0}$, %	B	$\frac{B}{C_{O_2}} \cdot 10^{-3}$
5.0	373.8	22.47	4.48
13.2	690.6	61.51	4.68
39.5	1366.7	214.10	5.42
59.2	1542.9	269.0	4.54
81.5	1831.3	372.0	4.56
113.1	2380.0	615.0	5.44
210.4	2967.0	940.0	4.47
394.5	4190.0	1838.0	4.66
1578.0	8333.0	7111.0	4.51

which relates the change in conductivity σ to concentration of dissolved oxygen $[O_2]_p$ as well as to the rate constants of competing processes of adsorption and chemical desorbtion expressed by the equations (3.23) and (3.24) [53].

Designating for the brevity sake the left hand side of the equation (3.25) as B for low occupation degrees of the film by adsorbed oxygen ($\Delta q\varphi \ll q\varphi_0 + U$) we obtain:

$$B = K_0[O_2]_p, \text{ where } K_0 = \frac{K_{ads}\Gamma}{K_{ds}[nM]}\exp\left[-\left(q\varphi_0 + U\right)/kT\right]. \qquad (3.28)$$

The validity of the above equation for different solvents is demonstrated by the Table containing experimental data which show that the value K_0 ($K_0 = B/[O_2]_p$) is fairly constant for each solvent over a wide concentration range (up to several orders of magnitude) of dissolved oxygen, according to the expression (3.28), but it increases with reduction of ε in the solvent. Evidently, the latter is mostly related to reduction of the value K_{ds}, i.e. to decrease in chemical activity of oxide complexes containing solvent polar molecules caused by the drop in the ε value of solvent (see Table 3.6). However, the explicit dependence of the value K_{ds} on ε does not follow from expression (3.28) at all. To trace the above dependence it is apparently necessary to use the relationships which describe homogeneous reactions in solutions and associate the rate constants of interacting charged and polar particles with ε value of the solvent assuming that these are applicable to adsorbed layer – solvent interface. If the above assumption is correct, the process in the heterogeneous system can be described as either a surface reaction of superoxide ion-radicals O_2 with polar molecules nM in polar medium, or as an interaction of a short ion pair of ($Zn^{\bullet+}...O_2^-$)-complex with solvent dipoles:

$$\left[Zn_s^{\bullet+}...O_2^-\right]_{(solv)} + nM \xrightarrow{+e_s} H\dot{O}_2 + \dot{M} + Zn_{s\,(solv)}^\bullet$$
$$ZnO$$

Assuming that $\varepsilon = D$ in the normal state, according to [60] for the dipole-ion interaction (the first case) the sought-for equation is:

$$\ln K_\varepsilon = \ln K_\infty + Y_q d_{MN} / kT r_{max}^3 \varepsilon, \qquad (3.29)$$

where Y_q is the ion charge; d_{MN} is the dipole moment of a polar molecule; r_{max} is the largest distance between the particles at which they start interacting; K_ε is the rate constant of reaction oxygen ion-radical with a solvent molecule. It is noteworthy that the second term of the above equation is negative for negative ions $\left(Y_q < 0\right)$. For a dipole-dipole interaction along a straight line (the second case), i.e. for oriented dipoles, the Amis equation [60] can be expressed as:

$$\ln K_\varepsilon = \ln K_\infty - 2d_A d_D \, / \, kT r_{\max}^3 \varepsilon, \qquad (3.30)$$

where d_A and d_D are the dipole moments of interacting molecules. Note that the contacting ion pair can be considered as a polar molecule possessing a large dipole moment. Figure 3.21 shows the dependence of reaction rate of oxygen adsorbed on ZnO with solvent molecules in water-dioxane medium with different concentration, i.e. for different ε value. It follows from the figure that the Amis equation is satisfied fairly well. This conclusion is supported by calculation of the $r_{\max}$ value using the tangent of experimentally obtained straight line plotted for case $d_{MN} \approx 2D$. The calculation yields that the value of $r_{\max}$ is quite reasonable. The result of the calculation of the same value $r_{\max}$ based on equation (3.30) under assumption dipole moments $d_{MN} = 2D$ and $d\left(Zn^+...O_2^-\right) \approx 10D$ gives a substantially underrated value $r_{\max}$ $(< 1 \overset{\circ}{A})$.

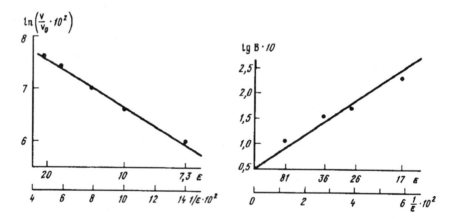

Fig. 3.21. Rate of oxygen "removal" from a ZnO film (a.u.) as a function of ε of water-dioxane mixtures (verification of the equation (3.29)).

Fig. 3.22. Value B plotted as a function of dielectric constant ε for various solvents (water, alcohols) for ZnO film in presence of oxygen $\left(C_{O_2} = 10^{-3} \text{ vol.-\%}\right)$ (verification of the equations (3.29) and (3.31)).

In case of weaker dependence of the rate of oxygen adsorption on ZnO K_{ds} on the value ε of solvent, compared to that of K_{ads} as a function of ε (which is supported by experimental results), we can deduce the following equation on the basis of the expressions (3.28) and (3.29):

$$\lg B \approx const - Y_q d_{MN} / kT r_{max}^2 \varepsilon , \tag{3.31}$$

A straight line plotted in coordinates $\lg B - 1/\varepsilon$ for a negatively charged ion $O_2^- (Yq < 0)$ must have a positive slope which is consistent with the experimental data (see Fig. 3.22). The value r_{max} has been also evaluated for this particular case and has been found to constitute 2.0 $\overset{\circ}{A}$, i.e. quite reasonable.

Figure 3.22 shows the result of verification of equations (3.28) and (3.29) under conditions of interaction of oxygen adsorbed on ZnO film with molecules of polar and chemically analogous solvents for the same concentration of dissolved oxygen. The values ε of these solvents vary from 17 to 81. Experimental points are fairly satisfactorily plotted against the straight line $\lg B - 1/\varepsilon$ which is in consistency with the requirements of equations (3.28) and (3.29).

Thus, the experimental data do not contradict above assumption of electrolytic dissociation of surface complexes formed by adsorbed oxygen and metal oxides (in particular, by ZnO), as well as allowing for chemical interaction of solvated superoxide ion-radicals with solvents molecules developing on the surface at room temperature in the dark. Experiments [53] with aproton solvents like dimethylphormamyde, dimethylsulphoxide, piridine, acetonitrile, acetone, etc. revealed that similar effects are exhibited in these cases and the dependencies deduced for proton-donor solvents are valid. In the event of thoroughly desiccated and pure aproton solvents with a reasonably high ε, there is a fairly active interaction between the solvent molecules and oxygen adsorbed on ZnO, developing as a first order reaction in relation with respect to concentration of dissolved oxygen, and zero order reaction in relation to the solvent.

It follows from the equation (3.28) that the liquid technique to determine concentration of dissolved oxygen or oxygen of extraneous gases being analyzed (nitrogen, hydrogen, inert gases, hydrocarbons, etc.) which are in equilibrium with liquid polar solvent (water, alcohols, etc.), is applicable as well. This means that relation

$$\sigma = const / \sqrt{[O_2]_p} , \tag{3.32}$$

holds which is quite a transparent conclusion for both cases judging by suggested equations to describe chemisorbtion and assumptions made with regard to the nature of chemisorbtion centres and formation of centre-bound chemisorbed charged oxygen.

Similar expression is applicable to describe chemisorbtion of other acceptor particles (free atoms and radicals) [52]. Major distinctions between the "liquid" and "dry" interfaces of sensors when detecting the above particles, provided that the experiment lasts long enough are:

a) attaining the stationary value of conductivity of a sensor in the first case and equilibrium value in the second;

b) reversibility of sensor readings for the first case at any temperature (within reasonable limits), and only at elevated (over 200°C) temperatures for the second. In the latter case the kinetic restrictions enable one to do measurements only within a sufficiently short period of time.

References

1   K. Gen, Formation and Stabilisation of Free Radicals, A. Bass and G. Boyd (eds.), Foreign Literature Publ., Moscow, 1962

2   A. Callir and J. Lambert, Excited Particles in Chemical Kinetics, A.A. Borisov (ed.), Mir Publ., Moscow, 1973

3   D. Lichtman, Methods of Surface Analysis, A. Zandera (ed.), Mir Publ., Moscow, 1979

4   N.E. Lobashina, N.N. Savvin, I.A. Myasnikov et al., Zhurn. Fiz. Khimii 56 (1982) 1719 - 1723

5   I.A. Myasnikov, National Conference on Radiation Chemistry, USSR Ac. of Sci. Publ., Moscow, 1962

6   I.N. Pospelova and I.A. Myasnikov, Zhurn. Fiz. Khimii, 41 (1967) 1990 - 1997

7   I.N. Pospelova and I.A. Myasnikov, Kinetika i Kataliz, 10 (1969) 1097 - 1105

8   I.N. Pospelova and I.A. Myasnikov, Zavod. Lab., 10 (1971) 1202 - 1208

9   I.A. Myasnikov and G.V. Malinova, DAN SSSR, 159 (1964) 894 - 901

10   A.M. Panesh and I.A. Myasnikov, Zhurn. Fiz. Khimii, 39 (1965) 2326 - 2329

11   A.M. Panesh and I.A. Myasnikov, Zhurn. Fiz. Khimii, 42 (1968) 2100 - 2102

12   I.A. Myasnikov, Zhurn. Fiz. Khimii, 62 (1988) 2770 - 2781

13 N.E. Lobashina, N.N. Savvin and I.A. Myasnikov, DAN SSSR, 268 (1983) 1434 - 1439

14 E.I Grigoriev, I.A. Myasnikov and V.I. Tsivenko, Zhurn. Fiz. Khimii, 56 (1982) 1748 - 1756

15 M.V. Yakunichev, I.A. Myasnikov and V.I. Tsivenko, Zhurn. Fiz. Khimii, 59 (1985) 2584 - 2588

16 I.A. Myasnikov, Zhurn. Fiz. Khimii, 32 (1958) 841 - 853

17 G. Heinland, Ztschr. Phys., 148 (1957) 15 - 22

18 I.A. Myasnikov, DAN SSSR, 120 (1958) 1298 - 1307

19 I.A. Myasnikov, Zhurn. Fiz. Khimii, 31 (1957) 1721 - 1729

20 I.N. Pospelova and I.A. Myasnikov, DAN SSSR, 167 (1966) 625 - 634

21 G. Thomas and J.J. Lander, J. Chem. Phys., 6 (1956) 1136 - 1141

22 N.V. Ryltsev, E.E. Gutman and I.A. Myasnikov, Interaction of Atom Particles with Solids, Kharkov University Publ., Kharkov, 1976

23 N.V. Ryltsev, E.E. Gutman and I.A. Myasnikov, Zhurn. Fiz. Khimii, 52 (1978) 1796 - 1803

24 D. Brennan, Physical Chemistry of Heterogenous Catalysis, Mir Publ., Moscow, 1967

25 N. Ramsay, Molecular Beams, Foreign Literature Publ., Moscow, 1960

26 R. Herny, Electron Processes in Ion Crystals, A.F. Ioffe (ed.), Foreign Literature Publ., Moscow, 1950

27 G. Thomas, Semiconductors, B.F. Ormont (ed.), Foreign Literature Publ., Moscow, 1962

28 I.A. Myasnikov, E.V. Bolshun and V.S. Raida, Zhurn. Fiz. Khimii, 47 (1973) 2349 - 2360

29 B.S. Agayan, I.A. Myasnikov and V.I. Tsivenko, Zhurn. Fiz. Khimii, 47 (1973) 2904 - 2907

30 A.N. Nesmeyanov, Vapour Pressure of Chemical Elements, USSR Ac. of Sci. Publ., Moscow, 1961

31 I.A. Myasnikov, E.V. Bolshun and B.S. Agayan, DAN SSSR, 220 (1975) 1122 - 1129

32 E.V. Bolshun, I.A. Myasnikov and D.G. Tabatadze, Zhurn. Fiz. Khimii, 45 (1971) 2499 - 2507

33 I.A. Myasnikov and E.V. Bolshun, Kinetika i Kataliz, 8 (1967) 182 - 193

34 E.E. Gutman, I.A. Myasnikov, S.A. Kazakov et al., Khim Fizika, 5 (1986) 386 - 397

35 G.V. Malinova and I.A. Myasnikov, Kinetika i Kataliz, 10 (1969) 336 - 348

36 G.V. Malinova and I.A. Myasnikov, Kinetika i Kataliz, 11 (1970) 715 - 729

218

37   V.I. Tsivenko and I.A. Myasnikov, Kinetika i Kataliz, 11 (1970) 267 - 277

38   V.I. Tsivenko and I.A. Myasnikov, Khimiya Vysokikh Energii, 7 (1972) 180 - 186

39   V.I. Tsivenko, Studies of Adsorption Nitrogen Atoms and Nitrogen-Hydrogen Radicals on Oxide Adsorbents by Electroconductivity Technique, PhD (Chemistry) Thesis, Moscow, 1973

40   B.S. Agayan, I.A. Myasnikov and V.I. Tsivenko, Zhurn. Fiz. Khimii, 47 (1973) 1292 - 1296

41   I.A. Myasnikov, Electron Phenomena in Adsorption and Catalysis on Semiconductors, F.F. Volkenshtein (ed.), Mir Publ., Moscow, 1969

42   F. Paneth and W. Hofeditz, Berichte, 62 (1929) 1335 - 1352

43   V.I. Vedeneev, L.V. Gurvich and V.N. Kondratiev, Handbook. Energy of Chemical Bonds. Ionisation Potentials and Electron Affinity, V.N. Kondratiev (ed.), USSR Ac. of Sci. Publ., Moscow, 1962

44   I.A. Myasnikov, Vest. AN SSSR, 8 (1973) 40 - 52

45   I.A. Myasnikov, E.V. Bolshun and E.E. Gutman, Kinetika i Kataliz, 4 (1963) 867 - 875

46   I.N. Pospelova and I.A. Myasnikov, Zhurn. Fiz. Khimii, 46 (1972) 1016 - 1030

47   B.S. Agayan and I.A. Myasnikov, Zhurn. Fiz. Khimii, 60 (1977) 2965 - 2972

48   N.V. Ryltsev, E.E. Gutman and I.A. Myasnikov, Zhurn. Fiz. Khimii, 55 (1981) 986 - 998

49   I.A. Myasnikov and E.V. Bolshun, DAN SSSR, 135 (1960) 1164 - 1169

50   I.A. Myasnikov and I.N. Pospelova, Zhurn. Fiz. Khimii, 41 (1967) 1028 - 1037

51   I.A. Myasnikov, Zhurn. Fiz. Khimii, 55 (1981) 2059 - 2072

52   I.A. Myasnikov, Zhurn. Fiz. Khimii, 55 (1981) 1283 - 1291

53   I.A. Myasnikov, Zhurn. Fiz. Khimii, 55 (1981) 2053 - 2058

54   I.A. Myasnikov, Zhurn. Fiz. Khimii, 55 (1981) 1278 - 1283

55   D.M. Shub, A.A. Remnev and V.I. Veselovski, Elektrokhimiya, 10 (1974) 657 - 664

56   I.B. Afanasiev, Uspekhi Khimii, 48 (1979) 977 - 999

57   A.S. Morkovnik and O.Yu. Okhlobystin, Uspekhi Khimii, 48 (1979) 1968 - 1991

58   S.V. Morrison, Catalysis. Electron Phenomena, Foreign Literature Publ., Moscow, 1958

59   S.S. Shalt, Semiconductors in Science and Technology, USSR Ac. of Sci. Publ., Moscow, 1957

60   E. Amis, Effect of Solvents on Rate and Mechanism of Chemical Reactions, Mir Publ., Moscow, 1966

Semiconductor Sensors in Physico-Chemical Studies
L.Yu. Kupriyanov (Editor)
1996 Elsevier Science B.V.

219

Chapter 4

Application of semiconductor sensors in experimental investigation of physical - chemical processes

4.1. Recombination of Atoms and Radicals

Lifetimes of free atoms and radicals account for the degree of interaction of these particles with an ambient medium and with each other. Due to high reaction capability of active particles in gaseous and, especially, in liquid media, their lifetimes are rather small. In gaseous phase, at small pressures these lifetimes are determined by heterogeneous recombination of these particles on vessel walls and by interaction of these particles with an adsorbed layer. At high gas pressures, the lifetimes are determined by bulk recombination and chemical interaction with ambient molecules.

According to [1], in the case of recombination of atoms and radicals governed by the first-order kinetics, the radicals' concentration distribution over the height h in a cylindrical vessel can be written as

$$N / N_0 = \exp\left(-\delta\gamma^{1/2}h\right). \tag{4.1}$$

For surface recombination governed by the second-order kinetics, this quantity is given by

$$N / N_0 = \frac{1}{\left[1 + \left(N_0\alpha P / 6SD\right)^{1/2}h\right]^2}, \tag{4.2}$$

where h is the distance measured from a reference point (the shutter), N and N_0 are the radical concentrations corresponding to the level h and the reference zero level h_0, $\delta = (vP/4SD)^{1/2}$, v is the mean velocity of particles, P is the vessel perimeter, S is the vessel cross-section, D is the diffusion coefficient of particles, γ and α are the coefficients (probabilities) of recombination due to collisions of particles with the surface and with each other, respectively. These quantities are defined

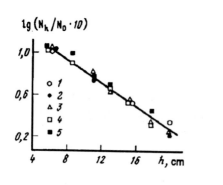 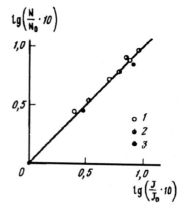

Fig.4.1. Dependence of the concentration of free CH_3 radicals in a quartz cylindrical cell on the distance between the film sensor and the point where the radicals are produced, for different temperatures of the vessel walls: 25°C (*1*), 140°C (*2*), 241°C (*3*), 269°C (*4*), and 300°C (*5*).

Fig.4.2. Relative concentration N/N_0 of hydrogen atoms in a reaction vessel as a function of relative discharge intensity (current) in the Wood tube J/J_0. The measurements were carried out by a semiconductor sensor (*1*), microcalorimeter (*2*), and by using the diffusion Wrede method (*3*).

as the ratios of the number of collisions resulting in recombination to the total number of collisions of the respective type. We note that the Smith method [1] suggests a uniform radial distribution of the concentration of active particles. This condition is met when

$$\lambda / 2 \ll r \ll \frac{2\lambda}{3\gamma} , \qquad (4.3)$$

where λ is the mean free path and r is the vessel radius.

As is seen from the above formulas, in the case of heterogeneous recombination of particles governed by the first-order kinetics, absolute values of γ are determined solely by the distribution of relative concentration of particles along the axis of the cylindrical vessel. For heterogeneous recombination of particles governed by the second-order kinetics, a knowledge of the absolute concentration of these particles for certain h is also required.

Numerous studies show that, usually, heterogeneous recombination of free atoms and radicals is governed by the first-order kinetics. Thus, to determine γ we need only to find the dependence $v = f(h)$. To do this, we may use the distribution of relative values $v = (d\sigma/dt)_{t \approx 0}$

along the direction in which the sensor is displaced. This distribution determines γ, because $v/v_0 = N/N_0$ (see formula (4.1)).

As an example the experimental results on heterogeneous recombination of CH_3 radicals on glass at different temperatures are plotted on Fig. 4.1. The experimental conditions in this case are chosen in such a way that inequality (4.3) is satisfied ($\lambda < 1$ cm, γ is about 10^{-4}, $r = 3$ cm). Thus, formula (4.1) holds in this experiment. This conclusion is supported by the fact that for all experimental series the results obtained at different temperatures of the reaction vessel walls are satisfactorily approximated by the same straight line. This means that methyl radicals on glass substrate undergo recombination governed by the first-order kinetics, and the activation energy is close to zero.

Similarly, we used the method of semiconductor sensors (SS) to study heterogeneous recombination of NH_2 and NH radicals [2, 3], as well as hydrogen [4], oxygen [5], and nitrogen [6] atoms.

Investigation of heterogeneous recombination of active particles on individual faces of single crystals can be also carried out by the method of semiconductor sensors SS. In this case, however, one should use collimated molecular beams reflected from the target faces [7]. These beams affect the SS film inserted in the reaction vessel in an appropriate manner. Note that in the case of heterogeneous recombination of hydrogen atoms [4], the results of the SS method were compared with the data obtained under identical experimental conditions by other (classical) methods, namely, microcalorimetry and the Wrede diffusion method [4]. This comparison shows that the results obtained by all three methods fit fairly well, although they are applicable in essentially different ranges of concentrations of hydrogen atoms. As compared with the two other methods, the SS technique can be used in the range of concentrations 6-8 orders of magnitude lower. In this range, the two other methods are insensitive to active particles (Fig.4.2). On the contrary, at large concentrations of hydrogen atoms sensors exhibit low sensitivity to these particles.

We note also that, due to its high sensitivity, the SS technique can be successfully used for studying homogeneous recombination of active particles in gaseous media under various pressures, including high pressures, provided the above specified conditions are satisfied, e.g., in adsorbent voids. When employing the SS method for detecting active particles, one should use the Smith formula written in the form

$$v_h / v_{h=0} = \frac{1}{\left[1 + \left(N_0 \alpha P / 6D\right)^{1/2} h\right]^2} , \tag{4.4}$$

222

where N_0 is the initial concentration of radicals, P is the gas pressure, α is the bulk recombination coefficient, $v = (d\sigma/dt)_{t\approx0}$ is the initial variation rate of electric conductivity of the sensor for different h.

4.2. Pyrolysis of Simple Molecules on Hot Filaments

Analysis of thermal decomposition of molecules on hot surfaces of solids is of considerable interest not only for investigation of mechanisms of heterogeneous decomposition of molecules into fragments which interact actively with solid surfaces. It is of importance also for clarifying the role of the chemical nature of a solid in this process. Furthermore, pyrolysis of molecules on hot filaments made of noble metals, tungsten, tantalum, etc., is a convenient experimental method for producing active particles. Note that it allows continuous adjustment of the intensity of the molecular flux by varying the temperature of the filament [8].

In a number of cases, the temperature of the filament and thermodynamic parameters allow one to calculate [9] the flux intensity of free atoms produced in dissociation of molecules. Specifically, in the case of dissociation of hydrogen, oxygen, and nitrogen molecules on hot metal filaments under pressures of molecular gases higher than 10^{-6} Torr, the flux intensity I_A of atoms A originating from A_2 molecules is given by

$$I_A = \tilde{A}\sqrt{\frac{P_{A_2}K}{2\pi M_A RT}} = \frac{B}{\sqrt{T}}e^{-\frac{E_0}{2RT}}, \qquad (4.5)$$

where $\tilde{A}$ is the Avogadro number, P_{A_2} is the pressure of a molecular gas, K is the equilibrium constant of the process $A_2 \rightleftarrows 2A$ for a given temperature of the filament T, M_A is the atomic mass, R is the universal gas constant, $B = \tilde{A}\sqrt{\frac{P_{A_2}K}{2\pi M_A R}}$, and E_0 is the dissociation energy of molecules A_2.

When using a semiconductor sensor for analysis of the flux of free radicals and atoms produced on the filament, one should rewrite expression (4.1) as

$$\sqrt{T}v = \text{const} \cdot \exp\left(-\frac{E_0}{2RT}\right) \qquad (4.6)$$

or

$$\ln\left(\sqrt{T}v\right) = \text{const} - \frac{E_0}{2R}\frac{1}{T} , \tag{4.7}$$

where $v = \left(\dfrac{d\sigma}{dt}\right)_{\theta\approx0}$ is the initial variation rate of the electric conductivity of the sensor (for $\theta \ll 1$), const is a value independent of the fila-
ment temperature. It is evident from formula (4.7) that temperature
dependence of the quantity $\ln\sqrt{T}$ is negligible as compared with $1/T$.
Consequently, we can plot experimental data for pyrolysis in a $(\ln v - 1/T)$ frame, with error thus introduced being negligible.

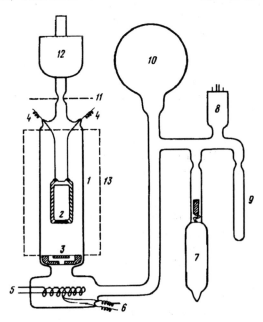

Fig.4.3. Experimental arrangement for investigation of pyrolysis of molecules by
the method of semiconductor sensors: *1* - reaction vessel, *2* - quartz slab with a
ZnO film (sensor), *3* - filter, *4* - contacts, *5* - incandescent filament, *6* - - thermocouple, *7* - cell with a substance, *8* - lamp - manometer, *9* - pin, *10* - flask,
11 - sealing bulkhead, *12* - trap, *13* - thermostat.

A simplest vessel used for experimental investigation of pyrolysis of
hydrogen, oxygen, and nitrogen, as well as other molecules is shown in
Fig.4.3. The pyrolysis filament (below) is separated from the adsorption
chamber (above) by a plane shutter driven by a magnet, which permits
the sensor to be exposed to the established atomic flux during a required

224

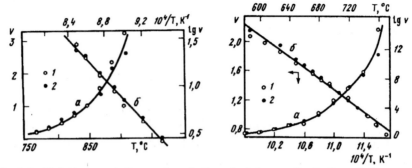

Fig.4.4. Relative variation rate of the electric conductivity of the sensor $v = (\partial\sigma/\partial t)_{t\approx0}$ as a function of the temperature of the pyrolysis filament, plotted in $v - T$ axes (*a*) and $\lg v - T^{-1}$ axes (*b*). The temperatures in the vessel are 370°C (*1*) and 380°C (*2*), the pressure of hydrogen $P_{H_2} = 10^{-2}$ Torr.

Fig.4.5. Relative variation rate of the electric conductivity of the sensor v as a function of the temperature of the pyrolysis filament, plotted in $v - T$ axes (*a*) and $\lg v - T^{-1}$ axes (*b*). The temperatures in the vessel are 323°C (*1*) and 350°C (*2*), the pressure of acetone $P_{ac} = 0.5$ Torr.

time interval. The sensor is located in the adsorption chamber and can be displaced along the cylindrical vessel. The exposure time and the distance between the movable sensor and the shutter are chosen in such a way as to minimize the area of the surface of the sensitive element exposed to adsorbed atoms. The higher is the intensity of the incident flux, the shorter is the exposure time (usually it ranges from several seconds to a minute) and the longer is the distance. Note, however, that, in this experiment, we do not deal with the beam technique, where the active particles do not collide with each other and with the vessel wall. Beam particles collide only with a target under study. In our case, we should take into account heterogeneous recombination of active particles on the vessel walls occurring at small pressures of molecular gases. We examine this phenomenon in detail in Section 2.4.3, where we show that the quantities v_h and v_0 are related by a linear expression. Hence, to find the energy at which, for example, the H–H bond breaks in an

H$_2$ molecule, we can plot a graph in a $(\ln v_h - \frac{1}{T})$ frame and then cal-

culate the sought-for quantity from the slope of the straight line (Fig.4.4). Thus we find that the energy at which the bond in H$_2$ breaks is equal to 101 kcal/mol. This value differs from the one obtained by spectroscopic measurements (104 kcal/mol) by only several percent. Similar results were obtained by the same method for pyrolysis of oxygen [5] and nitrogen [6]. Note that in experiments with hydrogen, the steady-state concentration of hydrogen atoms near the sensor was calculated to be only 10^7 atoms per cm^{-3} for the distance between the sensor and the filament of 15 cm and the filament temperature of 750°C. Such a small value of the concentration indicates high sensitivity of the sensors used in experiments.

TABLE 4.1

Substance	Type of bond	Active particle	E_D, kcal/mol	
			our results	reference data
Hydrogen	H–H	$\dot{H}$	100–102	104
Oxygen	O–O	$\dot{O}$	115	118
Nitrogen	N–N	$\dot{N}$	220	225
Carbon Dioxide	O–CO	$\dot{O}$	80	–
Acetone	$CH_3–COCH_3$	$\dot{C}H_3$	71	72
Propylbenzol	$C_2H_5–CH_2C_6H_5$	$\dot{C}_2H_5$	56	58
Benzylamine	$NH_2– CH_2C_6H_5$	$\dot{N}H_2$	60	59
Nitrogen-hydrogen acid	$NH_2–N$	$\dot{N}H$	9	9
Dibenzyl	$C_6H_5CH_2–CH_2C_6H_5$	$C_6H_5\dot{C}H_2$	50	48

Pyrolysis of more complex molecules proceeds via production of free radicals. Then formula (4.5) fails, because reactions of creation and recombination of radicals in these systems are irreversible. Therefore, the steady-state concentration of active particles in these systems depends on conditions of pyrolysis, determining the first or the second order of recombination of active particles, and is governed by the following equations [8]

$$[\dot{R}] = \sqrt{K_1[M]} \exp\left(-E_0/2RT\right) , \qquad (4.8)$$

$$[\dot{R}] = K_2 \exp\left(-E_0/2RT\right) , \qquad (4.9)$$

where K_1 and K_2 are the ratios of rate constants of the reactions of decay of molecules, recombination of free radicals, and interaction of free radicals with parent molecules, $[M]$ and $[\dot{R}]$ are the steady-state concentrations of molecules and radicals, and R is the universal gas constant.

The results on pyrolysis of acetone displayed in Fig. 4.5 are consistent with formula (4.8). Thus, variation of the concentration of free radicals near the sensor surface and, consequently, variation of the value $(dv/dt)_{t\approx0} = v$ as functions of the filament temperature are governed by relation (4.8). As the acetone pressure increases, this relation fails because of fast interaction of CH_3 radicals with acetone molecules.

Table 4.1 summarizes the energies corresponding to breaking of bonds for different molecules and compares the results obtained by the method of semiconductor sensors with reference data [10]. Obviously, these results agree with each other, which indicates their validity and applicability of the SS technique for these studies.

4.3. Photolysis in Gas Phase

Upon absorption of a light quantum, a particle is usually transferred to a short-lived excited state. From this state, the particle either relaxes to the initial (ground) state via radiative and radiationless transitions, or enters into chemical reactions. Furthermore, it may dissociate into fragments. The resulting free atoms and radicals may also interact with ambient molecules, as well as with each other. Experimental study of these processes in thermodynamic conditions using conventional light sources is hampered by low sensitivity of available techniques (without accumulation of a signal) for detecting free atoms and radicals. For this reason, in a number of cases, these experiments employ indirect measurements using various substances active with respect to free radicals (e.g., iodine). These substances are added to the reaction medium.

Such techniques imply analysis of chemical products of photolysis. Application of mass-spectrometers of various types is often hampered by a number of circumstances. These difficulties will be discussed later on. The EPR method, which is currently the most extensively employed technique, features low sensitivity and is usually used for analysis of primary fragments of photolysis. For this purpose, the radicals produced are frozen on the walls of a quartz pin and are thus accumulated inside the device. On one hand, this approach allows one to overcome the sensitivity threshold of the device. However, on the other hand, this excludes the possibility of direct kinetic measurements. The SS technique permits the use of weak light sources for detecting active particles under usual conditions. Furthermore, it allows determination of the concentration of particles in the process of photolysis not only in the gas phase, but in liquid media as well. Note that lifetimes of particles in liquid media are short, and, consequently, their concentrations are vanishingly low. In this case, the specified measurements can be performed in both kinetic and the steady-state regimes.

As an example of application of semiconductor sensors for this purpose, we consider photolysis of simplest olefines (ethylene, propylene, acetylene, etc.) occurring in the range of vacuum ultraviolet. It is well-known (e.g., see [11]) that photolysis of ethylene may result in detachment of either hydrogen molecules (detached in one act) or hydrogen atoms. Hydrogen atoms subsequently associate into molecules or interact with ethylene molecules. In what follows, we consider how this problem can be solved with the help of sensors.

It is well-known that hydrogen is detached from an ethylene molecule through the following reactions

$$C_2H_4 \xrightarrow{hv} C_2H_2 + H_2 \ ,$$
$$C_2H_4 \xrightarrow{hv} C_2H_2 + 2\dot{H} \ .$$

These reactions yield acetylene and hydrogen. However, it is evident that these processes result in the following secondary reactions

$$\dot{H} + C_2H_4 \longrightarrow (C_2H_2)^{\bullet} \ ,$$

and

$$(C_2H_5)^{\bullet} + (C_2H_5)^{\bullet} \longrightarrow C_4H_{10}$$

which yield ethyl radicals $(C_2H_5)^{\bullet}$ and valence-saturated molecules of aliphatic hydrocarbons (butane). Furthermore, free hydrogen atoms recombine. The above expressions show that photolysis of ethylene may result in production of intermediate active particles of two types, hydrogen atoms (primary particles) and ethyl radicals (secondary particles). In contrast to valence-saturated hydrocarbons and molecular hydrogen, these particles may be adsorbed on the surface of semiconductor sensors, and thus they may influence significantly the electric conductivity of the sensors. For such studies, one should choose the sensors with the minimum area of the sensitive surface (zinc oxide films), so that the sensor itself does not disturb essentially the steady-state distribution of active particles in the reaction vessel (Fig.4.6). We note that this distribution depends on the ethylene pressure, temperature, the intensity of the active light, and the activity of the reaction vessel walls (fused quartz) with ethylene adsorbed on them. In experiments [11], photolysis of ethylene was studied in a quartz cell shown in Fig.4.6. A resonance radiation from a quartz mercury lamp was used as a light source. The lamp was sealed in the lower section of the vessel separated from the upper part by a marblyte porous filter. This filter prevented the active light of the lamp from penetrating into the adsorption chamber. However, this filter was "transparent" for diffusion of active particles from the lower to the upper part of the vessel.

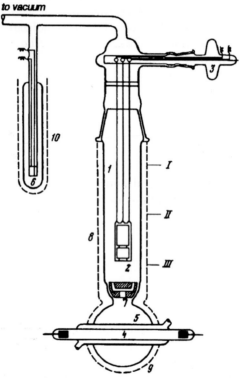

Fig.4.6. Schematic of the reactor: *1* – vessel, *2* – quartz slab with a ZnO film (sensor), *3* – movable T-handle, *4* – lamp, *5* – cooling jacket, *6* – auxiliary sensor, *7* – replaceable marblyte filter, *8* – *10* – thermostats.

To monitor a possible influence of molecular products of photolysis of ethylene (acetylene, ethane, and butane) on the sensor, a second sensor was positioned at a distance of 50 cm from the photolysis zone. The second sensor was designed to introduce corrections into the readings of the movable sensor. The specified distance was chosen so that atoms and radicals produced in the lower part of the vessel could not reach the upper sensor, because of their recombination and interaction with molecules. At the same time, these circumstances did not hinder diffusion of molecular products to the second sensor, where these molecules could be adsorbed.

Suppose that the movable sensor does not indicate any decrease or increase in its electric conductivity after switching on the active light source for ethylene pressures P ranging from 0.01 Torr to 0.1 Torr. This result would allow an unambiguous conclusion that, under the specified conditions, photolysis of ethylene proceeds only according to the molecular scheme. In other words, molecular hydrogen is detached from an ethylene molecule in a single act. However, the results of experiments [11] were essentially different. It was found out that photolysis of ethylene, occurring at room temperature under low pressures (below 0.01 Torr), resulted in a positive signal (an increase in the electric conductivity) of the movable sensor at every level in the reaction vessel. On the contrary, the upper (monitoring) sensor produced no signal.

This result means that the signal detected can be attributed to hydrogen atoms prevailing throughout the vessel. They are produced in the lower part of the vessel and easily reach its upper operating part under the specified pressure of ethylene. As the ethylene pressure increases, readings of the movable sensor change. At the ethylene pressure of 0.05 Torr, the electric conductivity of the sensor located below level II (see Fig.4.6) increases in the process of photolysis. On the contrary, all other conditions being the same, the sensor positioned above this level indicates a decrease in the electric conductivity, whereas being placed exactly at level II it shows no variation at all (neutral level!). As the ethylene pressure rises up to 0.07 Torr, the neutral level lowers down to mark III. Further increase of the ethylene pressure up to the value of 0.1 Torr and higher throughout the height of the vessel results in the decrease of the electric conductivity of the movable sensor, in contrast to the case where the pressure is 0.01 Torr. This effect is due to ethyl radicals (semiconductor electron acceptors) prevailing throughout the height of the vessel. Note that in experiments carried out under this pressure the monitoring sensor yields zero readings.

The results of these experiments can be accounted for by the fact that photolysis of ethylene occurring according to the atomic scheme at low pressures of ethylene yields hydrogen atoms in the lower section of the vessel. These atoms easily penetrate into all regions of the upper section of the vessel. Thus, under these conditions, the sensor produces a positive (hydrogen) signal in every point of the vessel. On the contrary, under high pressure of ethylene ($P = 0.1$ Torr), all hydrogen atoms enter into the specified reaction with ethylene already in the vicinity of the lamp, i.e., below filter 7 (see Fig.4.6). Thus the sensor produces a signal of the opposite sign at every level of the vessel, this fact being associated with prevailing adsorption of ethyl radicals on the sensor surface. These ethyl radicals diffuse from the lower part of the vessel into the upper one.

Under intermediate pressures of ethylene, hydrogen atoms prevail in the lower section of the vessel, whereas in the upper section, (C_2H_5) radicals prevail. At the neutral level hydrogen atoms and (C_2H_5) radicals give rise to signals equal in amplitude but opposite in sign. These signals are produced due to simultaneous adsorption of corresponding amounts of donor and acceptor particles on the sensor.

Now, we consider a question, whether it is possible to use sensors in experiments involving two types of active particles, to keep track of each type of particles. In a number of cases, such studies are needed to provide a deeper insight into the mechanism of photolysis. We can conclude that this technique can be employed if it is based on understanding of chemical properties and peculiarities of particles under investigation. Thus, it is well-known that hydrogen atoms actively recombine on palladium, platinum, and some other metals (with the coefficient of recombination γ close to unity). At the same time, alkali radicals exhibit similar activity with respect to other metals, such as lead, antimony, etc. Note that hydrogen atoms are rather passive with respect to the latter materials. Depositing specified metals on the inner surface of properly prepared glass porous filters, one can easily make filters selective with respect to transmission of various types of particles. These filters should be mounted above the optical filter (see Fig.4.6). This allows observation of a definite type of particles in photolysis. Furthermore, taking into account chemical properties of each type of particles, one can use two sensors in such experiments, each detecting particles of a definite type. This allows one to keep track of all types of particles produced in photolysis simultaneously, for example, ethylene and active particles. To manufacture these filters, one can use porous caps activated by corresponding metals. Another method is to activate the surface of a sensitive element. In our case of photolysis of ethylene, palladium allowed complete separation of $(C_2H_5)^\bullet$ radicals from hydrogen atoms, whereas the influence of lead was opposite.

Note that the same sensor technique was used in rather intriguing experiments on photolysis of acetone [12], ammonia [13], nitrogen [14], oxygen [15], and carbon dioxide [16].

When a $\lambda = 1849$ Å light acts on an ammonia molecule, the latter breaks into a hydrogen atom and an NH_2 radical [13]. At the zinc oxide surface the former particle is an electron donor, whereas the latter one is an acceptor. Experiments indicate that in photolysis of ammonia in a vessel shown in Fig. 4.6. only hydrogen atoms can be detected at every level of the vessel, starting from the source. This experimental result can be accounted for by the fact that, even in presence of such acceptors as

NH$_2$ radicals, hydrogen atoms adsorbed on the surface of a semiconductor sensor more actively affect the electric conductivity of the sensor.

If we use the same light (yet another technique) to irradiate a gaseous mixture NH$_3$ + He (1:10), the electric conductivity signal becomes negative in the vicinity of the lamp (at a distance of 1 – 5 cm) and positive at large distances (10 – 15 cm). In this case, the volume of the cylindrical reaction vessel can be divided into two zones in which sensors exhibit opposite variations in the electric conductivity. The results obtained can be accounted for by the fact that an addition of a rare gas slightly enhances bulk recombination of hydrogen atoms. This enhancement is due to triple collisions. At the same time, variation of the recombination rate of more complex particles, such as multiatomic radicals NH$_2$, caused by addition of the rare gas, is smaller. Consequently, the relative concentration of radicals increases in presence of neon (relative to concentration of hydrogen atoms). Similar effect occurs when hydrogen atoms are mixed with other multiatomic radicals that may feature recombination in absence of a third particle. Note that experiments on pyrolysis and photolysis of carbon dioxide [16] show that two sensors (one of them being located close to the sample, the other one being offset by 120 cm) allowed simultaneous study of two processes,

$$CO_2 \longrightarrow CO + O \ ,$$

and

$$CO_2 \longrightarrow CO + O_2 \ .$$

Oxygen atoms (acceptors) are the most active particles in the first process, which was detected by the nearest sensor. The second process, i.e., production of molecular oxygen, is detected by the remote sensor. The latter was affected only by molecular products (O$_2$ and CO), which diffused to this region. Oxygen atoms were entirely absent at this distance. Note that, similarly to O$_2$, CO molecules feature low electric activity, as compared with oxygen. Using these effects, we succeeded in determining the activation energy of atomic-molecular reaction of CO$_2$ with oxygen atoms by the sensor technique. It was found to be 5 kcal/mole, which is consistent with the available reference data [17].

4.4. Photolysis in Adsorbed Layers

Using the sensor technique for studying photolysis of adsorbed layers of cetene on metal oxides we observed the decay of adsorption layers under the influence of ultraviolet light. The reaction yields methylene radicals in the surface layer

$$\left(CH_2CO\right)_{ads} \xrightarrow{\ h\nu\ } \left(CH_2\right)_{ads} + \uparrow CO \ ,$$

This radicals do not escape from the surface (this is indicated by a semiconductor microdetector located near the adsorbent surface) undergoing chemisorption on the same semiconductor adsorbent film. Thus, they caused a decrease in the electric conductivity of the adsorbent sensor, similarly to the case where free radicals arrived to the film surface from the outside (for example, from the gas phase). Note that in these cases, the role of semiconductor oxide films is twofold. First, they play a part of adsorbents, and photoprocesses occur on their surfaces. Second, they are used as sensors of the active particles produced on the same surface through photolysis of the adsorbed molecular layer.

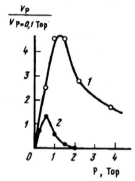

Fig.4.7. Photolysis of cetene in the layer adsorbed on a ZnO film (curve 1) and in the gas phase (curve 2) at room temperature.

Figure 4.7. shows the initial variation rate of the electric conductivity of a ZnO sensor film $v = (d\sigma/dt)_{t\approx 0}$ as a function of the pressure of a cetene vapor, measured in arbitrary units, at room temperature. The displayed dependencies govern photolysis of an adsorbed cetene layer by a light beam normal to the film surface (curve 1) and photolysis of cetene vapor by a light beam propagating parallel to the film at a distance of 1 cm (curve 2). As is seen from Fig. 4.7, in the initial stage (rising sections of curves 1 and 2) the increase of the cetene pressure and, consequently, the increase of the surface coverage of the ZnO film with adsorbed cetene molecules result in growth of v both in gas phase and on the surface. Consequently, the steady-state radical concentrations increase in both the gas phase and the adsorbed layer. On further increase of the cetene vapor pressure the v values reach their maxima and then rapidly decrease. In the gas phase, v decreases down to zero, whereas in the surface layer it reaches a limiting value independent of the cetene pressure. This limit is achieved when the film surface is completely covered with adsorbed molecules.

The specified decrease of the radical concentration in the gas phase near the film surface and in the layer adsorbed on the film is caused by the fact that interaction of these particles with cetene molecules becomes stronger as concentration of the latter increases. Another reason for the decrease of the radical concentration is the decrease of the diffusion coefficients of active particles in the gas and on the surface. This results in a growth of the time it takes for active particles from a gas phase to reach the film surface. Furthermore, it leads to an increase in the time it takes for active particles in the adsorption layer to reach the centers of chemisorption.

Intriguing results were obtained for photolysis of chemisorbed layers of methyl radicals on oxide films [18]. Experiments confirmed the hypothesis [19] concerning the nature of surface compounds produced in chemisorption of elementary radicals on metal oxides.

The SS technique has also proved to be informative for studying intermediate particles in radiolysis of organic mixtures in gaseous and liquid media [20].

Using semiconductor oxides doped with various metals (doping can be performed, for example, in molecular beams [21]) and measuring the electric conductivity of oxide targets, we found out that Zn, Sb, Cd, Pb, Ti, and other atoms used as over-stechiometric (dopant) elements provide the most active centers of adsorption of elementary alkali radicals. The above atoms and alkali radicals form surface metalloorganic complexes, which decompose at temperatures higher than 150 – 200°C. Below and in the range of room temperatures, chemisorption of alkali radicals on these oxides is characterized by low activation energies. Furthermore, as the temperature decreases, the chemisorption rate gradually slows down, and the surface coverage with adsorbed elements increases.

Thus, we considered a number of examples of application of the sensor technique in experiments on heterogeneous recombination of active particles, pyrolysis and photolysis of chemical compounds in gas phase and on the surface of solids, such as oxides of metals and glasses. The above examples prove that, in a number of cases, compact detectors of free atoms and radicals allow one to reveal essential elements of the mechanisms of the processes under consideration. Moreover, this technique provides new experimental data, which cannot be obtained by other methods. Sensors can be used for investigations in both gas phase and adsorbed layers. This technique can also be used for studying several types of active particles. It allows one to determine specific features of distribution of the active particles along the reaction vessel. The above experiments demonstrate inhomogeneity of the reaction mixture for the specified processes and, consequently, inhomogeneity of the thermal field. This means that in these cases the rate of the reaction varies from point to point. It is unlikely that this fact can be revealed by other methods (for example, by the EPR technique, mass-spectrometry, luminescence, and, finally, spectroscopic methods). Formally, in the considered cases, kinetic methods operate with only average concentrations of active particles. This disadvantage reduces reliability of these methods used for analysis of mechanisms of the specified reactions.

4.5. Examples of Elementary Processes in Heterogeneous Catalytic Reactions on Metal Oxides

In this Section, we consider examples of application of semiconductor sensors in investigation of heterogeneous catalytic reactions of dehydration of isopropyl alcohol and dissociation of hydrosine on zinc oxide.

It is well-known [22] that catalytic dehydration of isopropyl alcohol on zinc oxide yields predominantly hydrogen and acetone. However, the elementary mechanism of this process is still unclear. Obviously, this reaction proceeds either via molecular scheme, involving production of hydrogen molecules on the surface in a single act, or through the atomic scheme, where hydrogen atoms are detached from an alcohol molecule and then recombine into hydrogen molecules on the catalyzer surface. Furthermore, it is not clear in the latter case whether hydrogen atoms take part in dehydration of other adsorbed alcohol molecules, providing a kind of feedback, before they associate with each other. To answer these questions we carried out experiments using zinc oxide as a catalyzer of alcohol dehydration. In the presence of this catalyzer, the process under study is rather selective and features the efficiency of up to 96 – 98%. The role of zinc oxide in this experiment is twofold. It serves simultaneously as a heterogeneous catalyzer and as a high-sensitivity detector of atoms and radicals produced not only on the catalyzer surface, but in the gas phase as well, especially in the case, where the surface coverage of the catalyzer with the reagent is small. A small ZnO film was manufactured according to the standard technique, its contacts being available from the outside. The film was placed in the reaction vessel containing a doser of alcohol vapor and a thermostat. Throughout the whole process, the electric conductivity of the film was measured and recorded by a loop oscillograph. Sensitivity of the detector with respect to hydrogen atoms was provisionally inspected against pyrolysis on a filament of molecular hydrogen in the same vessel. In preliminary experiments, we verified that adsorption of hydrogen atoms on a zinc oxide film prepared for catalysis results in a sharp increase of the electric conductivity. The latter is manifested as an abrupt jump on a weak background corresponding to the signal due to molecular hydrogen. At room temperature this change in the electric conductivity is irreversible. However, the electric conductivity can be gradually reduced down to its initial level by increasing the temperature of the oxide film (the higher is the film temperature, the faster is the fall in the electric conductivity). The activation energy corresponding to the increase of the electric conductivity of zinc oxide caused by adsorption of hydrogen atoms was calculated to be 2 – 3 kcal/mol, whereas the decrease of the electric conductivity after switching-off the source of active particles corresponded to 30 kcal/mol [23].

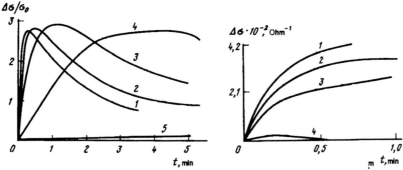

Fig.4.8. Oscilloscope traces of variation of the electric conductivity of a ZnO sensor upon admission of isopropyl alcohol vapor to the vessel (the initial vapor pressure is 0.01 Torr) at the temperature of 390°C (1), 370°C (2), 350°C (3), 320°C (4), and upon admission of H_2 at the temperature of 390°C (5).

Fig.4.9. Oscilloscope traces of temporal variation of the electric conductivity of a ZnO sensor for different initial pressures of the isopropyl alcohol vapor: $5.2 \cdot 10^{-2}$ Torr (1), $3.6.10^{-2}$ (2), $1.65 \cdot 10^{-1}$ (3), and $3.25 \cdot 10^{-1}$ Torr. The temperature of the ZnO film is 390°C.

Figure 4.8. displays oscillograms of evolution of the electric conductivity of the ZnO film in the process of catalytic dehydration of isopropyl alcohol at various temperatures of the catalyzer and equal portions of alcohol ($5 \cdot 10^{-2}$ Torr) admitted into the reaction cell. Experimental curves 1–4 are bell-shaped. We suppose that this fact is associated with two circumstances. On one hand, alcohol vapors dissociate on the oxide film producing hydrogen atoms. The jump in electric conductivity is caused by chemisorption of these hydrogen atoms on the film which plays a part of the sensor in this case. On the other hand, the drop in electric conductivity is caused by complete dissociation of the admitted portion of alcohol ("depletion" of the source of hydrogen atoms) and by surface recombination of chemisorbed hydrogen atoms produced in dehydration of alcohol. The influence of molecular hydrogen, released in the reaction, on electric conductivity of the film is negligible, as compared with hydrogen atoms. We verified this conclusion (see curve 5) by ensuring admission of molecular hydrogen into the reaction cell in amounts exceeding those produced by dehydration of the whole portion of alcohol admitted.

Estimations based on experimental data show that, in experiments with alcohol (at the left of the curves in Fig. 4.8), the activation energy corresponding to a change in the electric conductivity is about 45 kcal/mol, and the activation energy corresponding to a decrease in the electric conductivity (at the right of the curves) is about 30 kcal/mol. We note that the latter value coincides with the activation energy corresponding to the decrease in the electric conductivity of the film due to recombination of chemisorbed hydrogen atoms on the film surface in experiments on pyrolysis of hydrogen. This result unequivocally confirms that dehydration of alcohol on zinc oxide yields surface hydrogen atoms. The former value (about 45 kcal/mol) was found to be equal to the energy of activation of dehydration of isopropyl alcohol vapor on zinc oxide determined previously by other methods [24]. This coincidence indicates that the growth of electric conductivity of the film, on which alcohol molecules are dehydrated, is limited by dissociation of molecules. The effect of dissociation of molecules is stronger than that of the second stage of this process. The latter consists in chemisorption of hydrogen atoms and also leads to the increase in electric conductivity of the oxide film.

Further investigations of the above discussed effects show that, at fixed temperature of the oxide film (catalyst), the jump in the electric conductivity first increases in amplitude, as the portion of alcohol vapor admitted into the vessel increases. On further increase of the admitted portion of alcohol, the jump amplitude reduces (starting with the pressure of $3.6 \cdot 10^{-2}$ Torr). At the pressure of $3.2 \cdot 10^{-1}$ Torr, the jump in the electric conductivity of the zinc oxide film is less pronounced. Finally, at still higher pressures, it disappears (Fig.4.9). This effect is not unexpected. On our mind, it is associated with the fact that, as the concentration of alcohol vapor increases, the sum of the rate of interaction of the vapor with adsorbed hydrogen atoms and the rate of surface recombination of hydrogen atoms at the time instant of production becomes higher than the chemisorption rate of these atoms. The latter is responsible for the increase of the electric conductivity of the semiconductor oxide film via the reaction

$$H_{ads} \longrightarrow H^+_{ads} + e_S \ .$$

In conclusion, we note that the appearance of hydrogen atoms in the gas volume in catalytic reaction of dehydration of alcohol at low pressures observed in [25] by the sensor technique confirms that dehydration of alcohol on the surface of the zinc oxide catalyzer yields hydrogen atoms. In other words, this heterogeneous reaction does not result in production of hydrogen molecules through the process

$$CH_3CHOHCH_3 \longrightarrow CH_3COCH_3 + \uparrow H_2 \ .$$

The considered reaction yields hydrogen atoms, which subsequently recombine into molecules.

Experimental results clearly demonstrate that catalytic reaction of dehydration of alcohols on zinc oxide proceeds via formation of radicals. Emission of hydrogen atoms from the catalyzer surface may be associated with structure relaxation of the catalyzer surface excited during the reaction [26].

We used the sensor technique to analyze the mechanism of another heterogeneous catalytic reaction of dissociation of hydrosine on zinc oxide. The primary act of this reaction may consist in breaking of the N–N bond and production of $\dot{N}H_2$ radicals. Alternatively, this process may occur through breaking of the H–N bond accompanied by production of hydrogen atoms and $H_3\dot{N}_2$ radicals. The experiments [27] performed with the use of the sensor technique showed that the primary act of dissociation of hydrosine molecules on zinc oxide consisted in detachment of hydrogen atoms, i.e. in breaking of the H–N, but not N–N bond. Note that this result contradicts the experimental results and conclusions made in [28].

4.6. Evaporation of superstechiometric atoms of metals from metal oxide surfaces

It was shown in a number of works [29] that impurity conductivity of thin zinc oxide films are extremely sensitive to adsorption of atoms of various metals (see Chapters 2 and 3). Using this feature of oxide films, we first employed the sensor method to study evaporation of superstechiometric atoms of metals from metal oxide surfaces, zinc oxide in particular [30].

Quantitative studies of such processes are of great interest for understanding the mechanism of chemisorption and a number of heterogeneous catalytic reactions, because it is superstechiometric (admixture) atoms (ions) of metals become active centers of adsorption of different particles (radicals, molecules) on metal oxides, or centers of catalysis. Such

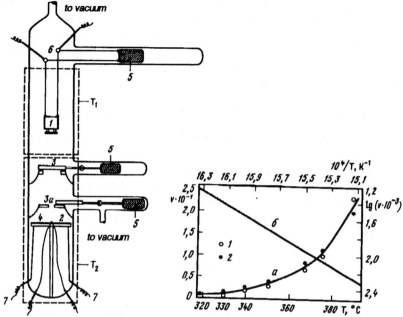

Fig.4.10. Reaction cell: 1 – quartz slab with a semiconductor sensor film; 2 – quartz plate with deposited metal oxide layer; $3, 3a$ – shutters; 4 – Pt/Pt-Rh thermocouple; 5 – iron rod sealed in a glass jacket; 6 – platinum ring contacts with Pt wires; T_1, T_2 – thermostats.

Fig.4.11. The initial rates v of increase and decrease of electric conductivity of the sensor (1) and adsorbent (2), as functions of the adsorbent temperature (a), and the Arrhenius plot (b) corresponding to curve (a). The sensor (ZnO) is kept at room temperature.

atoms appear on the surface either during the sample preparation, or as a result of doping from outside. The experiments based on the sensor technique were carried out in a cell shown in Fig. 4.10. In the lower part of the cell a quartz platform with electrodes was mounted, which served for deposition of a film of the oxide under study and for measuring the film conductivity with the help of electric leads brought in through the cell walls. The upper part of the cell separated from the lower part by two sliding shutters contained movable semiconductor film sensor 1. In a well-evacuated cell, oxide was heated up to required temperature by a thermostat T_2. Then, shutter 3 was opened to bring detector 1, which was heated by thermostat T_1 up to the temperature of 100-150°C, in thermal equilibrium with the evaporation chamber. After that, shutter $3a$ was opened, and the quantity $v = (d\sigma/dt)_{t\approx0}$ was measured. Thus, the value of v as a function of temperature of the oxide under consideration was obtained in a series of experiments. Moreover, the same series of experiments allowed one to measure simultaneously electric conductivity of both the zinc oxide film on the platform inside the evaporation chamber, and of movable sensor *1* in the upper adsorption chamber.

It was found in experiments that electric conductivity of lower oxide film *2* decreased in time. On the contrary, electric conductivity of upper sensor film *1* increased. This peculiarity can be attributed to transfer of superstechiometric atoms of zinc from the lower (evaporating) film to the upper (sensor) film. In other words, a double control was realized in this case.

It follows from calculations, that intensity of flux of zinc atoms incident upon the sensor film in these experiments amounted to $10^8 - 10^{10}$ atom$/$cm^2, on the average. It is seen from Fig. 4.11, that all experimental points depicting in arbitrary units the rates of increase (sensor) or decrease (evaporating film) of electric conductivity can be well approximated by the linear dependence in a ($\ln v - \frac{1}{T}$) plot.

The activation energy for evaporation of over-stechiometric zinc atoms Zn$^{\bullet}$ calculated from the tilt of this line is 32 kcal$/$gram$/$atom. This means that evaporation heat of superstechiometric zinc atoms (Zn$^{\bullet}$) from zinc oxide found in these experiments agrees well with the corresponding value found in [31] by mass-spectrometry.

Note that similar measurements were performed in [30] for the case of evaporation of defect atoms of lead from lead oxide, atoms of magnesium from magnesium oxide, and so on.

4.7. Surface and bulk diffusion of active particles

Many physical-chemical processes on surfaces of solids involve free atoms and radicals as intermediate particles. The latter diffuse along the adsorbent-catalyst surface and govern not only kinetics of catalytic, photocatalytic, or some heterogeneous radiative processes, but also creation of certain substances as a result of the reaction.

Heterogeneous recombination of active particles and their interaction with molecules of the adlayer are simplest processes of this type. The rates of such reactions as functions of surface coverage by the specified reagents are fully determined by the rate of their surface diffusion towards active centers. In a number of cases, the rate of lateral diffusion is determined not only by the type of diffusing particle, but also (sometimes, predominantly) by the composition and state of the solid substrate surface. Taking into account the role played by the composition of the active particles in an adsorbed layer, it is highly desirable to measure directly the coefficient of lateral diffusion of these particles and its dependence on a number of parameters characterizing the solid substrate.

The idea is to use for this purpose semiconductor sensors possessing very high detection capability with respect to active particles, such as atoms of hydrogen, nitrogen, oxygen, and simple radicals ($\dot{C}H_2$, $\dot{C}H_3$, $\dot{C}_2H_5$, $\dot{N}H$, $\dot{N}H_2$, $\dot{O}H$ and other). Moreover, semiconductor sensors are especially suitable for solving the above problem, because sensor films made of zinc oxide can be deposited upon substrate in the form of a small spot several tens of microns (or less) in diameter, or in the form of thin strips. Active particles diffusing along the substrate surface are expected to enter the area covered by the sensor film. However, it is not quite obvious, whether the supposition of a possibility for the particles under study to jump over the film boundary is true. Indeed, morphology and electronic state of the boundary between the substrate and the oxide film depend on a number of factors which are difficult to take into consideration. Nevertheless, the experiment made with our sensors in a "flat" configuration confirmed the validity of this supposition. The first experiment of this kind was carried out in [32], where surface diffusivity of hydrogen atoms (protium and deuterium) on polished quartz was measured. The measurement were made in a special vacuum chamber with a measuring unit shown in Fig. 4.12.

The rate of diffusion and flux of hydrogen atoms migrating along the surface were estimated from the time interval (induction period) between the beginning of the experiment (deposition of hydrogen atoms on quartz substrate) and the increase of the electric conductivity of a flat sensor (Fig. 4.13) caused by adsorption of hydrogen atoms migrating to the sensor along the quartz substrate. In their turn, hydrogen atoms were adsorbed on quartz from gas phase. Stationary concentration of hydrogen atoms near the surface of the substrate was estimated to be $7.7 \cdot 10^9$ particles/cm^3. Electric conductivity of the sensor was measured with a precision of 0.01%. Hydrogen atoms were generated from the gas phase by pyrolysis of hydrogen molecules on a hot iridium filament. This allowed smooth regulation of concentration of hydrogen atoms. During the measurement procedure concentration of hydrogen atoms in the gas phase and, consequently, the stationary hydrogen coverage of the quartz substrate remained constant. In order to exclude the influence on sensor conductivity of direct and reflected fluxes, as well as that of walking hydrogen atoms from the gas phase, the sensor made of zinc oxide was covered by a thin marblyte glass plate. A layer of titanium or palladium was preliminarily deposited (*in situ*) via sublimation on the surface of the plate exposed to sensor, to ensure more effective capture of hydrogen atoms incident on the plate surface.

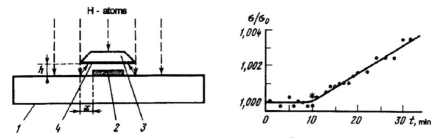

Fig.4.12. The sample construction: 1 − polished quartz plate; 2 − semiconductor sensor (ZnO); 3 − a strip of marblyte glass; 4 − a layer of titanium (palladium); $x = 0.027$ cm (the length of surface migration of H atoms; $h = 0.0025$ cm (the air gap between the quartz plate and the glass strip).

Fig.4.13. Relative variation of electric conductivity σ/σ_0 of the ZnO sensor under the influence of hydrogen atoms adsorbed from the gas phase and migrating toward the sensor. $T = 403$ K, $P_{H_2} = 6.7$ Pa, concentration of hydrogen atoms near the surface of the sample is $7 \cdot 7.10^9$ cm^{-3}.

As the working temperature of the substrate was increased, the induction period (the delay time) of increased conductivity decreased due to increased rate of lateral diffusion of hydrogen atoms towards the sensor. The activation energy for surface migration of particles along a SiO$_2$ substrate estimated from the tilt of the Arrhenius plot was found to be about 20 kJ/mol.

Similar experiments were conducted with deuterium. Comparing the results of these experiments with those obtained with protium, a conclusion was made that the ratio of calculated coefficients of surface diffusion of these particles is close to $\sqrt{2}$. In other words, this ratio is a manifestation of a usual isotopic shift for diffusivity given by $\sqrt{M_D / M_H} = \sqrt{2}$, as one might expect assuming that surface diffusion is the limiting stage of the three-stage process: surface migration of particles, adsorption of hydrogen atoms from quartz substrate on the surface of zinc oxide sensor, and, finally, surface ionization-chemisorption of adsorbed atoms on zinc oxide resulting in the appearance of a "jump" in the conductivity of sensor. To give support to this hypothesis, note that the activation energy of hydrogen atoms adsorbed on zinc oxide is only about 2 kcal/g-atom. As to the second stage of the process (there is no direct evidence), it seems that activation energy of this stage is not large, at least, at small hydrogen coverage of zinc oxide surface (the sticking coefficient is close to 1), as is seen from the heat of adsorption of hydrogen atoms on quartz substrate (about 20 kcal/mol) and on zinc oxide (about 30 kcal/mol). The above estimates are in agreement with the results of the experiments made with protium and deuterium.

The coefficient of diffusion for one-dimensional motion of particles is given by

$$D = \overline{x}^2 / 2\tau , \qquad (4.10)$$

where $\overline{x}^2$ is the mean square particle displacement per time interval τ.

For two-dimensional motion, (4.10) should be modified to give

$$D = \overline{z}^2 / 4\tau , \qquad (4.11)$$

$\overline{z}^2 = \overline{x}^2 + \overline{y}^2$, $\overline{x}^2$ and $\overline{y}^2$ being the mean square displacements of a particle in orthogonal directions. In our case mean value of D can be found from the relation (4.10).

From experimentally measured sensor signals as functions of distance x we estimated the coefficients of diffusion for protium and deuterium. At T = 345 K they are equal to $1.56 \cdot 10^{-11}$ and $1.00 \cdot 10^{-11}$ m/s^2, respectively.

Application of semiconductor sensors for measuring concentration of active particles in solids is of great interest for studies of peculiarities of the physical-chemical processes in real solids (for example, in polymers) involving free atoms and radicals.

Such a possibility has been pointed out for the first time in [13], where processes of diffusion and recombination of hydrogen atoms (protium and deuterium) have been studied in a water layer frozen on a semiconductor zinc oxide layer.

The authors showed that under bombardment of ice film by hydrogen atoms from gas phase, the sensor signal grows up. As the thickness of the water layer was increased, the time of the signal growth also increased. This fact has been attributed to slowing down of the process of onset of stationary concentration of hydrogen atoms diffusing through the water layer. In this case, the time of the signal growing up characterizes the time of diffusion of hydrogen atoms through the layer of water ice, because stationary concentration of hydrogen on the zinc oxide sensor surface is achieved, when the rates of particle arrival to the H$_2$O/ZnO interface due to diffusion and their chemisorption on ZnO surface become equal. The latter process takes rather short time to occur (the activation energy is about 3 kcal/mol). Consequently, variation of the stationary concentration of hydrogen atoms at the H$_2$O/ZnO interface as a function of time and thickness of the water layer should

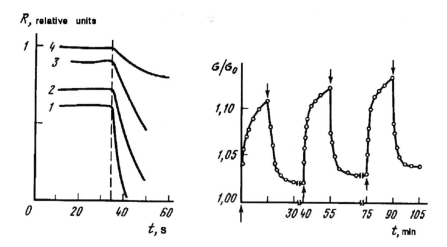

Fig.4.14. Kinetics of variation of resistance of pure ZnO sensor and of ZnO sensor covered with a layer of H_2O: *1,3* − interaction of H- and D-atoms with pure surface of ZnO; *2,4* − interaction of H- and D- atoms with surface of ZnO covered with layer of H_2O.

Fig.4.15. Kinetics of electrical conductivity of the Pd/ZnO sample in the processes of leaking-in (↑) and pumping-out (↓) of hydrogen at 298 K.

characterize the corresponding parameters of the process of diffusion, and not of chemisorption. In order to verify this supposition, the experiments were made with deuterium (D-atoms). In this case, a layer of water with a weight thickness of about $4 \cdot 10^{-3}$ cm was frozen upon the surface of ZnO sensor. Then, this layer was subjected to fluxes of H- and D-atoms of equal concentration. The characteristic results are shown in Fig. 4.14. Curves *1* and *3* describe variation of resistance of a pure ZnO film interacting with H- and D-atoms, respectively. Curves *2* and *4* describe interaction of these particles with a ZnO film covered with a layer of H_2O. It was found from that the ratio of relative rates of decrease of electric conductivity of a pure ZnO film and that covered with water layer under interaction with H- and D-atoms is about 1.53. This quantity is close to theoretically predicted value of the isotope effect in diffusion of H- and D-atoms in a solid matrix (≈ 1.41) [34]. The obtained results prove that the nature of interaction of an oxide sensor with H- and D-atoms from a gas phase and from a solid medium is the same. It is also shown that semiconductor sensors can be used successfully for studying physical-chemical processes in solids involving active particles.

4.8. Thermo- and photospillover of hydrogen atoms in multicomponent systems

A majority of publications available at the moment on the spillover effect, i.e., the effect in which active particles in heterogeneous systems flow from an activator (donor) to carrier (acceptor), is devoted to hydrogen. Among the systems considered are mainly metal-oxide ones, where this interesting effect has been observed for the first time [35].

The effect has attracted considerable interest, when the mechanism of heterogeneous thermo- and photoprocesses on the so-called deposited metal catalysts was studied. The catalysts usually consist of two components: an inert carrier (Al_2O_3, SiO_2, etc.) and an activator. The role of an activator is usually played by metals (Pt, Pd, Ni, etc.) dispersed on the surface of a carrier.

The effect is of rather complicated nature, and there are practically no descriptions of related direct observations or measurements in literature. Moreover, even theoretical concepts used for description of the effect are very cumbersome. Usually, the effect was observed indirectly, via its influence on isotope exchange, or the rate and completeness of some heterogeneous oxidation-reduction reactions in presence of metal oxide catalysts (like Al_2O_3, SiO_2, TiO_2, WO_3, and other) activated with transitional metals. The above effects could not be used for direct detection of active particles' migration, measurement of fluxes of these particles as functions of a number of reaction parameters, and obtaining experimental evidence to theoretical models of the nature of these particles and the phenomenon itself. The existing powerful methods of studying the surfaces of solids (Auger spectroscopy, electron spectroscopy, optical spectroscopy, etc.) do not allow one to observe this effect.

The idea to use oxide semiconductor sensors for this purpose has been suggested in [36]. The authors studied variation of electric conductivity of baked ZnO films activated with Pt and Pd as a result of their interaction with molecular hydrogen at room and elevated temperatures. It was noticed that in presence of an activator the increase of electric conductivity in the processes of leaking-in or pumping out of hydrogen from a cell containing samples was only partially reversible (Fig. 4.15), whereas in absence of an activator the effect was not detected at all. Obviously, these observations can be attributed to the hydrogen spillover effect. This effect, in a sense, "activates" zinc oxide via surface dissociation of hydrogen molecules allowed into the cell on an activator (Pt, Pd, etc.), with subsequent "creeping" of appearing hydrogen atoms from the activator to the surface of semiconductor, i.e., zinc oxide layer. Adsorption of hydrogen atoms on zinc oxide results in irreversible (at moderate temperatures of the oxide film) increase of electric conductivity of the semiconductor.

Note that this method enables one to observe variation of electric conductivity of a sample due to adsorption of hydrogen atoms appearing as a result of the spillover effect, no more. In a system based on this effect it is rather difficult to estimate the flux intensity of active particles between the two phases (an activator and a carrier). The intensity value obtained from such an experiment is always somewhat lower due to the interference of two opposite processes in such a sample, namely, "birth" of active particles on an activator and their recombination. When using such a complicated system as a semiconductor sensor of molecular hydrogen (in the case under consideration), one should properly choose both the carrier and the activator, and take care of optimal coverage of the carrier surface with metal globules and effect of their size [36].

In order to develop more informative and direct method of studying the spillover effect of active particles, the authors of [37] suggested to use the sensor method of detecting migrating particles based on separation of sensor and emitter (donor) of active particles. The latter consists of small metal globules, or clusters (with a diameter of about 20–30 Å) of Pt, Pd, Ni, etc. (activator) deposited on quartz or sapphire (Al_2O_3) plate in the form of a strip less than 1 cm wide. The sensor for detection of hydrogen atoms consisted of a zinc oxide strip (with a width of about 0.1 cm and thickness $\approx$100 nm) deposited on the same plate at a distance of 0.03 or 0.6 cm (two versions) from the inner boundaries of activator strips [38].

The scheme of the element is shown in Fig. 4.16. In order to increase variation of electric conductivity semiconductor film was deposited in the center of the plate, whereas activator was deposited at the plate edges at above specified distances through a mask. All stages of preparation were conducted in high vacuum ($\approx$10^{-8} Torr). Sensitivity of such sensors to adsorption of hydrogen atoms at room at lower temperatures was about $10^5 - 10^6$ at/cm^2, which corresponds to surface coverage of only $10^{-8} - 10^{-7}$% (!).

By varying the temperature at which the experiments were conducted and the distance between the activator and the sensor, the data were obtained (Fig. 4.17) which allowed us to calculate the activation energy of migration of hydrogen adatoms (protium and deuterium) along the carrier surface and coefficients of lateral diffusion of hydrogen atoms appearing due to the spillover effect (see Table 4.2).

Comparing these results with the analogous results obtained with hydrogen atoms adsorbed onto the surface of the same carrier directly from the gas phase (see previous paragraph), one concludes that the nature of the active particles formed under spillover of hydrogen is similar to that of hydrogen atoms adsorbed from the gas phase. In other words, particles migrating along the surface under the spillover of hydrogen have nothing to do, for example, with earlier predicted surface complexes $\left(H^+...e\right)_s$, $\left(H^+...H^-\right)_s$, $\left(H_3^+\right)_s$, $\left(H_2^{\cdot}\right)_s$ (vibrationally excited molecules), as well as with free or solvated protons [39].

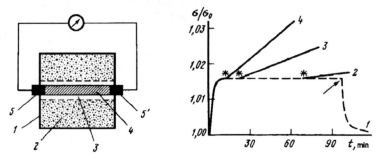

Fig. 4.16. The scheme of a working plate (element): 1 – a plate of SiO_2 (Al_2O_3); 2 – particles of deposited metal; 3 – strips of pure (not activated with metal) substrate (0.03 cm wide); 4 – ZnO film serving as a sensor of hydrogen atoms; $5,5'$ – platinum ohmic contacts.

Fig. 4.17. Kinetics of variation of electric conductivity of ZnO sensor following leaking-in H_2 into the reaction cell and subsequently pumping it out (indicated by pointer) before (curve 1) and after (2 – 4) activation of the working plate with palladium: $\theta_{Pd} = 5 \cdot 10^{15}$ cm^{-2}, $P_{H_2} = 6.7$ Pa, $T = 295$ K (2), 378 K (3), 413 K (4). Stars show the beginning of sharp rise of the sensor signal under the influence of H-atoms migrating toward it.

TABLE 4.2.

Migrating particle	System	Activation energy for migration, kJ/mol	Coefficient of surface diffusion at 345 K, m^2/s
H_s	Pd/SiO_2	18.8	$1.35 \cdot 10^{-11}$
H_s	Pd/Al_2O_3	33.5	$1.00 \cdot 10^{-11}$
D_s	Pd/SiO_2	–	$1.00 \cdot 10^{-11}$
H_v	Pd/SiO_2	19.7	$1.56 \cdot 10^{-11}$
D_v	Pd/SiO_2	–	$1.00 \cdot 10^{-11}$
H_s	$Pd/SiO_2 + UV$	16.6	$2.45 \cdot 10^{-11}$

However, it was established with the help of semiconductor sensors that hydrogen particles migrating under the spillover effect carry a small positive charge, as is seen from the results obtained when an electric field is applied between the activator strip and zinc oxide. The electric field (with the field strength of about several hundred volts per cm) directed from the activator to the sensor accelerated migration of active hydrogen particles, whereas oppositely directed field resulted in migration slowing down, as one might expect if the assumption is made that migrating hydrogen particles under the spillover effect carry an effective positive charge.

The results obtained under the thermal spillover effect of hydrogen in Pd/SiO$_2$ (Al$_2$O$_3$) systems put forward the question of whether light has an influence on this process, or not. Specially designed experiments based on application of semiconductor sensors of hydrogen atoms gave positive effect [40]. Thus, the effect has been called "the photospillover".

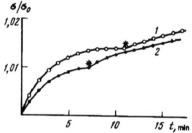

Fig. 4.18. Kinetics of variation of electric conductivity of the ZnO sensor on SiO$_2$ plate activated with Pd after leaking-in hydrogen: 1 − without illuminating the plate; 2 − during illumination with light at 313 nm from a mercury lamp with an additional water filter absorbing IR radiation. Stars show the beginning of sharp rise of electric conductivity.

The effect of photospillover of hydrogen consisting in variation of both the intensity and the direction of the thermal spillover under illumination of the phases' boundary (between an activator and a carrier) with light has been discovered in [40] (Fig. 4.18). Analyzing this phenomenon, the authors concluded that it was mainly due to light-induced additional charging and recharging of the metal-semiconductor interface, rather than due to light-induced acceleration of migration of ions along the surface of a wide-band semiconductors [41]. The above photoinduced effects cannot only considerably change the rate of spillover of charged hydrogen particles in metal-dielectric-semiconductor systems, but also change the direction of the thermal spillover effect.

Note that the effect of photospillover discovered with the help of semiconductor sensors may help a lot in understanding mechanisms of some important photocatalytic processes, such as photoheterogeneous decomposition of water and alcohols on surfaces of semiconductors (TiO_2, ZnO, etc.) doped with metal particles, which is accompanied by release of hydrogen, or photocatalytic oxidation of carbon on metal-activated semiconductor oxides, which is probably due to the spillover and photospillover of adsorbed atoms of oxygen.

4.9. Adsorption of atomic, molecular, and cluster particles on metal oxides

We studied the influence of adsorption of various particles of silver on electric conductivity of ZnO films. Silver was deposited on the semiconductor oxide target by using an atomic beam. The target consisted of thin baked film deposited on quartz substrate with preliminary formed platinum electrodes. The idea of the experiment was to follow variation of electric conductivity of the semiconductor sensor − target caused by surface aggregation of silver adatoms. It was assumed that molecular and larger (cluster) silver particles were less, if at all, active with respect to the influence produced by their adsorption on electric conductivity of the semiconductor [42], similar to recombination of hydrogen atoms adsorbed on a semiconductor surface, which is known to produce a small number of electrically active hydrogen molecules.

Atoms of metals are more interesting than hydrogen atoms, because they can form not only dimers Ag_2, but also particles with larger number of atoms. What are the electric properties of these particles on surfaces of solids? The answer to this question can be most easily obtained by using a semiconductor sensor which plays simultaneously the role of a sorbent target and is used as a detector of silver adatoms. The initial concentration of silver adatoms must be sufficiently small, so that growth of multiatomic aggregates of silver particles (clusters) could be traced by variation of an electric conductivity in time (after atomic beam was terminated), provided the assumption of small electric activity of clusters on a semiconductor surface [42] compared to that of atomic particles is true.

The experiments were conducted in a cell (Fig. 4.19) at residual gas pressure of less then 10^{-8} Torr kept constant during the measurements. The surface coverage in these experiments was only $10^{-4} - 10^{-2}$ %. In this case, after the atomic beam was terminated, relaxation of electric conductivity has not been observed even at elevated temperatures (100 − 180°C), when surface mobility of adatoms increased considerably. At larger coverages of the target surface with adatoms, or at higher surface temperatures electric conductivity relaxed to its initial value (before adsorption). As the temperature or surface coverage was increased, the relaxation rate also increased.

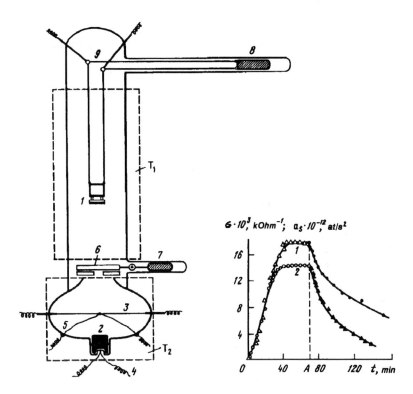

Fig. 4.19. Reaction cell: 1 – ZnO sensor; 2 – evaporator; 3 – band evaporator; 4, 5 – thermocouples; 6 – shutter; 7, 8 – iron rods sealed into a glass jacket; 9 – contact rings

Fig. 4.20. Variation of electric conductivity of ZnO film under bombardment with silver atoms (to the left from the dashed line) and after termination of the atomic beam (to the right from the dashed line). The film temperature is equal to: 1 – 94°C; 2 – 193°C; T_A = 780°C

Figure 4.20 shows variation of electric conductivity of ZnO film in time during bombardment with silver atoms and under conditions, when the atomic beam was terminated. We believe, that the obtained results (the behaviour of conductivity in time, its relaxation) can be explained by competition of adsorption and surface aggregation of adatoms leading to the appearance of larger particles of silver (dimers, trimers, and so on) on the surface. Such particles were assumed not to provide a considerable influence on electric conductivity of the film. Thus, the above two processes should shift the equilibrium of the reaction

$$Me_v \rightarrow Me_s \underset{\leftarrow}{\rightarrow} Me_s^+ + e_s \ ,$$

to the left. Here Me_v, Me_s are atoms of a metal in the gas phase and on the surface, respectively, Me_s^+ are metal ions in adlayer, and e_s denotes electrons in a conductivity band.

After the atomic beam is terminated, only surface processes caused by migration of adatoms and their aggregation into larger particles take place (evaporation from the surface at moderate temperatures does not occur). First, the process characterized by largest probability, that is, dimerisation (recombination) of adsorbed atoms takes place. This process is obviously the second – order process, provided migration of adsorbed atoms to centres of recombination does not impose any limitations. Later on, with the increase of the number of dimers and larger particles, the order of the process reduces down to the first order with respect to monomer surface particles (the probability for such particles to meet larger particles and adhere to them increases with the increase of surface coverage). Consequently, the second – order processes should obey the following rate equations:

$$da_s \ / \ dt = -K_s^2 \ , \quad 1 \ / \ a_s = Kt + 1 \ / \ a_s^0 \ , \tag{4.13}$$

where a_s is the surface concentration of adsorbed atoms, K is the rate constant of aggregation of adatoms, a_s^0 is the initial concentration of adatoms (immediately after termination of the atomic beam). At sufficiently small surface coverages of the semiconductor film, concentration of adatoms is directly proportional to variation of electric conductivity of the film (see Chapter 1). Thus, the values a can be calculated from known values of sticking coefficients and the coefficient of surface ionization of atoms by using the relations

$$\Delta\sigma_0 = \Delta[e] \cdot \mu \cdot 1.6 \cdot 10^{-19} \ , \quad \Delta\sigma = \frac{S}{l}\Delta\sigma_0 \ , \tag{4.14}$$

where $\Delta\sigma_0$ and $\Delta\sigma$ are variations of specific and total conductivity of the film, respectively, l is the film length, S is its cross – section, $[e]$ is the concentration of carriers, and $\mu = 10 \ cm^2/s \cdot V$ is their mobility.

Figure 4.21 illustrates the validity of an integral form of the rate equation governing aggregation of silver atoms at several temperatures. The activation energy of surface aggregation (second order) was found to be 3.2 kcal/g-atom. The rate constants at different temperatures can be found from the relation

$$K_{Ag}[\text{at} \cdot \text{cm}^2 \cdot \text{c}^{-1}] = 1.2 \cdot 10^{-15} \exp(-3200 / RT).$$ (4.15)

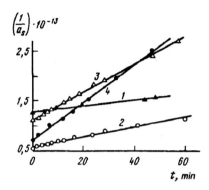

Fig. 4.21. Kinetics of surface aggregation of silver atoms on ZnO film (see Eq.(4.13)): *1* – 20°C; *2* – 94°C; *3* – 150°C; *4* – 193°C.

The above rate equations confirm the suggested explanation of dynamics of silver particles on the surface of zinc oxide. They account for their relatively fast migration and recombination, as well as formation of larger particles (clusters) not interacting with electronic subsystem of the semiconductor. Note, however, that at longer time intervals, the appearance of a new phase (formation of silver crystals on the surface) results in phase interactions, which are accompanied by the appearance of potential jumps influencing the electronic subsystem of a zinc oxide film. Such an interaction also modifies the adsorption capability of the areas of zinc oxide surface in the vicinity of electrodes [43].

In order to confirm the results on aggregation of silver adatoms under above experimental conditions, obtained with the help of the semiconductor sensor, and thus to check the validity of using semiconductor sensors for solving this kind of problems, the authors of Ref. [42] studied surface aggregation of silver atoms by a photographic method based on processing of photographic emulsions containing centres of hidden image (silver aggregates consisting of at least ten atoms). However, when studying aggregation of silver atoms on ZnO film, instead of usual procedure of processing the photosensitive layers, the well known method of physical development [44] was used, consisting of processing of ZnO film with silver adlayer in a solution containing only AgBr and metholhydrochinon. It was found in the experiments that in the case where preliminary $10^9 - 10^{11}$ at/cm^2 were adsorbed on ZnO surface (electric conductivity increased irreversibly, thus indicating the absence of aggregation of silver atoms), the film did not reveal any darkening after physical development. However, if surface concentration of adsorbed silver-atoms was as high as $10^{12} - 10^{14}$ at$/$cm^2, then the electric conductivity revealed relaxation (aggregation of silver atoms!), and a contrast spot was seen on the film surface after development. Note, that after physical development of adsorbents, with aggregates of silver atoms deposited on them, similar results were also obtained for glass and for Al_2O_3 film. This fact indicates that the nature of the substrate, on which silver aggregates are deposited, is important in the processes of film development.

The results of these experiments also confirm the conclusions made in the above paper dealing with the mechanism of aggregation. These conditions were based on the data obtained using the method of semiconductor sensors. However, the technique used in [42] was seemingly more sensitive, because it enabled observation of elementary surface processes, such as the appearance of centres of condensation of metal atoms on atomic scale.

The above results put forward the question, whether or not one can directly compare in experiment the influence produced by adsorbed atomic and molecular silver particles on electric conductivity of a zinc oxide film.

For this purpose, the authors of Ref. [45] conducted experiments with silver using the method of collimated atomic beams in a strongly inhomogeneous magnetic field. The experiment made use of the fact that in an inhomogeneous magnetic field paramagnetic atoms in atomic beam were deflected, whereas molecular particles retained their direction of motion. Thus, using the Stern-Gerlach effect, one can split a beam of silver particles into atomic and molecular components, and to subject the surface of a zinc oxide film to bombardment with atomic and molecular particles separately. For this purpose, an experimental arrangement was used (similar to that used by Stern and Gerlach), which is shown in Fig. 4.22. The strength of an inhomogeneous magnetic field was 15000 Oe, the length of a capillary in the field was 70 mm, its diameter was 2 mm, and the target diameter was 2 mm. The magnet was made of "Armco" iron. Alignment of the device was performed by using a laser beam.

In the experiment, target 1 (semiconductor ZnO film) was exposed to a beam of metal particles for a specified time interval by activating a shutter 3 (controlled by a magnetic device) installed in front of a diaphragm 4, with magnetic field on and magnetic field off. The rate of variation of an electric conductivity was measured. At small surface coverages, this rate is strictly directly proportional to the number of metal atoms incident on the film surface, i.e., $(d\sigma/dt)_{t\approx0} \sim I_A$, where I_A is the atomic beam intensity. The shorter was the time of exposition and the lower was the beam intensity in each experiment, the larger number of such experiments could be conducted without regenerating the film.

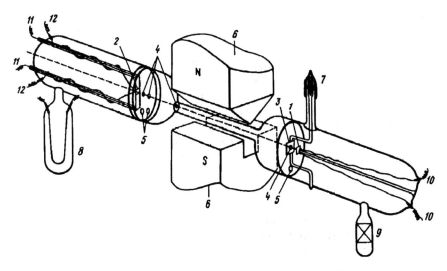

Fig. 4.22. Reaction cell: *1* − ZnO sensor; *2* − evaporator of silver atoms; *3* − shutter used to terminate the beam of Ag atoms; *4* − collimating apertures; *5* − an aperture used for pumping the cell out; *6* − magnet; *7* − magnetic drive for a shutter; *8* − getter; *9* − vacuum-measuring tube; *10, 11* − electrodes; *12* − thermocouple.

The experiments with a beam of silver particles were conducted at room temperature. The energy of dissociation of diatomic molecules of silver is 1.78 eV, the heat of evaporation of silver molecules is 95 kcal/mol [46], and the heat of evaporation of an uniatomic silver is 64 kcal/mol. Mass-spectrometric studies [46] of silver vapour above a metallic silver showed that the ratio of number densities of ions Ag^+/Ag_2^+ is equal to two. In other studies [47], a considerably larger value of this ratio was found. At 1037 − 1147°C molecular mass of silver particles in vapours was found to be 278 ± 90 [46], i.e., an average number of atoms in a molecule of silver is 2.56.

We found that the increase of film conductivity observed without magnetic field disappeared immediately after the field turn-on and did not appear again until the filed was turned off again. This result seems to confirm the earlier supposition of electric activity being present only if the surface of a semiconductor film is covered with silver atoms. We used the method of physical development of an oxide film alter sufficiently long exposition of the film to the beam of silver particles, with the magnetic field turned on, in order to be sure that particles of silver reach the surface of a target (when silver atoms are incident upon the surface, this fact can be verified by the signal of electric conductivity). The degree of darkening of the film after physical development is approximately directly proportional to the amount of silver deposited on the film.

From the positive results of these experiments we concluded that the behaviour of atomic and molecular particles of silver with respect to their influence on electrophysical properties of oxide films is similar to that of atoms and molecules of nonmetals, with the only difference that metal-atom interstitials behave similar to hydrogen-like donors of electrons, independent of the kind of a metal. As to metal molecules, at low temperatures of a semiconductor film, when their surface dissociation does not occur, they do not reveal considerable activity with respect to electrophysical properties of the film.

The above results once again make clear the reason for relaxation of impurity conductivity of a zinc oxide film alter termination of a beam of metal particles used for doping the surface of these films. Additionally, these results contain earlier suppositions of molecular particles of metals being electrically inactive.

4.10. Measurement of concentration profile of oxygen in the lower thermosphere of the Earth with the help of semiconductor sensors

The method of semiconductor sensors allows one to determine the flux of atoms, to which the sensor was exposed, from electric conductivity measurements (provided coefficients of ionization and reflection of oxygen atoms from zinc oxide films are known). In other words, the sensor technique can be used in this case as an absolute method [21]. Indeed, variation of electric conductivity of a semiconductor film $\Delta\sigma_{film}$ due to adsorption is known to be caused by variation of carrier concentration Δn in the film, rather than by variation of their mobility μ [21]:

$$\Delta\sigma_{film} = \Delta\sigma^0 \frac{S}{l} = \Delta n^0 q\mu \frac{S}{l} = \left(\frac{\Delta n}{lS}\right) q\mu \frac{S}{l} = \frac{\Delta nq\mu}{l^2}, \tag{4.16}$$

where σ^0 is the specific electric conductivity of the film, n^0 is the carrier concentration per unit volume of the film, q is the charge of an electron, S is the visible cross-section of the film, l is the distance between the electrodes, and μ is the mobility of electrons.

If coefficients of ionization ν and reflection ω of atoms for zinc oxide film are known, we have

$$\Delta n \, / \, \Delta t = [I_A(1 - \omega)] \, \nu \, , \qquad (4.17)$$

and

$$I_A = \frac{\Delta n \, / \, \Delta t}{\nu \, (1 - \omega)} = \frac{(\Delta \sigma_{film} \, / \, \Delta t) \, l^2}{q\mu\nu \, (1 - \omega)} = \frac{v_0 \, l^2}{q\mu\nu \, (1 - \omega)} \, , \qquad (4.18)$$

where I_A is the flux of atoms incident on the sensor, v_0 is the rate of variarion of electric conductivity of the sensor, ν is the coefficient of ionization of adatoms, ω is the coefficient of reflection of atoms from the sensor film. The values of ν and ω for the case of interaction of atoms of oxygen and hydrogen with zinc oxide films were determined in [21]. The time constant of a semiconductor sensor is less than 10^{-2} s [48], which ensures high temporal and spatial resolution when measuring concentration of atoms with a semiconductor sensor. Sensitivity of a semiconductor sensor with respect to adsorbed atoms is $10^9 - 10^{10}$ at·cm^{-2}·s^{-1}. Such a high sensitivity with respect to active particles is retained at pressures up to atmospheric one.

The above properties of semiconductor sensors allow one to use them for measuring the contents of oxygen and hydrogen in different parts of the atmosphere. An important problem in this kind of measurements is to ensure selectivity of measuring certain components. This problem was solved for the first time in Ref. [49]. The selective semiconductor sensor developed by the authors enables detection of oxygen atoms in presence of hydrogen atoms. A detailed description of the analyzer and its operation principles can be found in Ref. [49]. Note, however, that the known effect of critical capacity of a semiconductor sensor restricts the time of exposition of semiconductor films to intense fluxes of atoms. Thus a possibility of performing a sufficient number of measurements with the help of a sensor in the range of linearity, where intensity of flux of incident particles is directly proportional to the rate of variation of the sensor film electric conductivity in each measurement, is limited. The effect of critical capacity is one of the main factors determining the design of semiconductor analyzers. Thus, in the case of an "open" construction of a semiconductor sensor, where the stream of atmospheric gases directly interacts with the sensor (assuming that high sensitivity of such an analyzer is ensured), the time of continuous measurements in the range of linearity at atomic flux intensities of $10^{10} - 10^{11}$ at·cm^{-2}·s^{-1} is only $10^2 - 10^3$ s. However, in the lower thermosphere, especially, in the vicinity of maximum concentration of atomic oxygen equal to 10^{11} −

10^{12} at·cm³, where the time of measurements in the range of linearity of a semiconductor sensor is small, the flux of atoms incident of the sensor surface should be automatically restricted and terminated, in order to prolong the time of measurements in the range of linearity. This situation demonstrates the necessity of using a "closed" construction of the analyzer. It this case, the sensor is installed in a cavity connected with the atmosphere by a channel with a valve.

The analyzer contains three sensitive elements (thin films of zinc oxide): two elements are used in the measurements, whereas the third one is isolated from the atmosphere and serves to compensate for time and temperature drifts and to exclude the influence from molecular components of the environment.

In order to develop an effective sensor for detection of atomic oxygen, it would be interesting to estimate possible contributions from different atmospheric constituents to the value of signal attributed to oxygen atoms and detected in accordance with the above procedure by a semiconductor sensor analyzer based on zinc oxide.

Molecular nitrogen, noble gases, simple hydrocarbons, like CH_4, C_2H_6, etc., as well as ozone, nitrogen oxides, like N_2O, NO, NO_2, have practically no influence on the detector conductivity at temperatures below $50 - 100°C$. The effect of atomic oxygen on zinc oxide semiconductor film exceeds similar effect from O_2 by several orders of magnitude [15]. At the ratio of concentrations of molecular and atomic oxygen $[O_2]/[O] = 10^6$, contribution of O_2 to the sensor signal (zinc oxide sensor) amounts to 50%, provided no special measures are taken to reduce its influence. The compensating detector reduces this value to $5 - 7\%$ (compensation is incomplete because the sensor semiconductor films are not identical). In this case, the error introduced by molecular oxygen in measured concentration of atomic oxygen at the altitude of 100 km (the ratio $[O_2]/[O]$ is about 10) is only $(5 \div 7) \cdot 10^{-5}\%$ and decreases with the increase of altitude. This situation is typical of "open" detectors. However, usually the detectors are installed in a special isolated cavity connected by a channel with the atmosphere. It should be noted that the gas in the cavity is enriched with molecular oxygen, compared to the surrounding atmosphere, because atoms recombine on walls of the cavity and the chamber. This fact results in increased error due to contribution of molecular oxygen. Thus, for the analyzer of concentration used in the experiment, concentration of atoms in the cavity is four orders of magnitude lower than that in the atmosphere, which makes the error at an altitude of 100 km as large as $0.5 - 0.7\%$.

The influence of molecular oxygen in an excited state $O_2(^1\Delta_g)$ (singlet oxygen) is much larger. For example, measured electric conductivities of a ZnO film exposed to $O_2(^1\Delta_g)$ and that exposed to atomic oxygen are equal at the ratio of concentrations $[O_2(^1\Delta_g)]/[O] =$ 3.5·10. The error due to contribution of singlet oxygen at an altitude of 100 km ($[O_2(^1\Delta_g)]/[O] \approx 10^{-2}$) is about 3·10^{-2}%.

The influence of other active components, such as $\dot{N}, O\dot{H}, \dot{H}$ on a semiconductor sensor, with other conditions being the same, is comparable with the influence of atomic oxygen [50]. Contribution of $\dot{N}$ and $O\dot{H}$ is proportional to their relative contents (compared to that of atomic oxygen) in the atmosphere and may become essential at altitudes lower than 60 − 70 km. The use of selective detectors excludes the influence of atomic hydrogen. Studies of adsorption of water vapours on ZnO films [50] show that their influence is negligibly small at the film temperatures below 100°C. Variations of electric conductivity of the films under the influence of water vapours and of an atomic oxygen are comparable at the ratio of their concentrations $[H_2O]/[O] = 10^{-9}$.

Measurements of concentration of atomic oxygen with the help of a sensor with a semiconductor sensitive elements installed on a meteorological rocket MR-12 were carried out on December 27, 1979, at the test site "Volgograd". The measurements were made on the rising side of the rocket trajectory from an altitude of 96 km till the ultimate point of the trajectory at an altitude of 162 km. The rocket was launched at 18:11, local time. Geomagnetic and solar activity during the experiment was characterized by the following parameters: flux of solar radiation of the radiofrequency range $F_{10,7} = 164·10^{-22}$ W/m²·Hz, mean value of this flux during a three-month period $\overline{F}_{10,7} = 219·10^{-22}$ W/m²·Hz, the index $K_p = 3$. The sensor was installed in the rocket head with its longitudinal axis being parallel to the rocket axis. Since the rocket was stabilized in space using fast rotation (about 5 revolutions per second), on the rising part of the trajectory the sensor input aperture was directed forward, along the rocket velocity. When processing the primary experimental data, we used the relation between concentration of atomic oxygen in a free atmosphere and its concentration in the device consisting of a spherical cavity connected with the atmosphere by a tube, which was found in Ref. [48] for the case of a free molecular flow at zero angle of attack, with proper accounting for recombination of atoms.

Figure 4.23 shows the results of measuring the electric conductivity of the semiconductor sensor obtained by remote control means from board of the rocket MR-12, along with the data obtained in our experiments and the data of model calculations by other authors. Also shown are the experimental results of similar measurements obtained by other

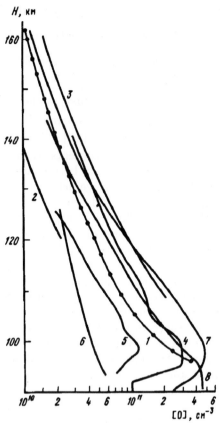

Fig. 4.23. Distribution of concentration of oxygen in upper atmosphere: *1* – obtained with a semiconductor sensor on December 27, 1979 near observation site "Volgograd" (present investigation); *2, 3* – the data of mass-spectrometric measurements; *4, 5* – the data of a resonance spectroscopy; *6* – silver films; *7, 8* – model calculations

methods in various moments of time and under different experimental conditions. This circumstance does not allow direct comparison of the data. However, in general, the data obtained by the semiconductor sensor technique [51] are in agreement with the available experimental data. Thus, the obtained results are close to those found in model calculations [52] for winter weather conditions. Note that the use of semiconductor sensors in this kind of measurements is a pioneering work. We believe, that this technique will be developing fast and successfully in the near future, because semiconductor sensors are characterized by small size, weight and low energy consumption. Moreover, the semiconductor sensors are cheap and simple to use.

4.11. Measurement of small concentrations of oxygen in various buffer gases at atmospheric pressure and room temperature

The problem of measuring small concentration of oxygen in a buffer gas can be solved by using the semiconductor sensor with a sensitive element consisting of a zinc oxide film immersed in a polar or, better, a protodonor liquid (see Section 3.4).

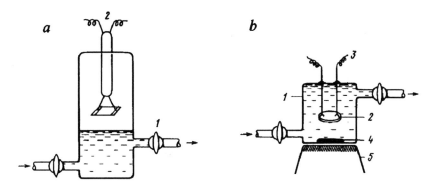

Fig. 4.24. Schemes of gas-analytic cells for vapour (*a*) and liquid (*b*) phases: *1* — the cell; *2* — sensor (ZnO); *3* — contacts; *4* — iron rod sealed in a glass tube; *5* — magnetic mixer.

The scheme of a laboratory gas analyzer for detection of oxygen is shown in Fig. 4.24. The gas being analyzed is passed for some time over the liquid (the gas can be passed through the liquid, as well) and is then "locked" in the gas analyzer. An equilibrium is usually established between oxygen contained in a carrier gas, for example, hydrogen, and that dissolved in a solvent. The larger is the dielectric constant ε of the protodonor solvent (water, alcohols, aqueous-dioxane mixtures, etc.), the more rapidly stationary concentration of oxygen adsorbed on ZnO film is established. Simultaneously, electric conductivity of the sensor attains a stationary value, which is equal to concentration of oxygen in a carrier gas multiplied by the Bunsen coefficient, i.e.,

$$N_{O_2}^r = N_{O_2}^g \beta \; , \tag{4.19}$$

where N denotes concentration, superscripts r and g correspond to solvent and carrier gas, respectively, β is the Bunsen coefficient (equal to the ratio of equilibrium volume concentration in the liquid and the gas over the liquid).

Note that stationary concentration of oxygen adsorbed on a ZnO film immersed in a solvent with large dielectric constant ε is established rapidly only in case where a magnetic mixer is installed under a semiconductor film deposited on the rear surface of a quartz plate (diskshaped). The mixer consists of a glass tube $3 - 4$ mm in diameter and $15 - 20$ mm long with an iron rod sealed inside. The rod rotates at a distance of about 2 cm from the quartz plate with sufficiently high angular rate (about 500 rpm). Mixing solutions eliminates diffusive slowing-down in delivering oxygen to the film and helps to remove the products of the reaction between adsorbed oxygen and molecules of solvent (see Chapter 3). Moreover, agitation of the solution determines the absolute value and reproducibility of the effect of establishing a stationary electric conductivity of a sensor.

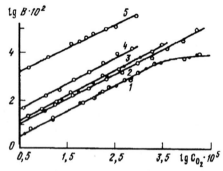

Fig. 4.25. The results of verification of the equation (3.29) in the case of ZnO film immersed in various solvents in a hydrogen atmosphere at room temperature $(1 - 4)$ or in vacuum at 250°C (5): 1 – water (liquid ($\varepsilon = 78$) and saturated vapour); 2 – methanol ($\varepsilon = 36$); 3 – ethanol (liquid ($\varepsilon = 26$) and saturated vapour); 4 – butanol ($\varepsilon = 17$).

The minimum required angular rate of the mixer is different for various solvents. This fact can be attributed to different values of the rate constant of adsorption of the reaction of adsorbed oxygen with solvent (see Chapter 3). Thus, a weak dependence of the rate of establishing the stationary conductivity of the sensor on the rotation rate of magnetic mixer was found in the case where acetonitrile was used for detection of oxygen in the atmosphere of hydrogen. Consequently, using this solvent one can detect trace concentrations of oxygen in various buffer gasses using the above solvent at room temperature without mixing the solvent. This circumstance greatly simplifies the analysis. The results of experiments are shown in Table 4.3 and in Fig. 4.25. The data show reasonable reproducibility of the signals of variation of electric conductivity of the ZnO sensor from one measurement to another and for various concentrations of oxygen diluted in a solvent. Moreover, satisfactory reproducibility was obtained in experiments conducted several days or even months later without replacing the film or a solvent. This circumstance is clearly seen from the data shown in Fig. 4.25 and Table 4.3.

Note, however, that in the case under consideration the analysis of oxygen in the atmosphere of hydrogen was performed in a liquid in presence of saturated vapour, in which equilibrium volume concentration of oxygen was higher than that in the liquid phase, as predicted by the Bunsen coefficient. This fact makes the analysis of oxygen in a carrier gas from measurements of its concentration in a liquid phase less precise, compared to analysis made in a gas – vapour phase under the same conditions. Consequently, detection of oxygen in hydrogen, nitrogen, and other gases (especially, in experiments with acetonitrile and similar liquids) should be done in saturated acetonitrile vapours (see Chapter 3). The drawbacks of this method noted in Ref. [53] become less pronounced when using acetonitrile as a solvent, because in this case the results of measurements are practically insensitive to possible variation of the thickness of the liquid layer on the surface of the sensitive element (zinc oxide).

Table 4.3.

	Acetonitrile						
	$\Delta R / R_0$, %						
$C_{O_2} \cdot 10^5$, vol.%	exp. No.89	exp. No.90	exp. No.91	exp. No.92	exp. No.93	mean value	$\pm \Delta X$, %
3.9	7.3	5.8	6.6	5.9	–	6.4	11.0
11.7	15.4	12.6	13.4	12.9	–	13.6	10.0
23.4	22.5	18.6	19.8	20.7	21.7	20.7	10.0
46.8	31.2	26.3	28.2	32.0	30.1	29.6	10.0
93.6	42.1	36.0	38.5	43.2	42.0	40.4	8.7
187.2	55.3	48.0	51.2	63.2	56.7	54.9	14.0
374.0	83.4	77.0	79.2	78.5	75.3	78.7	5.1

In addition, when using semiconductor sensors for measuring concentration of oxygen in carrier gases, or in liquid media, one can also determine the relative contents of oxygen in a liquid and in saturated vapour over its surface (the Bunsen coefficient).

For this purpose, one should measure variation of electric conductivity of one and the same movable sensor in the saturated vapour-gas phase and in a liquid, caused by the presence of any given concentration of oxygen in a carrier gas (hydrogen, nitrogen, noble gas, etc.). From the results of these measurements the Bunsen coefficient β can be found in accordance with the relation (see Chapter 3, Section 4)

$$\beta = \left[\frac{(\sigma_0 / \sigma)_r^2 - 1}{(\sigma_0 / \sigma)_g^2 - 1} \right] , \qquad (4.20)$$

where σ_0 and σ are the values of equilibrium electric conductivity in absence and in presence of oxygen in the corresponding phases. If $(\sigma_0)_r \approx (\sigma_0)_g$ and $\sigma_0 \gg \sigma$ (which is usually the case), relation (4.20) reduces to

$$\beta = \left(\frac{\sigma_g}{\sigma_r} \right)^2 . \qquad (4.21)$$

The latter formula enables one to estimate the Bunsen coefficient using a semiconductor sensor.

It should be noted that in a vapour phase the liquid layer on the surface of a sensitive element of the sensor (zinc oxide) must be sufficiently thin, so that it would not produce any influence on the diffusion flux of oxygen through this layer. Possible lack of the film continuity (the presence of voids) does not prevent determination of concentration of oxygen in the bulk of the cell by the vapour – gas method. In this case, one deals with a "semi-dry" method. On the contrary, the presence of a thick liquid layer causes considerable errors in measuring σ, because of different distribution of oxygen in a system gas – liquid layer – semiconductor film (this distribution is close to that in the system semiconductor film – liquid), in addition to substantial slowing down of oxygen diffusion in such systems.

Note in conclusion, that similar problem can he solved at temperatures of the liquid lower than room temperature. At lower temperatures one may expect longer times of establishing stationary electric conductivity of the sensor, in the first place due to lower rate constant of the reaction between adsorbed oxygen and solvent. Consequently, at lower temperatures one should use a liquid with larger ε (see Section 3.4). Obviously, the larger is concentration of oxygen to be detected in a carrier gas, the less is the necessity of mixing the liquid medium. Similar statement can be made regarding the choice of dielectric constant of the liquid: the lower is the dielectric constant, the less important becomes mixing of the liquid, when detecting oxygen in a liquid medium (one of the examples is acetonitrile). In other words, with the decrease of ε, the effect of mixing ($\Delta\sigma = \sigma_w - \sigma_m$, where subscripts w and m stand for "without mixing", or mixing , respectively) reveals itself at lower concentrations of oxygen The effect of mixing is zero for any liquid containing no oxygen, or obtaining it in large concentration (in other words, in this case stationary value of variation of the semiconductor sensor conductivity is independent of mixing).

Maximum effect of mixing in different liquids is observed at different concentrations of diluted oxygen. For example, in methanol this concentration is equal to $2 \cdot 10^{-2}$ volume % (in other solvents this value may be different). These conecentrations are called the critical concentrations and are denoted as $N_{O_2}^{cr}$. At oxygen concentrations above the critical one, the effect of mixing as a function of oxygen contents in the liquid becomes less pronounced.

The effect of mixing $\Delta\sigma$ for various liquids with different values of the dielectric constant ε is seen from Fig. 4.26. At lower values of ε of the solvent, critical concentration $N_{O_2}^{cr}$ also decreases. Moreover, at substantially low values of ε, in addition to decrease of $N_{O_2}^{cr}$, the effect of mixing also decreases. Finally, for benzene, hexane, and some other nonpolar liquids, the effect of mixing is zero at any concentration of diluted oxygen.

Note that in weakly polar ($\varepsilon < 5$) and nonpolar solvents, it is impossible to blow off oxygen adsorbed on ZnO film with an inert gas hydrogen, nitrogen, etc.), similar to the case of gas or saturated vapour phase (polar liquid at any ε) at room temperature, i.e., under the conditions, where $\varepsilon = 1$, and no liquid layer is condensed on the film.

At room temperature, unsaturated vapours of the above specified polar and nonpolar liquids do not influence considerably the rate of adsorption and chemical activity of not only adsorbed oxygen layers, but also of acceptors of semiconductor electrons of another type, namely, of alkyl radicals [54]. This is seen from the electric conductivity of ZnO films with adsorbed alkyl radicals or oxygen being invariable in the atmosphere of the saturated vapours of the above specified solvents. In the case of oxygen, this can be also seen from the fact that the oxygen concentration features no decrease.

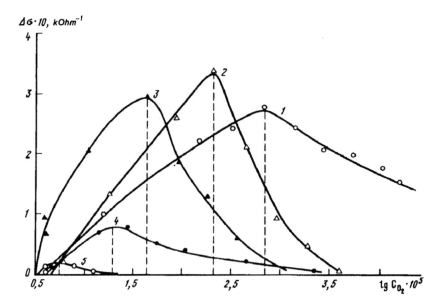

Fig. 4.26. The effect of a mixer for water and organic solvents at various concentrations of oxygen: $1 - H_2O$ $(\varepsilon = 78)$; $2 - CH_3OH$ $(\varepsilon = 36)$; $3 - C_4H_9OH$ $(\varepsilon = 17)$; $4 - C_6H_5Cl$ $(\varepsilon = 7)$; $5 - C_2H_5OOC_2H_5$ $(\varepsilon = 4)$; $6 - C_6H_{12}$ $(\varepsilon = 1,9)$; curve 6 coincides with abscissa.

However, the situation is completely different if saturated vapours of polar solvents from a liquid layer on the surface of a semiconductor film. In this case, variation of electric conductivity of ZnO films, as well as less pronounced variation of electric conductivity of TiO_2 films, is caused by adsorption of oxygen or free alkyl radicals $\dot{C}H_3$, $(C_2H_5)\dot{}$, $(C_3H_7)\dot{}$ is reversible even at room and lower temperatures, as seen from signals of the semiconductor sensor [54].

The new phenomenon discovered in these experiments consists in different chemical activity revealed by one and the same kind of adsorbed particles in contact with one and the same kind of molecules of the medium, but at different nature of the interface: either interface of a solid (ZnO film) with a polar liquid or interface of the solid with vapours of the polar liquid. This difference is caused by the fact that in the case of contact of the film with an adsorbed layer (oxygen, alkyl radicals) with a polar liquid, the solvated ion-radicals O_2^- chemically interact with molecules of the solvent (see Chapter 3, Section 3.4). In the case where alkyl radicals are adsorbed on ZnO film, one can assume, by analogy with the case of adsorbed oxygen, that in the process of adsorption on ZnO, simple alkyl radicals from metalloorganic complexes of the type

$CH_3^{-\delta} - Zn^{\bullet+\delta}$ with superstoichiometric (defect) zinc atoms ($Zn^{\bullet}$-impurity centres of conductivity). The larger is the electric positivity of the metal in these complexes, the larger is the ionicity of the carbon-metal bond, carbon being at the negative end of the dipole. Thus, in the case of C – K bond, ionicity amounts to 51%, whereas for C – Mg and C – Zn bonds ionicity amounts to 35% and 18%, respectively [55]. Consequently, metalloorganic compounds are characterized by only partially covalent metal-carbon bonds (except for mercury compounds).

Metallo-organic compounds possess high reactivity to water, oxygen and nearly all organic solvents, except hydrocarbons and ethers. Chemical properties of the suggested surface complexes of the type $(Zn^{\bullet+\delta} - M^{-\delta})_s$ in presence of various solvents are unknown. However, by analogy with the oxygen complexes of the type $(Zn^{\bullet+\delta} - O_2^{-\delta})_s$, one may expect considerable reactivity of these complexes in solvents possessing different polarity due to solvation of the complex components. In this case, short-lived solvated surface carboanions may emerge.

In order to verify the properties of alkyl radicals adsorbed on semiconductor ZnO film, we conducted experiments, in which methyl radicals were adsorbed from the vapour phase of acetone at room temperature. Under these conditions, the radicals were formed via photolysis of acetone vapours (for details, see Chapter 3) at small pressure of acetone (of about 0.5 Torr).

Electric conductivity of the ZnO film in the process of adsorption is known do decrease substantially down to some stationary value determined by processes of adsorption and subsequent recombination of adsorbed active particles with each other and with free radicals approaching the surface (see Fig. 4.27, curve 3).

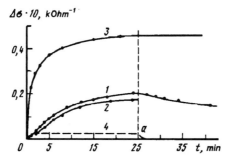

Fig. 4.27. Variation of electric conductivity of ZnO film under the influence of adsorption of CH_3 radicals at room temperature for various pressures of acetone vapours: *1, 2* – 200 Torr; *3* – 1 Torr; *1, 2* – before and after immersion of the film in liquid acetone; *4* – the film covered with a liquid layer.

After that, in the same cell the movable ZnO film (*in situ*) was immersed for some time in liquid acetone. Then, the film was taken out of the liquid and dried. Radicals were adsorbed from a gas phase, other conditions being the same. In the second experiment, we obtained the values of variation of electric conductivity of the ZnO film, caused by adsorption of alkyl radicals, close to those obtained in the first experiment.

Using the above technique, the experiments could be repeated unlimited number of times, whereas in the atmosphere of acetone under low pressure (0.1 – 0.5 Torr) the experiment could not be repeated without immersing the ZnO film with adsorbed radicals into liquid acetone, because the layer adsorbed in the first experiment remained intact. The latter experiment with immersing the ZnO film into liquid can be conducted in a somewhat different way. It is sufficient to gradually increase the pressure of acetone vapours from 0.5 – 1 Torr to 200 Torr (the pressure of saturated acetone vapours at room temperature) alter deposition of radicals on zinc oxide film. Immediately following the formation of a liquid layer of acetone on the surface of the oxide semiconductor, its electric conductivity rapidly attains its initial value, existed before adsorption of radicals (see Fig. 4.27, curve *1* to the right of dashed line). After that, the experiment could be repeated, and the obtained results were close to those obtained in the first experiment (see curve *2*). Note that at low pressure of acetone vapours, no variation of electric conductivity were observed during a long period of time (curve *3*) after generation of radicals by light was terminated (to the right of dashed line).

The results obtained in above experiments confirm the removal of chemisorbed particles in the process of immersion of the film with preliminary chemisorbed radicals in a liquid acetone. Note that at low pressures of acetone, the CH_3-radicals absorbed on ZnO film could be removed only by heating the film to the temperature of 200 – 250°C. Moreover, if the film with adsorbed radicals is immersed in a nonpolar liquid (hexane, benzene, dioxane), or vapours of such a liquid are condensed on the surface of the film, then the effect of removal of chemisorbed radicals does not take place, as is seen from the absence of variation of electric conductivity of the ZnO film after it is immersed in liquid and methyl radicals are adsorbed anew onto its surface. We explain the null effect in this case by suggesting that the radicals adsorbed on the surface of the ZnO film in the first experiment remained intact after immersion in a nonpolar liquid and blocked all surface activity of the adsorbent (zinc oxide).

The above results on detection of trace concentrations of oxygen by sine oxide films (and titanium oxide films, to a lesser degree), as well as the results on detection of alkyl radicals, which are acceptors of semiconductor electrons, show that the behaviour and electric properties of these particles in various media (gas, vapour phase, condensed medium) are very much alike as functions of the dielectric constant of the medium. The same parameters considered as functions of chemical properties of the medium reveal lesser similarity.

Moreover, quantitative relations of stationary concentrations of these particles with variation of the electric conductivity of a semiconductor sensor (based on zinc oxide) caused by their adsorption are also similar.

Thus, it is quite natural to consider the properties of other acceptor particles, for example, atoms of nitrogen, aminoradicals, hydroxyl radicals, and many others, adsorbed on oxide semiconductors. However, the properties of these particles are not studied yet. As to adsorbed donor particles, it was found in our experiments that liquid media with different values of the dielectric constant do not have any influence on the properties of adsorbed atoms of hydrogen.

4.12. Application of Semiconductor Sensors in Investigation of Radiation and Plasma Chemical Processes

Active particles, free atoms, radicals, ions, electrons, and excited particles produced in the field of high-energy radiation or in electric discharge interact with various chemical compounds. This interaction is accompanied by a variety of homogeneous and heterogeneous physical-chemical processes.

Mechanisms of these processes and their specific features are usually studied by analysis of final products of reactions. This method is especially extensively used, when studying radiolysis in condensed media at normal temperatures. Under these conditions, physical-chemical methods of detecting intermediate particles, such as mass-spectrometry, optical spectroscopy, etc., are not applicable because of their low sensitivity. From our point of view, kinetic measurements are of a particular interest in such studies. These measurements provide data concerning temporal evolution of these processes. For example, they allow one to keep track of production of free radicals and atoms at various moments of time corresponding to different stages of irradiation of gaseous and liquid hydrocarbons, and other compounds. We note also that, under the specified conditions, the active particles are produced in extremely small amounts. Thus, the methods used for detecting and measuring concentration of these particles should provide the required data without additional manipulations, such as accumulation of radicals by freezing. The main requirements imposed on this technique are high sensitivity (it should be capable of detecting concentrations not higher than $10^8 - 10^{10}$

particles per cm^3) and a short response time. Moreover, it should be easy to use and provide an opportunity to carry out measurement in any point of the medium under investigation. The SCS technique seems to be best suited for these purposes.

Now, we consider some examples of physical-chemical experiments carried out in this field and show applicability of sensors to investigation of gaseous and condensed media. First, we consider the case of a low-pressure gaseous media. In experiments [56], semiconductor sensors were used for detecting and measuring concentration of nitrogen atoms in active nitrogen. In the discharge region, active nitrogen consists of nitrogen atoms, ions, and excited atoms and molecules. Various physical-chemical methods are used to detect and analyze each type of articles. However, sensitivity of these methods should be rather high, permitting detection of $10^{14} - 10^{15}$ atoms per cubic centimetre. In [57] we showed applicability of zinc oxide SCS for measuring small concentrations of nitrogen ($10^8 - 10^{10}$ atoms per cm^3) produced, for example, in pyrolysis of molecular nitrogen on incandescent tungsten filament. Due to its high sensitivity, the SCS technique allows one to study not only intense discharges (glow, high-frequency, etc.) frequently used for production of nitrogen atoms, but also low-voltage discharge with an incandescent cathode, characterized by lower power as well. The advantage of this type of discharge is that it allows continuous variation of concentration of produced nitrogen atoms in a wide range, by changing the discharge parameters. These experiments were performed in the reaction cell described in [58]. A discharge device was mounted in the lower section of the vessel. It consisted of a cathode, grid, and collector of ions. To prevent charged particles reaching the sensor, electric and magnetic filters were placed between the discharge zone and the semiconductor sensor. An important result of this experiment is identification of significant decrease in electric conductivity σ of the zinc oxide film used as a sensor in a discharge produced in nitrogen at the pressure of $0.01 - 0.5$ Torr. This decrease depends on the discharge intensity and indicates production of nitrogen atoms under the specified conditions [58]. Further experiments confirmed that sensors provided an opportunity of detecting nitrogen atoms. Moreover, this technique allowed comparison of the concentration of nitrogen atoms with the magnitude of the ionic current. This comparison revealed similarity of the dependencies of these quantities on different parameters of the discharge and the experimental conditions.

Figure 4.28 shows these quantities as functions of the grid potential. It is seen that both the ion current and concentration of hydrogen atoms exhibit significant increase at the same threshold value of the grid potential. We note that s-like behaviour of the ionic current as a function of the grid potential is typical of low-voltage discharge with incandes-

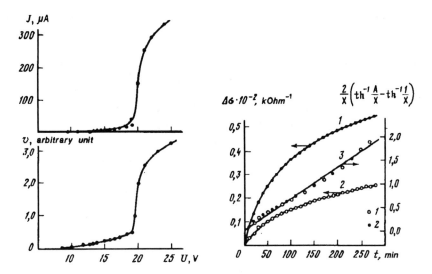

Fig. 4.28. Ionic current J and concentration of nitrogen atoms [N] as functions of the grid potential of the discharge device (the source of nitrogen atoms).

Fig. 4.29. γ- effect in butane at room temperature (20°C) under different pressures: P_{but} = 572 Torr (1), 250 Torr (2). Curve 3 displays the results of calculation based on equation (4.26) under the given experimental conditions (see curves 1 and 2).

cent cathode. These results indicate that, in a wide range of the discharge intensities, the following expression holds

$$v = \left(\frac{d\sigma}{dt}\right)_{t \approx 0} = KJ \ , \qquad (4.22)$$

where K the coefficient of proportionality, J is the current, v is the rate of initial variation of electric conductivity of the film, proportional to the steady-state concentration of nitrogen atoms [58]. Consequently, in discharge in nitrogen, the steady-state concentration of nitrogen atoms [N] is a linear function of the ionic current, i.e.,

$$[N] = \text{const} \cdot J \ , \qquad (4.23)$$

where const is the coefficient of proportionality independent of the discharge characteristics.

Note that similar dependence was observed in analogous experiments for hydrogen atoms produced in the discharge [59]. In this work the dependence determined by the SCS technique was compared with the results obtained by the Wrede diffusion method and with the data on microcalorimetry. The results obtained by these classical methods confirm the data of the SCS technique (see Fig. 4.2). Experiments [60] demonstrate applicability of the SCS method for investigation of radiolysis of vapours and condensed phase of organic components of aliphatic and olefine types. These experiments dealt with butane and isobutylene in gas phase and with liquefied hydrocarbons. Investigations were performed in sealed cells. Zinc oxide film sensors were placed inside the cells. A cobalt source of γ- radiation was mounted in a specially equipped room aided by automatic control devices designed to insert the cells in the "bottle-neck" of the source and to remove them from the source. The power of the source amounted to $\approx 10^{-2}$ roentgen per hour.

Figure 4.29 illustrates the increase in electric conductivity of a semiconductor sensor placed inside the cell in the process of irradiation of butane. This figure also displays the kinetics of electric conductivity of the sensor calculated according to the equation governing the presumed scheme of this complex process. The main intermediate stage of this process consists in production of hydrogen atoms detached from hydrocarbon molecules. Under irradiation of hydrocarbon in the presence of zinc oxide film adsorbent, the specified process mainly occurs in the adsorbed layer. This behaviour can be attributed to both secondary (X-ray, etc.) emission and the fact that the density of the matter in the adsorbed layer is essentially higher than that in the gas phase. This process is governed by the reaction

$$M_v \rightarrow M_s \xrightarrow[\gamma]{} \dot{R}_s + H_s \rightarrow \dot{R}_s + H_s^+ + e_s \ ,$$

where M_v and M_s are hydrocarbon molecules in the gas phase and in the adsorption layer, respectively, $\dot{R}_s$, H_s and H_s^+ are the radicals, atoms, and ions of hydrogen in the adsorption layer on zinc oxide, e_s is the conductivity electron in the semiconductor. The activation energies of chemisorption of atoms H_s and radicals $\dot{R}_s$ on zinc oxide are equal to 2 and 8 kcal/mol, respectively. At moderate temperatures, this leads to prevailing chemisorption of hydrogen atoms on zinc oxide, which is responsible for the increase of electric conductivity of zinc oxide sensor, because under these conditions acceptor properties of large radicals are rather weak. The processes competitive with adsorption of hydrogen atoms in the adlayer result in a decrease of concentration of H_s atoms. These processes are recombination of atoms and interaction of atoms with hydrocarbon molecules in the adsorbed layer. The specified reactions can be written as

$$H_s + H_s \rightarrow (H_2)_s \quad \text{and} \quad H_s + M_s \rightarrow \dot{R}_s^1 + H\dot{R} \ .$$

Another competitive process is associated with the cell effect in the adsorption layer

$$H_s + \dot{R}_s \rightarrow M_s \ .$$

The efficiency of this reaction is not high because of high total mobility of H_s particles, as compared with that of $\dot{R}_s$. At the same time, for large radicals the cell effect is significant due to low mobility of these radicals. Production of H_s atoms in the adsorption layer under the influence of γ radiation proceeds practically without activation energy. Along with fast competitive reactions induced by radiation, production of hydrogen atoms results in rapid establishment of steady-state concentration of H_s particles at the oxide surface. As a consequence, at the initial stage of irradiation, the rate of variation of electric conductivity becomes constant

$$v = \left(\frac{d\sigma}{dt}\right) = \alpha[H_s](1 - \theta) \ . \tag{4.24}$$

This variation of electric conductivity is caused by the reaction $H_s \rightarrow H_s^+ + e_s$, where α is the coefficient of proportionality, θ is the relative surface coverage with H_s^+ ions. The variation rate is constant as long as the adsorbent surface coverage with chemisorbed hydrogen atoms (the value $[H_s^+]$) is small. At large values of $[H_s^+]$ the rate v rapidly decreases and approaches zero. Note that in these processes the value q remains small for quite a long time. This is associated with the fact that, as the rate of chemisorption of H_s particles increases, the rate of recombination via reaction of H_s^+ ions with H_s atoms increases in direct proportion to $[H_s^+]^2$. At the same time, the rate of recombination due to reaction of H_s^+ ions with adsorbed hydrocarbon molecules increases proportionally to $[H_s^+]$. Note, however, that the rate of these processes is considerable only for high-temperature radiolysis.

From the above consideration we suppose that the increase of electric conductivity σ of the zinc oxide film is proportional to $[H_s]$, as is the case when hydrogen atoms produced by pyrolysis or discharge in the gas phase are adsorbed on the zinc oxide film. Reverse (competitive) changes of electric conductivity are proportional to the σ value of the semiconductor film and to $[H_s^+]$. Thus, taking into account both chemisorption of hydrogen atoms and recombination of H_s^+ particles, we can write the kinetic equation governing this process as

$$\frac{d\sigma}{dt} = K_1[H_s] - K_2[H_s^+][e_s^-] \; , \tag{4.25}$$

where K_1 and K_2 are the temperature dependent kinetic constants, $[e_s^-]$ is the concentration of conduction electrons. Note that, in the process of radiolysis, the value $[H_s]$ is constant.

Now, taking into account the conditions $[H_s] = \text{const}$ and $\sigma = \sigma_0$ at $t = 0$, we can transform and integrate equation (4.25). Thus, in the case $A < x$, we find

$$2x\left[\text{th}^{-1}\left(\frac{A}{x}\right) - \text{th}^{-1}\left(\frac{1}{x}\right) \right] = K_2' t \; , \tag{4.26}$$

where $A = (2\Delta\sigma / \sigma_0 + 1)$; $x = \sqrt{1 + 4K}$; $K = K_1' / K_2'$; $\Delta\sigma = \sigma - \sigma_0$; $K_1' = K_1 / [e_s^-]_0$, $K_2' = K_2 / [e_s^-]_0$; σ_0 and σ ($\sigma = [e_s^-] \cdot q\mu$) are electric conductivities of the semiconductor film before and after radiation, q and μ are the charge and mobility of carriers, and t is time.

The value of K_1', characteristic of the rate of the direct process $H_s \rightarrow H_s^+ + e_s^-$ depends on the efficiency of adsorption of H_s particles and the radiation intensity. The K_2', value characteristic of the reverse process depends only on the temperature. To verify the validity of expression (4.26), we calculated the values using the experimental data and plotted graphs. The slope of the resulting straight lines determined the K_2' values. The K_1' values were calculated from K and K_2'.

Figure 4.29 demonstrates validity of equation (4.26) in the case of butane at various pressures. It is seen that the dots representing two experimental series fit the same straight line. This means (according to the equation) that K_2' values are equal and independent of the pressure of butane. In liquefied butane, the K_1' value increases significantly. This indicates that adsorption involves not only hydrogen atoms produced in the adsorption layer, but those produced in distant layers of the liquid adjacent to the surface.

Similarly, validity of equation (4.26) is verified for γ- irradiated isobutylene. However, in this case the K_1' value is a weak function of the pressure and aggregate state. This result may be associated with the fact that in all experiments with isobutylene the adsorbent surface was covered with at least a single layer of isobutylene. Apparently, in this case, this layer plays a major part in delivering active particles to the surface. This is particularly true for hydrogen atoms, because olefines are active receptors for them.

Along with atom particles and radicals, ions and electrons play an important role in radiation and plasma chemical processes. Ions and electrons are being produced and interact actively with irradiated matter both in gases and, especially, at the surfaces of solids (vessel walls, adsorbents, etc.).

Studying interaction of charged particles with the adsorption layer requires knowledge of the data concerning elementary interaction of ions with the surface of a solid body. In earlier experiments [61], we studied interaction of a beam of H_2^+ ions with the energy 10 – 50 eV with thin baked polycrystalline zinc oxide films. This films simultaneously served as sensors of active particles (for example, hydrogen atoms) produced at the surface. Design of the source of ions provides beams of H_2^+ ions with the intensity up to $8 \cdot 10^{11}$ ion/cm^2·s and the energy from 10 to 200 eV. The electron energy was equal to 20 eV. This ensured the absence of H^+ ions in the beam. The experiments were performed at the film temperature equal to 150°C and the pressure of hydrogen in the cell equal to 10^{-4} Torr. As hydrogen was admitted into the vessel with heated filament cathode, a slight increase in electric conductivity of the ZnO film was observed. This was associated with adsorption of hydrogen atoms on the film. The hydrogen atoms were produced in pyrolysis of H_2 on the incandescent filament. An amount of hydrogen atoms adsorbed on the ZnO film can be measured by the increase of electric conductivity of the film. Estimates show that, under the specified conditions, this value amounts to $7 \cdot 10^{13}$ cm^{-2}. Thus, the relative surface coverage with hydrogen atoms does not exceed several percent. Next, the source of ions was switched on. Upon bombardment of the film surface by the beam of H_2^+ ions, we observed an increase in electric conductivity of the zinc oxide film for all energies of the ion beam (10, 20, 30, 40, and 50 eV). Therewith, the higher was the energy, the greater was the change of electric conductivity $\Delta\sigma$ in time. After the switching off the beam the electric conductivity of the film does not completely restore to its initial (before bombardment) value. In other words, bombardment of the ZnO film with the adsorbed layer of H_2 with H_2^+ ions results in partially irreversible increase in electric conductivity. This change is completely removed by heating the film in vacuum at 350°C.

Note that, as the temperature of the film increases, the change of σ induced by bombardment of the film surface becomes a weak function of the ion beam energy. The experimental results show that the effect can be due to two reasons. First, bombardment of the adsorbed layer of H_2 with H_2^+ ions induces a partial dissociation of molecules resulting in formation of adsorbed hydrogen atoms. The effect observed may be caused by chemisorption of these atoms. The second possible mechanisms consists in dissociation of H_2^+ ions caused by a collision of these ions with the film surface. This results in production of the hydrogen atoms in the adsorbed layer. Then, the process follows the same pattern as in the first case.

Investigation of interaction of electrons of different energies with a solid material in plasma processes may be even more intriguing and important, especially in the case of an adsorbed layer of materials contained in the reaction vessel. Provided thin semiconductor films deposited on the walls of the reaction vessel are used as solid targets, these films can be simultaneously used as targets and semiconductor sensors. This is also the case when such films are deposited on the specially manufactured quartz plates with electrodes accessible from the outside of the vessel. These sensors can be placed in any point of the vessel.

One of intriguing problems can be solved by using semiconductor sensors, for example, zinc oxide films (as well as TiO_2, CdO, and others) deposited on a quartz (or glass ceramics) substrate, and an electron beam. This problem is associated with investigation of adsorption mechanism of different molecular components on metal oxides. In particular, such data is required for understanding mechanisms of various catalytic reactions proceeding through chemisorption of reagents. Note that chemisorption is an obligatory initial stage of every heterogeneous process. In experiments [62], this problem was solved for molecular hydrogen adsorbed on zinc oxide films at moderate, as well as high, temperatures.

For this purpose, the authors used a special vacuum cell with a controlled focused electron beam incident on a zinc oxide film target. In these experiments, the role of the film was twofold. It served as an adsorbent and as a high-sensitivity detector of hydrogen atoms (10^6 at/cm^3). Hydrogein atoms were produced due to surface dissociation of adsorbed molecular hydrogen. This process was induced by heating or bombardment of the adsorbed layer by an electron beam.

Experimental technique was as follows. Vacuum system was evacuated to the pressure of 10^{-8} Torr. At room temperature electric conductivity σ of the zinc oxide film did not change upon bombardment of the film by electrons with energies up to 80 eV (Fig. 4.30, curve 1). Next, hydrogen was adsorbed on zinc oxide at room temperature, with the electron beam being switched off. After admission of hydrogen and exposure of the film to hydrogen during 5, 30, and 120 min, at the same pressure of hydrogen the system was reevacuated to the pressure of

10^{-8} Torr. Electric conductivity of the film σ did not change during these procedures. Experiments showed that, under the influence of electrons with the energy of 20 eV, electric conductivity σ of the film with adsorbed hydrogen increased. Note that the action of electron beam did not change the temperature of the surface of the zinc oxide film, i.e. additional thermal chemisorption of hydrogen molecules did not occur.

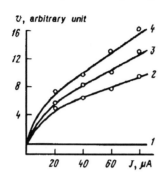

Fig. 4.30. Variation rate of the electric conductivity of a ZnO film as a function of the intensity of the electron beam bombarding the film, for different times of preliminary exposure of the film to molecular hydrogen at room temperature: (1) – ZnO surface free of hydrogen, (2) – the exposure time is 5 min, (3) – the exposure time is 30 min, (4) – the exposure time is 120 min.

Figure 4.30 displays the experimental results. It is seen that, as the time of interaction of hydrogen and zinc oxide was increased, other conditions being the same, the concentration of adsorbed hydrogen atoms increased. These atoms are produced under the influence of bombardment of the adsorption layer at the surface of zinc oxide with electrons. Based on these data, we conclude that, since adsorption of hydrogen on zinc oxide at room temperature does not change its electric conductivity σ, adsorption of hydrogen molecules is not accompanied by their considerable dissociation into atoms at this temperature. In other words, in this case we deal with associative adsorption of hydrogen. This contradicts classical results of Eischens and Pliskin, who showed by IR-spectroscopic measurements that adsorption of hydrogen on zinc oxide under these conditions leads to dissociation of hydrogen [63].

Experiments with electron beam show that, if molecular hydrogen is adsorbed at temperatures higher than 200°C, bombardment of the adsorbed layer does not lead to a change in electric conductivity of the film, with other conditions being the same. From our point of view, the above data indicate that, under these conditions ($T_{ZnO} \gg 200$°C), adsorbed hydrogen completely dissociates, contrary to the case of room temperature.

To account for the contradiction between the data obtained at room temperature in experiments [63] and our results, we assume that, at low temperatures of adsorbent, hydrogen atoms produced in the dissociative adsorption of H_2 molecules differ from those produced in adsorption of free hydrogen atoms proceeding directly from gas phase (for small covered areas of the surface). The difference is that in the inner case interaction of adsorbed atoms is absolutely impossible, because of large distances between them. (We note that mobility of atoms chemisorbed on oxides at room temperatures is low, because the chemisorption heat is about 30 kcal/mol). Consequently, these atoms can be chemisorbed as charged particles, thus increasing electric conductivity σ of semiconductor oxide. This effect is observed in experiments. At the same time, chemisorption of H_2 molecules at room temperatures (as well as at lower temperatures) does not change electric conductivity σ of zinc oxide. Thus, H_s atoms produced in this case possess different properties and exhibit no capability for chemisorption in a charged form. This may be due to the well known "flat cell effect" in the adsorption layer. This phenomenon implies that these atoms remain "coupled" with each other. Therefore, even after dissociative chemisorption of H_2 molecules, these atoms cannot be ionized yielding two adjacent and equally charged particles, because this requires a significantly higher activation energy than that corresponding to chemisorption of charged particles at large distance.

Note that, unfortunately, the authors of [63] did not study adsorption of hydrogen atoms from gas phase by IR spectroscopy. However, this investigation was carried out in [64]. The results obtained in these experiments confirm the above notion of the process.

Now, we consider H_s atoms produced from hydrogen molecules adsorbed on zinc oxide under the influence of electron (ion) impact. We suppose that in this case the energy released in interaction of an electron (ion) with an adsorbed molecule is enough to break any bond between hydrogen atoms. As a consequence, H_s atoms bounce apart over the surface. Hydrogen atoms produced in this case are similar to H_s atoms adsorbed on the oxide surface from the gas phase at small surface coverages. In other words, they can be chemisorbed as charged particles and thus may influence electric conductivity of zinc oxide. This conclusion is consistent with the experimental results.

Excited particles (molecules, atoms, and ions) also play an important role in plasma and radiation chemical reactions. These particles interact actively with components of a gas (liquid) phase and with the adsorbed layer. These processes are discussed in detail in the next Chapter.

References

1   M.V. Smith, J. Chem. Phys., 2 (1943) 110.
2   V.I. Tsivenko and I.A. Myasnikov, Zh. Fiz. Khimii, 45 (1971) 1814.
3   V.I. Tsivenko and I.A. Myasnikov, Zh. Fiz. Khimi, 45 (1971) 2609.
4   l.N. Pospelova and l.A. Myasnikov, Kinetika i Kataliz, 7 (1966) 190.
5   G.V. Malinova, Investigation of Adsorption of Oxygen Atoms on Semiconductor Oxides of Metals, Doctorate thesis (Chemistry), Moscow, 1971.
6   V.I. Tsivenko and I.A. Myasnikov, Kinetika i Kataliz, 11 (1970) 267.
7   N.V. Ryl'tsev, E.A. Gutman, and l.A. Myasnikov, Interaction of Atomic Particles with Solids, Khar'kov State University, Khar'kov:1976, (in Russian).
8   I.A. Myasnikov and I.N. Pospelova, Zh. Fiz. Khimii, 41 (1967) 1028.
9   D. Brennan, Catalysis: Physics and Chemistry of Heterogeneous Catalysis, Mir Publ., Moscow, 1967 (translation into Russian).
10 V.I. Vedeneev, L.V. Gurvich, V.N. Kondratyev, et al., Handbook of Energies of Breaking of Chemical Bonds. Ionization Potential and Electron Attachment Energies, Kondratyev,V.N. (ed.), USSR Acad. of Sci. Publ., Moscow, 1962 (in Russian).
11 V.I. Tsivenko and I.A. Myasnikov, Zh. Fiz. Khimii, 39 (1965) 2376.
12 I.A. Myasnikov and E.V. Bol'shun, Doklady Akad. Nauk SSSR, 135 (1960) 1164.
13 V.I. Tsivenko and I.A. Myasnikov, Zh. Fiz. Khimii, 41 (1967) 43.
14 V.I. Tsiveiko and I.A. Myasnikov, Khimiya Vysokikh Energii, 7 (1972) 377.
15 G.V. Malinova and I.A. Myasnikov, Kinetika i Kataliz, 10 (1969) 328.
16 G.V. Malinova and I.A. Myasnikov, Kinetika i Kataliz, 10 (1969) 336.
17 P.J. Warneck, Chem. Phys., 41 (1964) 3435.
18 I.A. Myasnikov, Chemisorption and Photocatalytic Phenomena in Heterogeneous Systems, Novosibirsk, 1974 (in Russian).
19 I.A. Myasnikov, Electron Phenomena in Adsorption and Catalysis on Semicoductors, F.F. Volkenstein, (ed.), Mir Publ., Moscow, 1969 (in Russian).
20 I.A. Myasnikov, Doklady Akad. Nauk SSSR, 155 (1964)1407.

278

21 N.V. Ryl'tsev, E.E. Gutman, and I.A. Myasnikov, Zh. Fiz. Khimii, 52 (1978) 1769.
22 O.V. Krylov, Nonmetal Catalysis, Khimiya Publ., Moscow, 1967 (in Russian).
23 I.A. Myastiikov, Zh. Fiz. Khimii, 33 (1959) 2564.
24 I.A. Myasnikov and S.Ya. Pshezhetskii, Problems of Kinetics and Catalysis, Publ. of USSR Acad. of Sci., Moscow, 8 (1955) 175.
25 I.A. Myasnikov and E.V. Bol'shun, Methods of investigation of Catalysts and Catalytic Reactions, Publ. of Sib. Div. of the USSR Acad. of Sci., Novosibirsk, v.1, 1965,(in Russian).
26 S.A. Zavyalov, I.A. Myasnikov, and E.E. Gutman, Doklady Akad. Nauk SSSR, 236 (1977) 375.
27 I.A. Myasnikov, Zh. Vses. Khim. Obschehestva im. D.I. Mendeleeva, 20 (1975) 19.
28 O.V. Krylov, Doklady Akad. Nauk SSSR, 130 (1960) 1063.
29 E.V. Bol'shun and I.A. Myasnikov, Zh. Fiz. Khimii, 47 (1973) 878.
30 I.A. Myasnikov and E.V. Bol'shun, Kinetika i Kataliz, 8 (1967) 182.
31 M.I. Danchevskaya, O.P. Panasy'uk, and N.I. Kobozev, Zh. Fiz. Khimii, 42 (1968) 2843.
32 I.A. Myasnkov, Zh. Fiz. Khimii, 62 (1988) 2770.
33 L.Yu. Kupriyanov and I.A. Myasnikov, Zh. Fiz. Khimii, 63 (1989) 1960.
34 Ch. Kittel, Introduction to Physics of Solids, Nauka Publ., Moscow, 1978, (translation into Russian).
35 P.A. Sermon and G.C. Bond, Catal. Rev., 8 (1973) 211.
36 N.E. Lobashina, N.N. Savvin, and I.A. Myasnkov, Doklady Akad. Nauk SSSR, 268 (1983) 1434.
36 N.E. Lobashina, N.N. Savvin, and I.A. Myasnkov, Kinetika i Kataliz, 24 (1983) 747.
39 R. Kramer and M. Nadre, J. Catal., 68 (1981) 411.
40 A.Yu. Graifer and I.A. Myasnkov, Doklady Akad. Nauk SSSR, 291 (1986) 604.
41 V.Ya. Sukharev and I.A. Myasnikov, Kinetika i Kataliz, 24 (1985) 51.
42 I.A. Myasnikov, E.V. Bol'shun, and V.S. Raida, Zh. Fiz. Khimii, 43 (1973) 2349.
43 I.V. Miloserdov and I.A. Myasnikov, Doklady Akad. Nauk SSSR, 224 (1975) 1352.
44 K. Miz, Theory of Photographic Processes, Gosizdat Publ., Moscow, 1949 (in Russian).
45 I.A. Myasnikov and E.V. Bol'shun, Zh. Fiz. Khimii, 63 (1979) 1619.

46 A. Searcy, K. Freeman, and M.C. Michel, J. Amer. Chem. Soc., 76 (1954) 4050.

47 G.M. Martynovich, Vestnik Moskovskogo Universiteta, Seriya Matem., Mekhanika, Khimiya, N2 (1958) 151; Ibid, N5, 67.

48 A.I. Lifshitz, E.E. Gutman, and I.A Myasnikov, Zh. Fiz. Khimii, 53 (1979) 775.

49 A.I. Lifshitz, I.A. Myasnikov, and E.E. Gutman, Zh. Fiz. Khimii, 59 (1978) 2953.

50 N.N. Savvin, E.E.Gutman, and I.A. Myasnikov, Zh. Fiz. Khirnii, 48 (1974) 2107.

51 A.I. Litthitz, E.E. Gutman, and I.A. Myasnikov, Kosmich. Issled., 19 (1981) 415.

52 K. Fukuyama, J. Atoms and Terr. Phys., 36 (1974) 1297.

53 I.A. Myasnikov, Zh. Fiz. Khimii, 55 (1981) 2053.

54 I.A. Myasnikov, Zh. Fiz. Khimii, 55 (1981) 1283.

55 J. Roberts and M. Casserio, Fundamentals of Organic Chemistry, A.N. Nesmeyanov (ed.), Mir Publ., Moscow, 1968 (in Rusian).

56 V.I. Tsivenko and I.A. Myasnikov, Khimiya Vysokikh Energii, 7 (1972) 180.

57 V.I. Tsivenko and I.A. Myasnikov, Kinetika i Kataliz, 11 (1970) 267.

58 V.I. Tsivenko, Investigation of Chemisorption of Nitrogen Atoms on Semiconductor Oxides of Metals, Doctorate thesis (Chemistry), Moscow, 1971.

59 I.N. Pospelova and I.A. Myasnikov, Kinetika i Katatiz, 7 (1966) 196.

60 I.A. Myasnikov, Doklady Akad Nauk SSSR, 155 (1964) 1407.

61 A.P. Kascheev, A.M. Panesh, and I.A. Myasnikov, Zh. Fiz. Khimii, 49 (1975) 1022.

62 A.M. Panesh and I.A. Myasnikov, Problems of Kinetics and Catalysis, Nauka Publ., Moscow, 14 (1970) 172.

63 R.P. Eischens, V.A. Pliskin, and T.D. Low, J. Catal., 1 (1962) 180.

64 E.E. Gutman, A.I. Lifshitz, I.A. Myasnikov, and N.N. Savvin, Zh. Fiz. Khimii, 55 (1981) 995.

Semiconductor Sensors in Physico-Chemical Studies
L.Yu. Kupriyanov (Editor)

280

Chapter 5.

Interaction electron-excited particles of gaseous phase with solid surface

5.1. Electron-Excited State of Atoms and Molecules

Development of the present-day technology brings to the fore the study of nonequilibrium and nonstationary phenomena observed in the physical chemistry of heterogeneous systems. The problem of material and energy exchange between an excited gaseous phase and a solid surface is of importance from this standpoint. Determination of the nature of these processes plays a crucial role in understanding of such essential application phenomena as effects of beams of charged particles and laser radiation on the gas – solid interface, plasma – wall interaction, radiation- and photo-induced chemical reactions in heterogeneous systems, heterogeneous-homogeneous processes, etc.

An excited gaseous phase, as a rule, contains a whole "spectrum" of active particles. Examples are provided by atoms, radicals, electronically, vibrationally and rotationally excited molecules, or their fragments. The particular nature of excitation of particles and their concentration dependend on the chemical composition of the gas, the method of its excitation, and the energy delivery rate into the gaseous phase. Therefore, we examine a very complex system comprised by species that are frequently interacting with one another. In order to obtain a good understanding of the mechanism accounting for the energy exchange between an excited gaseous phase and a solid surface, one should know behaviour of each type of active particles at the interface. The previous chapter addressed the interaction of atoms and radicals with the surface of oxides and application of sensors to study developing processes. The present chapter is dedicated to the study of also important constituent of gases, i.e. electronically excited atoms and molecules. We shall use particular examples in attempting to show how semiconductor sensors can give an insight into heterogeneous processes involving electron-excited particles, and describe the probable mechanisms of changes in the structural and electron properties of solids during interaction with an excited gaseous phase.

A distinguishing feature of electronically excited atoms and molecules is that they have one or a few excited orbitals of an electron. The principal properties of these particles are represented by a high internal energy potential localized on the excited orbitals and the structure of electron shell essentially different from the electron ground state.

The lifetime of electron-excited particles in a gaseous phase is dependent on their radiative lifetime determined by the probability of spontaneous photon emission with particle transition to the ground state, and also by processes of nonradiative de-exciting due to collisions in the gaseous phase and with the walls. The excited states, for which transitions to the ground state through emitting a dipole photon are feasible, are termed resonance-excited. The resonance-excited states are highly efficiently formed either when the gaseous phase is light-excited, or when a current flows in the gas. However, these states rapidly decay due to photon emission. The lifetime of these states lasts from 10^{-7} to 10^{-8} s [1]. It is easy to compute that during its life an electron-excited particle will fly due to its thermal velocity some distance of 10^{-2} to 10^{-4} cm, i.e. at a gas pressure less than 1 Torr, the particle will be most likely radiation-deactivated without undergoing any collision with other molecules and walls. This circumstance allows us not to consider herein the phenomena the resonance-excited atoms and molecules participate in. Participation of these particles in homogeneous reactions is considered in detail by a series of monographs devoted to photochemistry [1 – 4].

Electron-excited states that are spin-prohibited to return into their ground state with dipole photon emission are termed metastable. These are, for example, $Is2s^1S$- and ^3S-states of He, $np^5(n+1)S^13P_{0,2}$-states of He, Ar, Kr and Xe (n = 2, 3, 4, and 5 respectively), singlet states of O_2, triplet states of organic molecules, etc. The metastable states feature long radiative lifetimes, the values of which lie between $10^{-5} - 10^5$ s as dictated by the particular nature of the excited particle. Table 5.1 covers as some examples the states, energies and lifetimes of the metastable states of some atoms and molecules (references are included in brackets). One can see that the energy potentials of these particles is sufficient and their radiative lifetimes are fairly long to provide their active participation in collision interactions both in the gaseous phase and at the gas – solid body interface.

Long radiative lifetimes of metastable states support the high density of these particles in slightly ionized plasma, or in excited gas. Thus, according to Fugal and Pakhomov [18, 19] the density of metastable atoms of helium at pressure of the order of a few Torrs, at temperatures ranging from 4 to 300 K, is about two orders of magnitude above the density of electrons. The density of metastable atoms and molecules in

TABLE 5.1

Particle	State	Excitation energy, eV	Radiative lifetime, s	
H	$2^2S_{1/2}$	10.20 [5]	0.143	[5]
He	^1S	20.60 [6]	$1.95 \cdot 10^{-2}$	[7]
He	^3S	19.82 [6]	$7.9 \cdot 10^3$	[8]
Ne	$^3P_0$	16.7 [6]	430	[9]
Ne	$^3P_2$	16.6 [6]	24.4	[9]
Ar	$^3P_0$	11.7 [6]	44.9	[9]
Ar	$^3P_2$	11.5 [6]	55.9	[9]
Kr	$^3P_0$	10.5 [6]	0.49	[9]
Kr	$^3P_2$	9.9 [6]	85.1	[9]
Xe	$^3P_0$	9.4 [6]	0.078	[9]
Xe	$^3P_2$	8.3 [6]	149.5	[9]
C	$^1D_2$	1.26 [10]	3230	[10]
C	^1S	2.68 [10]	2.0	[10]
N	$^2D_{5/2}$	2.38 [10]	$1.4 \cdot 10^5$	[10]
O	^1D	1.97 [11]	140	[11]
O	^1S	4.19 [11]	0.8	[11]
O	^5S	3.15 [12]	$1.8 \cdot 10^{-4}$	[12]
N_2	$A^3\Sigma_u^+$	6.22 [11]	2.0	[11]
O_2	$a^1\Delta_g$	0.98 [13]	$2.7 \cdot 10^3$	[13]
O_2	$b^1\Sigma_g^+$	1.63 [14]	7.0	[14]
Benzene	$^3B_{1u}$	3.64 [15]	$2.6 \cdot 10^{-5}$	[15]
Benzophenone	$(n\pi^*)$	2.99 [16]	$3 \cdot 10^{-5}$	[16]
Naphthalene	$(\pi\pi^*)$	2.62 [17]	$5 \cdot 10^{-4}$	[17]
Anthracene	$(\pi\pi^*)$	1.85 [17]	$2 \cdot 10^{-4}$	[17]

gas lasers [20] and in the plasma of different types of gaseous discharges [21] is high. Singlet oxygen and triplet states of organic molecules play an important role in the photoprocesses occurring in biological systems, photodestruction of polymers and dyes, and the like [22. 23]. Molecules of $N_2(A^3\Sigma_u^+)$ in binding "active" nitrogen in heterogeneous systems [24]. The above list speaks by itself of the fact that one of the most active roles in the interaction of an excited gaseous phase with a surfaces is just by metastable states. In the further presentation main attention will be paid to these states, their obtaining, detecting, and studying of their behaviour in heterogeneous systems with the aid of semiconductor sensors. For the sake of brevity, we will call them electronically excited particles (EEP).

In order to investigate into the interaction of EEPs with a solid surface, one has to provide conditions for their generation in a gaseous phase and their transport to the surface the investigator is interested in. Strong inhibition of radiative transitions between ground and metastable states makes it impossible to obtain excited atoms and molecules by selective optical excitation. Therefore, the general method of obtaining metastable states in simple gases is by nonselective excitation of gas, say, by discharge or a beam of charged particles and subsequent time or spatial separation of the obtained mixture of excited atoms and molecules. The time separation is generally accomplished in static gaseous systems by the pulse-excitation of the gas. The exciting pulse produces a great number of atoms and molecules excited to different states. However, after the pulse has been turned off, the resonance-excited states are de-excited in some time of the order of 10^{-7} s, and the concentration of metastable states varies, but little, during this time. The spatial separation is effected by the method of channeling the gas through the excitation zone. As the distance from the excitation zone increases, short-lived states grow de-excited and at some distance from the source, a distance that is determined by the pressure, flux velocity and the chemical nature of gas, therein remain solely metastable states.

The use of discharges and the other nonselective methods of excitation are not fruitful when it is necessary to excite gases and vapours comprised by polyatomic molecules, say, organic molecules. In discharge plasma a great number of highly active radicals are formed and chemical reactions are intensive. In the final the gaseous composition changes. By virtue of these reasons, other methods are applied to the organic systems to obtain metastable states. If a substance under the study is characterized by a high yield of triplet formation, as a result of intercombination conversion, then metastable states are obtained by selective optical excitation of relevant singlet levels. If the yield of triplets under optical irradiation is small, then the principle method to obtain triplets is by the triplet-triplet energy transfer from some donor-molecules selected for the purpose and featuring high efficiency of intercombination conversion [17].

Being formed in the gaseous phase medium, the electronically excited particles (EEPs) reach the solid surface by diffusion. The diffusion coefficients of EEPs are, as a rule, smaller than the self-diffusion coefficients of parent gas, a factor that is associated with increasing of the EEP elastic scattering cross-section at parent molecules due to the redis-

284

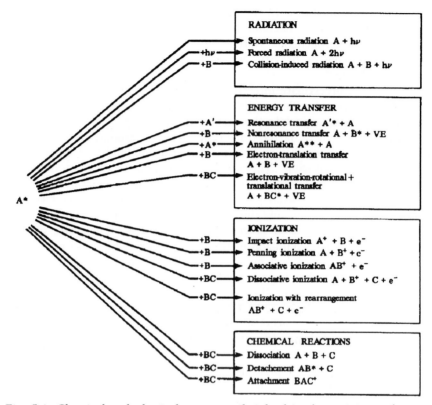

RADIATION
→ Spontaneous radiation A + hν
+hν → Forced radiation A + 2hν
+B → Collision-induced radiation A + B + hν

ENERGY TRANSFER
+A' → Resonance transfer A'* + A
+B → Nonresonance transfer A + B* + VE
+A* → Annihilation A** + A
+B → Electron-translation transfer
      A + B + VE
+BC → Electron-vibration-rotational +
      translational transfer
      A + BC* + VE

IONIZATION
+B → Impact ionization A⁺ + B + e⁻
+B → Penning ionization A + B⁺ + e⁻
+B → Associative ionization AB⁺ + e⁻
+BC → Dissociative ionization A + B⁺ + C + e⁻
+BC → Ionization with rearrangement
      AB⁺ + C + e⁻

CHEMICAL REACTIONS
+BC → Dissociation A + B + C
+BC → Detachement AB* + C
+BC → Attachment BAC⁺

A*

Fig. 5.1. Chemical and physical processes that lead to de-excitation of excited atoms in gaseous phase (taken from [28])

tribution of electron density in excited orbitals [25]. Besides, in the course of diffusion, the EEPs can enter various gaseous-phase interactions leading to excitation energy losses. For the diagram showing probable ways of deactivation of excited atoms, see Fig. 5.1. In addition to those listed in the figure, for molecules there exist still another series of gaseous-phase deactivation ways (channels) associated with intramolecular vibronic processes, dissociation, etc. The problems of diffusion and interactions of EEPs in a gaseous phase are considered in detail elsewhere, by a series of reviews and monographs [6, 26 – 29]. An inference valuable to us drawn from the listing of probable gaseous-phase processes with participation of EEPs is the necessity of selecting in each particular case an optimum gas-pressure region to provide conditions both for efficient gas excitation and EEP diffusion towards the surface under study. Another requirement is for the purity of gases involved and for the vacuum conditions in the experimental outfits, inasmuch as the presence of traces of substances known as energy acceptors may lead to essentially complete decay (death) of required EEPs in their way from the source to the surface. An example of such an energy dissipater is oxygen found in vapours of organic compounds. According to Stelling [30], the constant of de-excitation rate of benzene triplets in a gaseous phase is $5 \cdot 10^{-11} cm^3 \cdot mole^{-1} \cdot s^{-1}$. If the case is that nitrogen is excited by nitrogen electrical discharge, the formed N atoms de-excite metastable molecules of N_2 ($A^3 \Sigma_u^+$) [31] so efficiently that the energy transfer from metastable atoms of argon [32] is preferably used to obtain the latter molecules, rather than a discharge. The list of similar examples may be readily extended. Because of this, the experimental study of heterogeneous processes involving EEPs is very much in need of monitoring the gaseous-phase composition both when preparing the outfit for operation, and directly in the course of experiment.

5.2. Heterogeneous De-excitation of Electronically Excited Particles

The interaction of EEPs with solid surfaces is of complex, not yet fully studied nature. The complexity of the interaction processes is conditioned by a very large number of degrees of freedom featured by a solid body and, as a consequence, by a large number of probable ways of EEP energy relaxation on the surface. A diagram of fundamental physical and chemical processes leading to de-excitation of excited atoms on a solid surface appears in Fig. 5.2. In reality, the case is still more complex, since none of the relaxation ways shown in Fig. 5.2 probably exist in the pure state. Instead of this, a few possibilities are being realized at once. For example, an electron emission may occur through a stage of collective excitation of the solid-body electron subsystem, while an adsorption may be accompanied by origination of phonons or exitons, etc. If an EEP incident on the surface is a molecule, there appear new energy relaxation ways associated with the chemical, reactions, as well as with intramolecular vibronic processes taking place in the surface potential field.

The diversity of EEP reactions on a solid surface can be illustrated by the survey if interaction between excited atoms of mercury and zinc oxide [186]. When atoms of Hg^* get to an oxidized surface of ZnO at room temperature, an increase in the semiconductor electrical conductivity take place (Fig. 5.3, curve 2). The electrical conductivity change signal is irreversible, and in case of an increase in the temperature, after the Hg^* flux is disabled, an additional increase in the electrical conductivity (curves 3 and 4) takes place. One can logically suppose that we are dealing here with partial reduction of zinc oxide according to the scheme

286

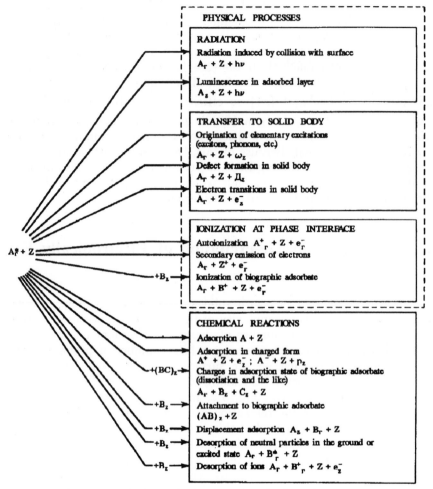

Fig. 5.2. Chemical and physical processes that lead to de-excitation of excited atoms during their interaction with a solid surface (Z – surface symbol)

$$ZnO + Hg_v^{\bullet} \rightarrow (HgO)_s + Zn_s \ . \tag{5.1}$$

The superstoichiometric zinc yielded as the reaction proceeds ionizes by giving an electron to the ZnO conduction band. Thus, the electrical conductivity of the semiconductor increases. A competing process that determines the beyond cutoff lengths of the curves in Fig. 5.3 is represented by the reaction

$$Zn_s + Hg_v^{\bullet} \rightarrow (HgZn)_s \ . \tag{5.2}$$

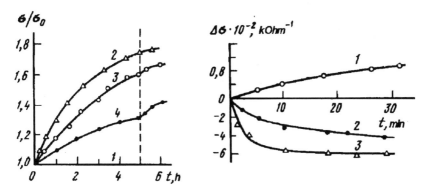

Fig. 5.3. Changes in electrical conductivity of ZnO film being acted upon by excited atoms of mercury [186]

1 – admission of unexcited mercury vapour (the straight line merges into the abscissa); 2 – film temperature is 25°C; 3 – 100°C; 4 – 200°C. The dashed line designates the instant the source of Hg* is switched off.

Fig. 5.4. Changes in electrical conductivity of pure (1) and zinc-doped (2, 3) films of ZnO under the action of Hg* [186]

Concentration of superstoichiometric zinc increases in series 2 < 3; films are at room temperature

Upon disabling the Hg* flux, the surface compounds of mercury decompose at elevated temperatures:

$$(ZnHg)_s \rightarrow Zn_s + Hg ,$$
$$(HgO)_s \rightarrow Hg_s + \frac{1}{2}O_2 .$$

(5.3)

The superstoichiometric zinc originated from causes an additional increase in the electrical conductivity of ZnO films after disabling the source of Hg* (see, Fig. 5.3, curves 3, 4). This scheme is confirmed by some experiments with introducing superstoichiometric zinc into ZnO. The curves illustrating variations of ZnO electrical conductivity on exposure to Hg*, depending upon the extent the film material is doped with zinc. From this figure we notice that for zinc-enriched samples one can see inversion of the electrical conductivity change sign, which fact points to an important role plaid by surplus zinc in de-exciting Hg* on a ZnO surface.

It is through selecting a proper level of doping zinc oxide with elementary zinc that one can completely counterbalance the processes of zinc reduction and fixation on the adsorbent's surface to passivate thus the ZnO film as to the interaction with Hg*. The film can be passivated as well by prolonged action of Hg[*] upon undoped film of ZnO. The latter films exhibit effects associated with the interaction between the excited mercury atoms and the adsorbed layers. Thus, Panesh et al. [187] showed that Hg[*] atoms actively interacted with adsorbed layers of H_2 and isopropanol on the ZnO surface. As this happens, the energy of electron excitation of Hg[*] is released in the adsorbed layer and becomes consumed in the course of the following chemical reaction

$$(H_2) + Hg_v^* \rightarrow H_s + H_s + Hg_s \ , \tag{5.4}$$

$$[(CH_3)_2 - CHOH]_s + Hg_v^* \rightarrow [(CH_3)_2 - CHO]_s + H_s + Hg_s \ . \tag{5.5}$$

The H-atoms originated during the reaction are ionized on the ZnO surface. As a result, the electrical conductivity of the adsorbent increases.

Therefore, by the examples of the Hg[*] – ZnO system, one can see that one and the same EEP can enter different particular processes on the surface as dictated by the biography of adsorbent and presence of adsorbed layers on it.

The diversity of probable ways of EEP heterogeneous deactivation leads to a situation, in which to define the deactivation mechanism for each EEP-surface pair requires a complex specific survey that includes taking measurements of primary EEP losses in the gaseous phase, probable emission of secondary particles and radiation, and also monitoring of the solid surface state. The complexity of carrying out these experiments is obvious, for which reason, the literature data on the heterogeneous deactivation of EEPs are scanty and generally scattered. Most of these data are systematized by reviews [33, 34]

Most of the investigations are devoted to the determining of the integral parameters of heterogeneous de-excitation of EEPs. These investigations are based on measuring the nature of EEP loss in the gaseous phase and yield information either about the reflection coefficients of EEP beams from various surfaces, or about coefficients of heterogeneous deactivation in the event of excited gas diffusion above a solid surface. The value of heterogeneous deactivation γ characterizes the probability of EEP decay on a single collision with the surface and can be obtained through measuring the spatial distribution of EEPs in a gaseous phase during the stationary diffusion of EEPs above the surface of the substance being studied. Smith and Semenov (35, 36) were the first to separately propose the diffusion techniques of evaluating γ, the techniques being then developed elsewhere [37, 38].

In case of low EEP concentration, when the EEP pair interactions in a gaseous phase may be neglected, and when there are no foreign gaseous-phase de-exciters, the equation of EEP steady-state diffusion in an infinite cylindrical tube takes the form

$$DVN(x, r_0) - (KN_1 + v)N(x,r) = 0 \qquad (5.6)$$

with boundary conditions $N(0, r) = N_0$; $N(\infty,r) = 0$; $\dfrac{dN}{dr}(x,0) = 0$;

$$\frac{dN}{dr}(x, r_0) + \frac{1}{\delta r_0} N(x, r_0) = 0 \; ; \; \delta = \frac{4D}{v r_0} \frac{2-\gamma}{2\gamma}.$$

The designations employed in equation (5.6) are as follows: D is the EEP diffusion coefficient in its own gas; ∇ is the Laplacian operator; N is the concentration of EEPs in a gaseous phase; N_1 is the concentration of parent gas; K is the rate constant of EEP de-excitation by own gas; v is the rate constant of EEP radiative de-excitation; r_0 is the cylinder radius; v is the heat velocity of EEPs; x, r are coordinates traveling along the cylinder axis and radius, respectively.

The general solution of equation (5.6) takes the form

$$\frac{N}{N_0} = 2\sum_{k=1}^{\infty} \frac{J_0(l_k r / r_0) \exp\left[-\left((KN_1 + v)/D + l_k^2/r_0^2\right)^{1/2} x \right]}{l_k\left(1 + \delta^2 l_k^2\right)^2 J_1(l_k)}, \qquad (5.7)$$

where J_0, J_1 are Bessel functions of the zero and first orders; l_k stands for the roots of equation $J_0(l_k) = \delta l_k J_1(l_k)$. In case of low activity of walls with regard to de-excitation of EEPs ($\delta \gg 1$), it will be enough to limit oneself to the first term of expansion (5.7) and to replace the Bessel functions with the relevant asymptotics. Then

$$\frac{N}{N_0} \approx \frac{2\delta}{1 + 2\delta} \exp\left[-\left(\frac{KN_1 + v}{D} + \frac{2\gamma}{2-\gamma} \frac{v}{2Dr_0} \right)^{1/2} x \right], \qquad (5.8)$$

Therefore, if the requirements of the described model are satisfied, the experimentally obtained distributions of EEP density along the cylinder should become rectified on coordinates $\ln\left(\dfrac{N}{N_0}\right) - x$, and the slope tangents of the resultant straight lines

$$k = \left(\frac{KN_1 + v}{D} + \frac{2\gamma}{2-\gamma} \frac{v}{2Dr_0} \right)^{1/2} \qquad (5.9)$$

should characterize the efficiency of EEP decay either inside the bulk, or on the solid surface. In order to determine the value of γ from expression (5.9), one has first to become convinced of the smallness of the homogeneous constituent of EEP losses in the gaseous phase. Let (5.9) be rewritten in the form

$$k^2 D = KN_1 + v + \frac{2\gamma}{2 - \gamma} \frac{v}{2r_0} \; . \tag{5.10}$$

The three summands found in the right-hand side of expression (5.10) correspond to the three major channels (ways) of EEP losses; the first summand characterizes the gaseous-phase de-excitation due to collisions, the second one stands for the gaseous-phase de-excitation on account of spontaneous radiation, and the third summand characterizes the heterogeneous decay of EEPs. A possible contribution of the radiative term to the value of $k^2 D$ can be done a priori. With the radiative time of EEP lifetime τ_{rad} known from the spectroscopy, one can easily estimate (by the formula of Einstein) the diffusion length over which the radiative decay of EEP will be perceptible:

$$\langle x_{rad} \rangle = \sqrt{2D\tau_{rad}}$$

If the $\langle x_{rad} \rangle$ far exceeds the cylinder length, over which experimental measurements of diffusion distribution of EEPs are taken, then the EEP radiative term found in expression (5.10) may be neglected. If such an approximation cannot be done, then the rate constant of radiative decay should be taken into consideration in processing the experimental data.

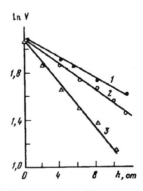

Fig. 5.5. Distribution of helium mnetastable atoms concentration along cylinder at different pressures of He[39] ($1 - 1.3 \cdot 10^{-2}$ Torr; $2 - 2.5 \cdot 10^{-2}$; $3 - 7 \cdot 10^{-2}$ Torr)

The second term of the right-hand side of expression (5.10), a term that describes the EEP decay in gaseous-phase collisions differs from the other terms of this expression in its dependence on pressure. This makes it possible to experimentally estimate the relative contribution of collision decay to EEP concentration losses in the gaseous phase. Taking measurements of EEP concentration distribution in the course of EEP diffusion along the cylinder versus the parent gas pressure, we can judge of the ratio between the homogeneous and heterogeneous constituents of the EEP decay. Let this thought be explained by a particular example. The concentration distribution of helium metastable atoms in their diffusion along the cylinder as a function of pressure [39] are exhibited in Fig. 5.5 in terms of a semilog plot. In this case, we may neglect the radiative decay of metastable atoms, since the characteristic diffusion length for this process is $<x_{rad}> \cong 10^4$ cm which is three orders of magnitude above the real degrees of He* concentration drop. Let us estimate the values of k for curves $1 - 3$ in Fig. 5.5 and analyze their dependence upon the helium pressure. If decay of metastable atoms is determined merely by heterogeneous processes, the value of $k^2D \sim k^2P^{-1}$ will be independent of pressure. If, on the contrary, the decisive role is plaid by gaseous-phase de-excitation, then the value of k^2P^{-1} should linearly increase as the pressure grows higher. For processing the data exhibited by Fig. 5.5, see Table 5.2. One can see that the value of k^2P^{-1} in this case is essentially independent of pressure, i.e. the loss of atoms is not determined merely by heterogeneous processes.

Therefore, for the example in question, formula (5.10) may be written as follows

$$\frac{\gamma}{2-\gamma} = \frac{k^2 D r_0}{v} \, , \tag{5.11}$$

hence, it is easy to compute the coefficient of heterogeneous deactivation.

TABLE 5.2

P, Torr	k, cm^{-1}	k^2, cm^{-2}	k^2P^{-1} (Torr·cm^2)$^{-1}$
$1.3 \cdot 10^{-2}$	$4 \cdot 10^{-2}$	$1.6 \cdot 10^{-3}$	$1.23 \cdot 10^{-1}$
$2.5 \cdot 10^{-2}$	$5.5 \cdot 10^{-2}$	$3.03 \cdot 10^{-3}$	$1.21 \cdot 10^{-1}$
$7.0 \cdot 10^{-2}$	$9.25 \cdot 10^{-2}$	$8.56 \cdot 10^{-3}$	$1.22 \cdot 10^{-1}$

292

One more case of essential importance is de-excitation of EEPs on very active walls ($\gamma \approx 1$). In this situation all homogeneous processes of EEP deactivation may be neglected a priori, and one can describe the drop of EEP concentration along the cylinder by the first term of expression (5.7). If that is the case, the value of γ can be evaluated by the formula

$$\frac{2\gamma}{2-\gamma} = \frac{4Dk}{v}\frac{J_1(kr_0)}{J_0(kr_0)} , \qquad (5.12)$$

where J_0, J_1 are Bessel functions of the zero and first order.

The analysis carried out allows us to state the general conditions needed to properly determine the coefficient of heterogeneous deactivation by the diffusion method:
– initial concentration of EEPs in the gaseous phase should be small enough to allow one to neglect their pair interactions in the gaseous phase;
– one should see to it that there are, as far as possible, no admixtures-de-exciters of EEPs in the gaseous phase;
– dimensions of the diffusion cylinder should be selected so as to meet the approximations of formulas (5.11) and (5.12), namely $r_0 \ll l_0$;
$\frac{\lambda}{2} \ll r_0 \ll \frac{3}{2}\frac{\lambda}{\gamma}$ (l_0 is the cylinder length; λ is the length of EEP free path in gas);
– EEP detector should not introduce essential changes in the diffusion distribution of EEPs along the cylinder.

In some cases, when all above-listed conditions cannot be satisfied simultaneously, the system geometry can be modified accordingly in order to estimate the value of γ from the solution of the newly arising diffusion problem. Thus, in work [40] a diffusion system comprised by two coaxial cylinders was used to study deactivation of singlet oxygen on highly active surfaces. The outer cylinder is made of glass characterized by a small value of $\gamma \sim 10^{-5}$, while the material under study with $\gamma \sim 1$ was applied to the inner cylinder smaller in diameter. Dickens et al [41] have analyzed the situations when the detector introduce disturbances into the primary diffusion distribution of active particles. Besides, they analyzed the case of high concentrations of active particles in the gaseous phase. Note, however, that the diffusion method modifications of this kind add much to the complexity of processing the experimental data and generally call for application of the present-day computer equipment.

Being an integral magnitude, the heterogeneous deactivation coefficient does not allow us to judge of the particular mechanisms of EEP de-excitation on solid surfaces, but its value can be used to estimate the general nature of the energy-exchange dynamics between the gaseous-phase EEPs and the surface. To determine the value of γ is as a rule the first step of studying the mechanism of interaction between EEPs and a solid body, and also is of practical importance for the choice of materials that contact the excited gaseous phase under particular conditions.

5.3. Methods for Evaluating Concentration of EEPs in the Gas Phase

The previous section stressed the importance of measuring the concentration of EEPs in the gas phase when studying the heterogeneous processes with EEP participation. In this case, the method to be used for measuring concentrations of EEPs must feature high sensitivity, minimize disturbances introduced into the system under measurements, and it is desirable that the method be applicable under the diffusion conditions. Let us try to consider from this standpoint the fundamental principles of detecting EEPs in a gaseous phase and to estimate the probability of their application to surveying heterogeneous processes.

All existing techniques of EEP detection may be divided into several groups: spectral, calorimetric, chemical titration, electrical methods, and also a method of sensor detection.

The spectral methods employ the absorption spectroscopy, laser-induced fluorescence spectroscopy, exchange luminescence, and EPR (electron paramagnetic resonance) technique. The absorption spectroscopy is one of the most frequently used methods. Milatz and Ornshtein [42] were probably the first to use this method to evaluate the rate of forming metastable atoms of neon as a result of electron impacts. This method is based on the phenomenon of resonant radiation absorption that corresponds to EEP transition from the metastable state to the resonance-excited state lying higher. By the intensity of absorption, one can evaluate the concentration of EEPs in the gaseous phase.

The sensitivity of the absorption spectroscopy method is dependent upon the optical path length of the probing light beam in an excited gas and the sensitivity may reach $10^8 - 10^{10}$ particle/cm^3. This method is generally employed for the survey of discharge plasma or decaying plasma in the earlier stages of afterglow. The absorption spectroscopic potentialities for the study of the heterogeneous processes with participation of EEP are not high. For example, to evaluate the efficiency of de-exciting EEPs on the reactor walls, one has, in case of flow system, to know the drop curve and EEP concentration along the excited gas, and, in case of a static system, the EEP spatial distribution. In the former case, the small length of the optical path of the probing light beam brings about the necessity of working at comparatively high pressures, when the processes of the homogeneous de-excitation because of utmost importance, and it is quite a problem to evaluate the contribution of the heterogeneous constituent [43]. In the latter case [44], under continuos gas excitation conditions, one can survey spatial distribution of EEPs at an overall gas pressure of $10^{-2} - 10^{-1}$ Torr that is enough to investigate heterogeneous processes. However, under the continuous excitation conditions, when the reacting space includes distributed sources of EEPs, in order to evaluate the efficiency of heterogeneous decay, one has to know the profile of EEPs concentration in the immediate vicinity of reactor walls, a situation that is unattainable within the scope of the absorption spectroscopy technique due to the effects of wall reflections on the probing beam.

Somewhat better sensitivity and spatial resolution are featured by the technique of laser-induced fluorescence spectroscopy [45]. This technique is also based on the phenomenon of resonance light absorption by metastable EEPs. The laser radiation pulse transfers EEPs in the probing space from the metastable state to the resonance-excited state, and the EEP concentration can be evaluated against the intensity of excited gas fluorescence, as early as in the resolved region of the spectrum. The high intensity and spatial localization of laser beam account for the advantages of this technique compared with the absorption spectroscopy method. Note, however, the disturbing effect introduced into the system by the probing beam increase as well.

The method of exchange-luminescence [46, 47] is based on the phenomenon of energy transfer from the metastable levels of EEPs to the resonance levels of atoms and molecules of de-exciter. The EEP concentration in this case is evaluated by the intensity of de-exciter luminescence. This technique features sensitivity up to -10^7 particle/cm^3, but its application is limited by flow system having a high flow velocity, with which the counterdiffusion phenomenon may be neglected. Moreover, this technique permits EEP concentration to be estimated only at a fixed point of the setup, a factor that interferes much with the survey of heterogeneous processes associated with taking measurements of EEP spatial distribution.

There exist methods of spatial evaluation of EEP concentration against the intensity of excited gas luminescence in the spectrum regions corresponding to the prohibited single-photon transitions from the metastable state to the ground state [48, 49], or against the intensity of the luminescence that arises during homogeneous de-excitation of EEPs by the own gas [50, 51]. In case of paramagnetic EEPs, say, $O^2(^1\Delta_g)$, Ar ($^3P_2$), their concentration can be evaluated by the ERP method [52,

53]. The sensitivity of these methods, however, is low and does not allow operations at low gas pressures.

The calorimetric method of detecting EEPs is based on measurements of the thermal effect arising on surfaces featuring high efficiency of de-excitation. This technique was used for evaluating $O_2(^1\Delta_g)$ [31] of a series of vibrationally excited molecules [54]. The EEP sensitivity in this case may amount to 10^{10} particle/cm^3. Absence of selectivity may be classified as a disadvantage of this method, inasmuch as the thermal effect occurs not only during the deactivation of EEPs, but also during recombination of atoms and radicals that are often present in an excited gas, during radiation, adsorption, etc. More than that, the calorimetric detectors introduce heavy distortions into the spatial distribution of EEPs in the system.

The chemical titration method consists in a quantitative analysis of the products resulting from some specific reactions of EEPs with specially selected reagents in gaseous phase [55], and also the products that results from passing an excited gas through a reagent solution [56], or from a gas in contact with a solid reagent [57]. The chemical technique has a number of essential disadvantages. These are low sensitivity, the necessity of having of a reagent-substance in the system, and chiefly the ambiguity of interpreting the obtained results.

High sensitivity is featured by the electrical methods used to detect EEPs. These are based on measuring small currents that occur in the course of selective ionization of EEPs, or currents of secondary emission of electrons or ions knocked by EEPs out of the surface of solid targets (if such emission is taking place).

The selective ionization of EEPs is attained due to an ionization potential difference between the excited and ground states of an atom or a molecule, a difference that is equal to the excitation energy. Thus, the ionization potentials of oxygen states $^1\Delta_g$ and $^1\Sigma_g^+$ are 1.0 and 1.6 eV, respectively, lower than the ionization potential of the O_2 ground state (12.06 eV). Using the resonance radiation of Kr (10.64 eV) and Ar (11.72, 11.54 eV) as an ionization effect in work [58], they succeeded in selectively separating excited states of O_2 against the background of triplet oxygen. It should be noted, however, that the interpretation of mass-spectrometric measurements becomes very complicated in case of analyzing multiatomic molecular gases and a mixture of gases with close potentials of ionization.

The emission methods of EEP detection are based on the ability of some metastable particles ($N_2(A^3\Sigma_u^+$, metastable atoms of rare gases) to knock electrons out of the surface of metals [59]. The present-day technique of measuring small currents allows one to record any small fluxes of electrons knocked out of metallic targets by EEPs. Thus, in surveying the scattering of thermal bundles of rare gas metastable atoms [60], the EEP fluxes being measured make up ~ 10^7 particle/cm^2·s which corresponds to gaseous-phase concentration ~ 10^3 particle/cm^3. The major complexity of employing the EEP sensors lies in the operation instability associated with the instability of the quantum yield of secondary electrons at the metallic target electrodes. The value of the quantum yield of secondary electrons during deactivation of rare gas metastable atoms on a surface of metal is dependent both upon the method of surface preparation [61] and on the presence of adsorbed layers [47], for which reason the value can randomly vary in the operation of the detector. Dunning [62] overcame this difficulty by continuous metal precipitation on the emitter surface. The structural complexity of this detector, however, makes this method suitable, but little, for practical purposes.

A general disadvantage that limits the field of employing the electrical techniques of detecting EEPs is in the complexity of the hardware involved. As a rule, all electrical methods call for use of superhigh vacuum, the technique of excited particle bundles, the electron beam optics, and need high noise immunity.

From the above-given condensed review of the EEP detection methods one can infer that none of these methods can independently satisfy all the requirements specified for the study of heterogeneous processes involving the EEP participation. To our opinion, the application of semiconductor sensors for detection of EEPs can be provided by a combination of required qualities. The sensors are highly sensitive, miniature, can be operated within wide ranges of gas temperatures and pressures, and are made of simple devices. At the same time, a series of problems arise connected with the preliminary preparation of sensors and improving their selectivity. These and other questions of general nature will be considered in the section that follows.

5.4. Detection of EEPs with the Aid of Semiconductor Films

The sensor detection of EEPs is methodically more complicated than the detection of atoms and radicals. With atoms and radicals being adsorbed on the surface of semiconductor oxide films, their electrical conductivity varies merely due to the adsorption in the charged form. If the case is that EEPs interact with an oxide surface, at least two mechanisms of sensor electrical conductivity changes can take place. One mechanism is associated with the effects of charged adsorption and the other is connected with the excitation energy transfer to the electron subsystem of semiconductor. The concepts of energy transfer from the excited adsorbed molecules to a solid body were first developed in the classical works of Terenin School [63] that deal with sensitizing the photoconductivity of zinc oxide by adsorbed dyes and pigments. By some direct experiments Vintsenets et al. [64, 65] have shown that one of the ways by which the energy of triplet-excited erythrosine molecules adsorbed on an oxidized surface of germanium dissipates is due to emptying of the traps found in the oxide layer and electron ejection into the conduction band of semiconductor. These processes are highly probable in case of interaction between the surface of oxide semiconductors with EEPs flying on from the gaseous phase. Sometimes the signals that indicate changes in the sensor electrical conductivity due to chemisorption and energy transfer to the EEPs have different signs, a factor that hinders the interpretation of sensor readings. A dramatic example of this effect is the interaction of $O_2(^1\Delta_g)$ with the surface of a sintered polycrystalline ZnO-based sensor, a property addressed in [66]. This work surveys changes in the electrical conductivity of oxidized and reduced ZnO films when acted upon by singlet oxygen. In order to prevent influence of probable gas admixtures that affect the electrical conductivity of ZnO, diverse methods were used to obtain $O_2(^1\Delta_g)$ molecules. One method was by a SHF-discharge in oxygen ambient, and the other, with the aid of a chemical source on the basis of reaction

$$HgO + O \rightarrow Hg + O_2\left(^1\Delta_g\right). \tag{5.13}$$

For the results of measurements, see Figs. 5.6 and 5.7. One can see that the changes in the electrical conductivity is of nonmonotonic nature. Initially, after the source of $O_2(^1\Delta_g)$ has been switched on, the sensor resistance decreases and then starts to smoothly rise, the resistance growth being continued even after the source of singlet oxygen has been switched off. The results obtained are evidence that during the interaction of $O_2(^1\Delta_g)$ with the ZnO film, two processes are taking place, one of which is associated with the transfer of $O_2(^1\Delta_g)$ molecule energy to the electron subsystem of zinc oxide and accordingly with the growth of electrical conductivity, and the other process, with the chemisorption of singlet oxygen at the superstoichiometric atoms of Zn that are centers of O_2 adsorption. It is the competition between these two processes that accounts for the extremum nature of the signal caused by changes in the sensor's electrical conductivity. This interpretation is supported by the fact that at the reduced samples featuring a high chemisorption ability with regard to O_2, the acceptor signal prevails over the donor's signal.

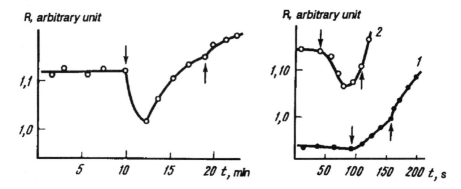

Fig. 5.6. Changes in electrical resistance of ZnO film under the action of singlet oxygen at its chemical generation [66]. Arrow-heads indicate on/off switching of $^1\Delta_g$ source.

Fig. 5.7. Changes in electrical resistance of reduced (1) and oxidized (2) films of ZnO under the action of singlet oxygen produced by UHF discharge [66]. Arrow-heads indicate on/off switching of discharge.

As for the energy transfer to the subsurface layers of zinc oxide from the singlet oxygen molecules, the transfer should lead to an increase in the electrical conductivity of semiconductor either due to ejection of electrons into the conduction band from shallow traps [67], or due to the injection of electrons into zinc oxide by excited particles [68]. Effects of this kind were observed in the interaction between a ZnO surface and excited pairs of benzophenone [70], and also in adsorption of singlet oxygen on the surface of ZnO monocrystal in electrolyte [69].

Therefore, the interaction of the EEPs with the surface of sensors is a complex process that, being dependent on the nature of the surface and the nature of the active particle, results either in chemical transformation (chemisorption, for instance), or in transfer of excitation energy to a solid body, the processes that proceed at different velocities.

The competition of unlike processes of EEP deactivation on the sensor surface much hamper the detection, since it does not allow quantitative evaluation of EEP concentration in a gaseous phase. For the proportional detection of EEPs, the sensor should be first treated to suppress certain ways of detection and to improve the energy dissipation efficiency by some other ways. With chemically active EEPs, say, of singlet oxygen, the strategy of detector preparation is aimed at increasing the chemisorption activity of the surface, a property that allows one to obtain a monotonic change in the ZnO electrical conductivity and proportional detection of $O_2(^1\Delta_g)$ [71]. With EEPs of low chemical activity, the preliminary treatment of the sensor should be generally aimed at surface passivation with respect to EEP chemisorption and to surface sensitization with additives that facilitate the energy transfer to the electron subsystem of semiconductor. An example is work [72] that succeeded in the proportional detection of triplet molecule of benzene and naphthalene by preliminary adsorption of magnesium phthalocyanine molecules on the surface of ZnO films.

Other aspects of the sensor detection of EEPs are represented by the problems of sensor calibration and improving of sensor selectivity. Low sensitivity of sensors is a common disadvantage, a property that shows itself in detection both of EEPs and radical particles. Modification of sensor surface with various additives that improve sensor selectivity is widely used in the gas analysis. This technique, however does not apply to the detection of EEPs, since no one can practically find additives that provide simultaneously good selectivity and conditions required for proportional detection of excited particles. A more promising approach is represented by preliminary separation of active particles in an excited gas. Separation like this may be obtained by placing suitable filters that absorb undesirable active particles and pass EEPs to be measured installed between the gas excitation zone and the detector. In the case of measuring concentration of singlet oxygen in desintegrating plasma of SHF-discharge in O_2 ambient, a chamber having its walls plated with freshly atomized silver. The silver film actively absorbs O-atoms [73]. The most efficient filters to separate EEPs and N-atoms in "active nitrogen" are films of metallic nickel. While diffusing over the Ni surface, the excited gas undergoes N-atom combining due to recombination, and the $N_2(A^3\Sigma_u^+)$ molecules add to the concentration of EEPs in the gaseous phase [74].

Compared with the sensors for atoms and radicals, the calibration of EEP sensors is also somewhat specific. To calibrate detectors of atomic particles, it will be generally enough to determine (on the basis of sensor measurements) one of the literature-known constants, say, the energy of parent gas dissociation on a hot filament. For the detection of EEPs when nonselective excitation of gas is taking place, in order to calibrate a sensor use should be made of some other selective methods detecting EEPs. The calibration method may be optical spectroscopy, chemical and optic titration, emission measurements, etc.

Therefore, we see that the development of EEP semiconductor detectors is a challenge that expects knowing of diverse experimental equipment, and, what is more, carrying out some particular investigations. The experimental difficulties, however, are compensated for by the advantages of the sensor detection of EEPs. Fairly high sensitivity, small size and energy capacity, a wide range operating temperatures and gas pressures featured by the semiconductor sensors allow important parameters of interaction between EEPs and solid surfaces to be estimated by means of experiments. More than that, the EEP-formed signal caused by a change in the sensor electrical conductivity contains information not only about the concentration of EEPs in the gaseous phase, but also about the relaxation of electron excitation energy on the surface. Varying the sensor's modes of operation, the investigator can obtain unique data on the actual mechanism of EEP heterogeneous de-excitation and ways of energy transfer to the electron subsystem of the semiconductor.

By way of illustration, the sections that follow will analyze in detail particular studies of deactivating the single oxygen and metastable atoms of rare gases on a surface of metal oxides. The singlet oxygen is a classical example of a chemically active EEP, while the metastable atoms of rare gases feature an extremely low chemical activity and their energy dissociates in the course of physical processes. These active particles may be used as examples to trace the methods of preparing and calibrating the EEP sensors, and also their application to study the mechanism of energy exchange between a gaseous phase and a solid surface.

5.5. Deactivation of Singlet Oxygen on Surface of Oxides

5.5.1. Physicochemical Properties of Singlet Oxygen

The term "singlet oxygen" designates the lower electron-excited states of O_2. These states are designated as $O_2(^1\Delta_g)$ and $O_2(^1\Sigma_g^+)$ and characterized by excitation energies of 0.98 and 1.63 eV, respectively. The radiative lifetime of $O_2(^1\Delta_g)$ is $2.7 \cdot 10^3$ s and $O_2(^1\Sigma_g^+)$, 7 s. Molecules of singlet oxygen are widely spread in the nature. They participate in a series of important heterogeneous and photoheterogeneous processes. These are photodestruction of polymers, fading of photographic films, oxidizing of petroleum films on water surface and a number of biological processes. There is reason to think that the high selectivity of some heterogeneous reactions of catalytic oxidation underwent by hydrocarbons is associated with participation of single oxygen in these reactions. In technology the singlet oxygen is used for water laser pumping, obtaining thin films of oxide on metal surfaces, oxidizing hydrocarbon contamination on the surface of film structures.

The physicochemical properties of singlet oxygen are studied fairly well and set forth in a series of reviews and monographs [22, 23, 75 – 77]. In this section we shall not dwell in detail upon the properties of singlet oxygen that show themselves in the gaseous-phase interactions. We shall make an attempt to give a concise review of the data on singlet-oxygen deactivation on surfaces of diverse chemical nature.

In most cases the studies of heterogeneous deactivation of singlet oxygen are carried out in flow-type set-ups, and are aimed at determining the coefficient of heterogeneous deactivation γ. The singlet oxygen is generally produced by UHF-discharge in O_2 ambient. The concentration of excited molecules in the reactor is estimated with the aid of spectroscopic or calorimetric techniques. It should be noted that in most cases the studies deal with $O_2(^1\Delta_g)$, since the molecules of $O_2(^1\Sigma_g^+)$ decay, as a rule, in their way from the generation source to the measuring instrumentation, owing to their shorter radiative lifetime and higher rate of de-excitation due to foreign gas admixtures contained in the deystaff [23]. Presence of uncontrollable admixtures and active particles (O-atoms, ozone, etc.) is the major disadvantage of this type methods. Seemingly, it is this disadvantage that accounts for the ambiguity of the conclusions drawn by the works that concern the mechanism of heterogeneous deactivation of $O_2(^1\Delta_g)$ and sometimes for the nonreproductivity of results.

For example, in works [78 – 81] the decay of $O_2(^1\Delta_g)$ at the Pyrex varied from $7.6 \cdot 10^{-6}$ to $5 \cdot 10^{-5}$. In works [82 – 86] the constant of heterogeneous decay of $O_2(^1\Delta_g)$ at the Pyrex varied from $9.3 \cdot 10^{-3}$ to $2.6 \cdot 10^{-2}$. The results show that the coefficient γ is essentially influenced by the nature of the surface and the method of preliminary treatment of the walls. This inference is also supported by works [87, 88] the authors of which observed changes in the γ as the walls were treated in the course of experiment. Ryskin and Shub [89] took measurements of the γ for deactivation of $O_2(^1\Delta_g)$ on glass, silicon, and some metals at room temperature. The value of γ was $4.4 \cdot 10^{-5}$ on glass, $5.5 \cdot 10^{-5}$ on aluminum, $6.5 \cdot 10^{-5}$ on titanium, $8.0 \cdot 10^{-5}$ on molybdenum, $1.2 \cdot 10^{-4}$ on niobium, $4.0 \cdot 10^{-4}$ on platinum, $7.3 \cdot 10^{-4}$ on silicon, $8.5 \cdot 10^{-4}$ on copper, $2.7 \cdot 10^{-3}$ on nickel, $4.4 \cdot 10^{-3}$ on iron, and $1.1 \cdot 10^{-2}$ on silver. The constant of $O_2(^1\Delta_g)$ decay on silver and copper varied from experiment to experiment by 50%; with silicon the constant was slowly increasing with time. It was supposed that the decay of singlet oxygen on metals might occur due to relaxation of the $O_2({}^1\Delta_g)$ electron excitation energy on free electrons of a metal.

Ryskin and Chernish [90] studied deactivation of $O_2({}^1\Delta_g)$ on a quartz surface coated by frozen layers of H_2O, D_2O, CO_2, C_2H_2, and SF_6. The values of γ were determined and it was shown that with a decrease in the temperature, the efficiency of heterogeneous decay rises. At temperatures below 150 K, the value of γ as functions of temperature for all surfaces studied can be presented by the relationship $\gamma = \gamma_0 \exp(Q/RT)$, where Q is the heat of singlet oxygen adsorption.

Works [40, 91] surveyed γ versus temperature for deactivation of $O_2({}^1\Delta_g)$ on quartz at 350 – 900 K. The obtained temperature dependencies were in the Arrhenius form with the activation energy of 18.5 kJ/mole. A conclusion was drawn up about the chemisorption mechanism of singlet oxygen deactivation on quartz surface. A similar inference was arrived at by the authors of work [92] relative to $O_2({}^1\Delta_g)$ deactivation (on a surface of oxygen-annealed gold).

Analyzing the above-mentioned data, one can come to the following conclusions:
– study of the heterogeneous decay of singlet oxygen on a solid surface should be accomplished under most "pure" conditions, i.e. in absence of foreign substances, lubricant vapours, water, etc.;
– it is good practice to use low concentrations of singlet oxygen, inasmuch as in its interaction with a surface, the latter surface undergoes modification;
– that the mechanisms of heterogeneous decay of singlet oxygen are vague is mainly traced to the disadvantages of the conventional experimental methods of deactivating $O_2({}^1\Delta_g)$.

Elimination of the above-mentioned difficulties can be facilitated by the application of sensor methods of detecting singlet oxygen.

5.5.2. Detection of Singlet Oxygen with the Aid of ZnO Sensors

Development of singlet-oxygen sensors faces two methodical complexities. First, one has to perform detection of singlet oxygen against the background of conventional triplet oxygen, which, as you know, in itself produces effects on the electrical conductivity of semiconductor sensors. Second, in case of interaction of singlet oxygen with a semiconductor surface, the above-described competition of physical and chemical ways of energy relaxation may take place. This leads to the extremal nature of changes in the sensor electrical conductivity under the action of EEPs.

According to experimental studies [71], singlet oxygen is a stronger oxidizer than molecules of triplet oxygen. This makes it possible to detect singlet oxygen with the aid of sensors based on metallic or semiconductor films against the background of ordinary oxygen. Finding a proper sensor material and by the procedure of sensor preparation for operation, one can suppress physical processes of energy relaxation and obtain high sensitivity and proportionality of the sensors.

An experiment has shown that thin films prepared of various materials (Co, Co_3O_4, Ag, TiO_2, ZnO) feature $O_2(^1\Delta_g)$ sensitivity. The highest sensitivity, 2 to 3 orders of magnitude above the sensitivity of calorimetric methods, is featured by the film prepared of partially reduced zinc oxide. During interaction with such films, the $O_2(^1\Delta_g)$ behave as acceptors of electrons more actively than an excited oxygen.

A ZnO film being placed in an oxygen ambient changes its electrical conductivity due to adsorption of O_2 triplet molecules on the film surface. However, sometimes later, new electrical conductivity sets in, the value of which is dependent upon the O_2 pressure and sensor temperature. If one now switches on the source of singlet oxygen, then an additional signal will be observed indicating a decrease in the sensor electrical conductivity associated with the interaction between the sensor and the singlet oxygen. The signal turns out to be in proportion to the concentration of singlet oxygen in the gaseous phase. This temporary separation allows one to detect singlet oxygen against the background of triplet oxygen.

With nonselective excitation of oxygen, a problem arises of distinguishing between the $O_2(^1\Delta_g)$ and other active particles, say, O-atoms or ozone molecules. All these particles cause a superequilibrium change in the electrical conductivity of ZnO-sensors, for which reason they cannot be distinguished by the nature and magnitude of the signal.

There are a few ways of solving this problem. One way is connected with the variation of the sensor operating temperature. Figure 5.8 shows how the sensitivity of a ZnO-sensor to O-atoms and singlet oxygen varies with temperature. One can see that the sensor sensitivity to $^1O_2$ decreases with a temperature drop more abruptly than the same parameter for O-atoms. This allows one to choose a range of operating temperature, within which the sensor signal caused by $^1O_2$ will be negligible compared with the signal caused by the same concentration of O-atoms. In practice, the temperature of 77 K is most suitable for realization of a situation like this. In the similar manner one can select a temperature range within which a stronger influence of singlet oxygen molecules upon the sensor electrical conductivity will be seen, whereas the effect of triplet oxygen will be materially lower [188]. Usually this is a temperature of the order 170 – 190 K.

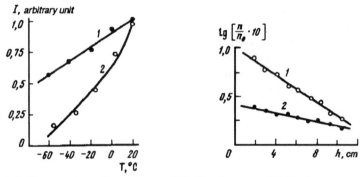

Fig. 5.8. Temperature dependence of ZnO sensor sensitivity to comparable concentrations of $^1O_2$ (1) and O(2)

Fig. 5.9. Distribution of oxygen atoms (1) and $^1O_2$ molecules (2) concentration during their diffusion along glass tube [71]

One can also arrive at a grounded conclusion about the nature of an active particle, having measured the spatial distribution of the concentration of particles during their diffusion along the tube. The appropriate concentration distribution for O-atoms and $O_2(^1\Delta_g)$ along a glass tube at room temperature appears in Fig. 5.9. From this figure we notice that the decline of O-atoms concentration proceeds at a higher rate than the concentration decline of singlet oxygen. Still lower rates of spatial distribution should be characteristic of ozone molecules. Therefore, the nature of heterogeneous decay permits us to judge what product of those resulting from O_2 excitation is being detected by the sensor. this result applies to the other inner surface linings of the tube (zinc oxide, titanium oxide, etc.) as well as to the glass, a property that additionally helps in the identification of singlet oxygen molecules.

A fairly efficient method of selective detection of active forms of oxygen by means of sensors is the preliminary separation of their mixture with the aid of filters designed for the purpose, filters that vigorously de-excite this or that form of oxygen. It has been mentioned in Section 5.4 that freshly atomized films of Ag efficiently absorb O-atoms from the gaseous phase, mildly de-excite $^1O_2$ molecules. To de-excite the singlet oxygen molecules proper, use should be made of filters with

Co_3O_4 coatings [190, 191]. As an experiment shows [101], the signals caused by $^1O_2$ are relaxed in this case at least twentyfold. When Co_3O_4 is used as a de-exciter, it is good to utilize an oxide calcined in air or oxygen ambient at 700°C. The calcined oxide is thus oxygen depleted and de-excites $^1O_2$ at an efficiency close to unity [101, 189]. Calcination at 400°C enriches the oxide in oxygen and reduces its activity as a de-exciter of singlet oxygen.

Nevertheless, in the sensor study of heterogeneous reactions of singlet oxygen, it is better to employ selective sources that do not generate other active particles and produce though low, but stable concentrations of $O_2(^1\Delta_g)$. These sources can be photochemical and chemical formations of $O_2(^1\Delta_g)$.

Photogeneration of singlet oxygen takes place at illumination of an O_2-surrounded quartz surface with molecule-dispersed vanadium pentoxide applied to the surface [93, 94]. A $10^8 - 10^9$ molecule/cm^3 concentration of $O_2(^1\Delta_g)$ in the gaseous phase can be obtained above a quartz surface employing a source of this type. A similar effect can be reached by illuminating quartz with molecules of organic dyes (trypaflavine, rose bengale, methylene blue) [95, 96]. The processes of heterogeneous photogeneration of $O_2(^1\Delta_g)$ will be described later, in the chapter dedicated to emission of active particles from a solid surface.

A typical example of a chemical generator is a device shown in Fig. 5.10. It consists of current-incandesced platinum strip *1* (a pyrolytic generator of O-atoms), and hole filter *2* coated with mercury oxide. The oxygen atoms formed through this pyrolysis interact with HgO by the known reaction [97 – 99]

$$HgO + O \rightarrow Hg + O_2(^1\Delta_g),$$
$$Hg + O \rightarrow HgO,$$

(5.14)

and the singlet oxygen molecules find their way into the internal space of the experimental set-up. Varying the temperature of the platinum strip, one can change the concentration of O-atoms to control thus the concentration of singlet oxygen within the range $10^4 - 10^{11}$ cm^{-3}. An advantage of this method is in that the mercury oxide very efficiently absorbs atoms of oxygen and ozone [188].

The use of low-capacity generators of singlet oxygen when working with sensors is justified from the standpoint that the sensors allow long series of measurements, remaining within the range of linear relationship between the signal of electrical conductivity measurement of the sensor and the concentration of active particles in the gaseous phase. The signal linearity region of a ZnO-detector relative to the $O_2(^1\Delta_g)$ concentration is dependent upon the temperature and degree of the zinc-oxide surface reduction. For the change in the ZnO film electrical conductivity against the time of treating with singlet oxygen, see Fig. 5.11. One can see that the dynamic range of linear detection makes it up $2-5\%$ of the initial value of sensor electrical conductivity. If the case is that the use is made of strongly reduced ZnO films, the proportionality of the initial change rate of sensor electrical conductivity and concentration of $O_2(^1\Delta_g)$ in the gaseous phase remains unchanged within a far more wide (up to 50%) range of the film electrical conductivity change [101].

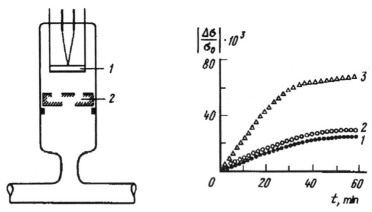

Fig. 5.10. Diagram of singlet oxygen chemical generator [100]
1 – platinum pyrolysis filament; *2* – hole filter

Fig. 5.11. Kinetic curves of changes in relative electrical conductivity of ZnO film against time of treating by singlet oxygen [100]
1 – 23°C; *2* – 45°C; *3* – 150°C

When the surface of a sensor is saturated with adsorbed active particles, the sensor loses its sensitivity and needs regeneration. The singlet-oxygen regeneration of ZnO sensors is obtained by heating in a hydrogen ambient at $500-600$ K.

To calibrate sources of low concentration of singlet oxygen, use may be made of any one of the traditional methods possessing integral sensitivity. One of these methods is chemical titration. This method is most likely of the qualitative nature. Some gas containing $^1O_2$ is passed (bubbled) through a 1,3-diphenyl izobenzofuran solution in n-hexadecane [192]. The extent of transformation underwent by this reagent can be evaluated either by the change in the optical density of the solution, or by the change in the luminescence intensity in the 480 nm band. The following equation gives a good fit to decolorization of reagent

$$\frac{d[A]}{dt} = \frac{1}{\tau}\left[{}^1O_2\right] + K\,[A]\left[{}^1O_2\right],\tag{5.15}$$

where [A] is the concentration of 1,3-diphenyl izobenzofuran in the solution (usually $10^{-3} - 10^{-4}$ mole/l; τ is the lifetime of singlet oxygen in n-hexadecane; $[{}^1O_2]$ is the concentration of singlet oxygen in solution. With the solubility of ${}^1O_2$ in n-hexadecane known, one can evaluate the concentration of singlet oxygen in the gaseous phase.

The calibration of a sensor with a view to evaluating absolute concentration of ${}^1O_2$ is quite complicated a problem. As of now, there seemingly exist only one method [23] of evaluating absolute concentrations of these particles, i.e. by an isothermic-shell calorimeter of thermal resistance with lining of Co_3O_4 [97]. However, concentration evaluations obtained by this technique should be cautiously interpreted. As it has been noted, the coefficient of heterogeneous decay of singlet oxygen on Co_3O_4 is dependent upon the condition of preliminary prepared oxide [189]. In addition, it is generally supposed in analyzing calorimeter readings that the thermal effect of deactivating a singlet-oxygen molecule equals the energy of electron excitation. In actuality, however, as shown elsewhere [97, 110], during the interaction between a mixture comprised by singlet and triplet oxygen and a Co_3O_4 surface, one can observe perceptible dissolution of oxygen in the oxide lattice, so that Zavyalova [111] even failed to single out the specific effect caused by chemisorption of ${}^1O_2$. It is obvious that with allowing for the thermal effect of oxygen dissolution, the overall thermal effect will differ from the energy of electron excitation. Therefore, it is believed that this calorimeter can be used merely for evaluating concentrations within an order of magnitude.

5.5.3. Deactivation of Singlet Oxygen on a Glass or Quartz Surface

Application of semiconductor sensors in the survey of heterogeneous deactivation of singlet oxygen allows one to accomplish experiments under the most pure conditions, including oil-free evacuation, use of static diffusion regime, use of highly purified gases and low concentrations of EEPs that do not modify the surface under survey in the course of deactivation. Evaluation of the coefficient γ of $O_2({}^1\Delta_g)$ on various surfaces is methodically most simple. The ZnO films are very good detectors for this type of measurements. They are small-scale, readily move inside the vacuum facilities, and disturb, but little, the initial distribution of active particles concentration. The wide range of operating temperatures and pressures featured by ZnO-sensors allows one to obtain values of γ as functions of the experiment conditions and using the analysis of these values as a basis to draw up some conclusions about the mechanism of heterogeneous deactivation.

Glass and fused quartz form one of the most inert surfaces with respect to heterogeneous de-excitation of $O_2(^1\Delta_g)$. The coefficient of heterogeneous decay of γ in this event lies within $10^{-5} - 10^{-6}$, depending upon the experiment conditions. As the temperature of glass walls increases, the γ decrease by the law

$$\gamma = \gamma_0 \exp(E / RT),$$

where the activation energy $E \approx 6$ kcal/mole [102]. Moreover, in case of glass, the γ is independent of the triplet oxygen pressure in the experimental set-up. The data obtained may be interpreted in terms of the mechanism of physical adsorption of oxygen with subsequent excitation energy transfer to a solid body, like it was done in [103, 104] in the event of vibrational relaxation underwent by vibration-excited molecules of nitrogen on silver and polytetrafluoroethylene (Teflon) at low temperatures. The following scheme gives a good fit to the mechanism in question:

$$O_2(^1\Delta_g) + Z \underset{K_{-1}}{\overset{K_1}{\rightleftarrows}} O_2(^1\Delta_g)Z,$$

$$O_2(^1\Delta_g)Z \overset{K_2}{\rightarrow} O_2(^3\Sigma_g^-)Z,$$
(5.16)

where Z is the symbol standing for the centre of adsorption.

Assuming $[O_2(^1\Delta_g)Z]$ to be stationary, we can obtain the following expression by the method of quasistationary concentrations

$$\gamma = \frac{4 K_1 H_2(1 - \theta)}{(K_2 + K_{-1}) v} \frac{S}{V} [Z_0],$$
(5.17)

where v is the mean heat velocity of $O_2(^1\Delta_g)$ molecules; θ is the surface coverage with singlet oxygen; S is the area of reactor surface; V is the reactor volume; $[Z_0]$ is the number of adsorption centres.

If the rate of singlet oxygen deactivation is far less than the rate of desorption, i.e. $K_2 \ll K_{-1}$, then

$$\gamma = \frac{4\,K_1 K_2 (1-\theta)}{K_{-1}\,v} \frac{S}{V} [Z_0]. \tag{5.18}$$

In this assumption, the singlet oxygen adsorption on a surface may be considered a quasistationary process and described by the Langmuir Adsorption Isotherm. Then the equilibrium constant is

$$K = \frac{K_1}{K_{-1}} = \frac{\theta}{(1-\theta)[O_2(^1\Delta_g)]}. \tag{5.19}$$

Since the concentration of singlet oxygen in the experiment described is low, one may consider the adsorption to take place in the Henry region. Then

$$\theta = \Gamma\,[O_2(^1\Delta_g)]\,kT\,/\,[Z_0], \tag{5.20}$$

where Γ is the Henry constant; k is the Boltzmann's Constant; T is the temperature.

Substituting θ into the expression for γ, gives us

$$\gamma = 4\,K_2\,\Gamma kTS\,/\,vV. \tag{5.21}$$

Hence, one can see that the energy of γ change activation may be interpreted as heat of singlet-oxygen adsorption on glass.

Henry constant Γ can be evaluated by the method of Tsvitering and Krevelyn (see [105]) based on the concept of mean lifetime of a molecule absorbed on a surface proposed by De Boer [106]. The number of molecules adsorbed on a surface is

$$N_s = \tilde{n}\tau,$$

where $\tilde{n}$ is the number of molecule – surface collisions; τ is the characteristic time a molecule stay on the surface. In accordance with the Frenkel formula (see [106]

$$\tau = \tau_0 \exp(Q / RT),$$

where τ_0 is the vibration period of adsorbed molecules in the direction perpendicular to the surface; Q is the adsorption heat. The number of molecule – surface collisions is

$$\tilde{n} = \tilde{A}P / (2\pi MRT)^{1/2},$$

where $\tilde{A}$ is Avogadro's Number; P is the pressure of triplet oxygen; M is the molecular mass of the gas; R is the unversed gas constant;

$$\Gamma = \frac{\tilde{A}}{(2\pi MRT)^{1/2}} \tau_0 \exp(Q / RT). \quad (5.22)$$

Using the experimentally obtained coefficient of heterogeneous de-activation of $O_2(^1\Delta_g)$ on glass (10^{-5}) and the adsorption heat (6 kcal/mole), one can evaluate the relaxation time of singlet excitation on a surface:

$$\tau_{rel} = K_2^{-1} = 4\Gamma kT / \gamma\upsilon. \quad (5.23)$$

At $\tau_0 = 10^{-12}$ to 10^{-13} s, we have $\tau_{rel} = 10^{-3}$ to 10^{-4} s. The magnitude essentially exceeds the relaxation times of vibration-excited nitrogen on glass (10^{-8} s [107]), a fact that points to the complicated nature of $O_2(^1\Delta_g)$ deactivation on pure glass surface. As the de-excitation of singlet oxygen is concerned, a surface of pure fused quartz behaves in a manner similar to the pure glass surface. However, modification of a quartz surface with adsorbed layers may lead not only to changes in the value of γ, but also to replacement of the deactivation mechanism. An example may be work [108], a work that investigated deactivation of $O_2(^1\Delta_g)$ on a quartz surface that contains adsorbed V_2O_5 complexes.

Experiments were carried out both for a strongly oxidized surface and for a surface partially reduced by calcination in a hydrogen ambient.

For an oxidized surface, the value of γ is $10^{-4} - 1.7\cdot10^{-5}$ and it decreases with increasing the experimental temperature. In this case the activation energy of a change in γ is 2.1 kcal/mole. From these data it can be inferred that the heterogeneous de-excitation of singlet oxygen proceeds in terms of the physical adsorption mechanism similar to that described for glass.

A different situation arises with a preliminary reduced surface. In this case the measured value of γ is within $10^{-3} - 10^{-2}$, and as the temperature increases, the γ grows by the Arrhenius' Law (Equation) with the activation energy of 5.2 kcal/mole. In addition, there is dependence of γ upon the triplet oxygen pressure in the set-up, though the experiment conditions allow us to neglect a priori the impact of homogeneous processes on the spatial distribution of $O_2(^1\Delta_g)$ molecules. Prolonged treatment of the surface being surveyed by a flow of singlet oxygen leads to an irreversible increase in the value of γ.

The above data allow us to infer that in case of a reduced V_2O_5/SiO_2 surface, a chemisorption mechanism of $O_2(^1\Delta_g)$-molecule deactivation is realized. As this happens, the value of γ activation energy may be attributed to the activation energy of oxygen desorption from the oxide surface, and the dependence of γ on the pressure can be accounted for by the competition of singlet and triplet oxygen at the reduced ions of vanadium. The chemisorption mechanism of deactivating singlet oxygen will be considered in more detail in the next section.

5.5.4. Deactivation of Singlet Oxygen on a Surface of Zinc Oxide

The survey of de-exciting singlet oxygen on a zinc oxide surface is of utmost interest in principles, as well as in terms of understanding the operation of $O_2(^1\Delta_g)$ sensors. A concurrent measurement of heterogeneous deactivation and response of ZnO electrical conductivity to the interaction with singlet oxygen in conjunction with the present day sensitive techniques of investigating into adsorption make it possible to form a fairly true notion of the mechanism of de-exciting $O_2(^1\Delta_g)$ on a surface of oxide semiconductor.

The first step of surveying was represented by studying the nature of heterogeneous deactivation of singlet oxygen on a surface of oxidized and partially reduced zinc oxide [102] with the aid of semiconductor sensors. At room temperature and the pressure of $O_2 \approx 10^{-1}$ Torr, the value of γ for de-exciting $O_2(^1\Delta_g)$ in ZnO is equal or approximately equal to 10^{-4}. As the temperature rises, the value considerably increases, following the exponential law (Fig. 5.12). As this happens, the activation energy is 25 kcal/mole. The dependence of heterogeneous decay coefficient on the oxygen pressure is shown in Fig. 5.13. One can see that the γ coefficient decreases, as the pressure increases. This fact cannot be explained by homogeneous processes, since the $O_2(^1\Delta_g)$ pressure is rather low, and the de-excitation of singlet oxygen in pair collision may be neglected.

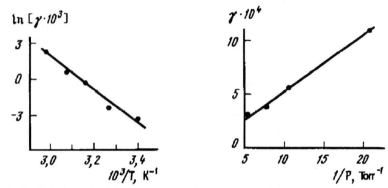

Fig. 5.12. Coefficient of heterogeneous deactivation of singlet oxygen versus the temperature of zinc oxide surface at $P_{O_2} = 0.1$ Torr [102]

Fig. 5.13. Coefficient of heterogeneous deactivation on zinc oxide surface versus oxygen pressure at 20°C [102]

The above-given results find an adequate explanation within the scope of the model supposing de-excitation of singlet oxygen due to its chemisorption on a surface of zinc oxide. A similar model was proposed in [107] to explain the heterogeneous de-excitation of vibration-excited nitrogen on a quartz surface at high temperatures. The proposed mechanism is based on the following kinetic scheme

$$O_2(^1\Delta_g) + Z \underset{K_{-1}}{\overset{K_1}{\rightleftarrows}} O_2(^1\Delta_g)Z,$$

$$O_2(^3\Sigma_g^-) + Z \underset{K_{-2}}{\overset{K_1}{\rightleftarrows}} O_2(^3\Sigma_g^-)Z, \qquad (5.24)$$

$$O_2(^1\Delta_g)Z \overset{K_3}{\rightarrow} O_2(^3\Sigma_g^-)Z.$$

According to this scheme, a loss of electron-excitation energy occurs as a result of a chemical coupling with the surface, i.e. due to chemisorption. It is supposed in this case that the lifetime of an absorbed molecule of singlet oxygen is significantly greater that the time of excitation energy loss, i.e. $K_1 \ll K_3$. By the method of steady-state concentrations, one can easily obtain the following expression

$$\gamma = \frac{4[Z]}{v} S_1 \frac{K_{-2}}{\left\{\frac{[O_2(^1\Delta_g)]}{[O_2(^3\Sigma_g^-)]} S_1 + S_2\right\}[O_2(^3\Sigma_g^-)] + 4K_{-2}\frac{[Z]}{v}} , \qquad (5.25)$$

where S_1, S_2 are the sticking coefficients of $O_2(^1\Delta_g)$ and $O_2(^3\Sigma_g^-)$, respectively; $[Z]$ is the surface density of chemisorption centres; v is the heat velocity of molecules.

It follows from the dependence of γ upon the pressure of O_2 that the degree of chemisorption centre coverage is close to unity. For this event, we obtain

$$\gamma = \frac{S_1}{\left(\frac{[O_2(^1\Delta_g)]}{[O_2(^1\Delta_g)]} S_1 + S_2\right)} \frac{4K_{-2}[Z]}{v[O_2(^3\Sigma_g^-)]} . \qquad (5.26)$$

It follows from the formula that the experimentally evaluated value of γ activation energy on the ZnO surface may be related to the activation energy of oxygen desorption from the zinc oxide surface. This value well agrees with the desorption activation energy measured with the aid of semiconductor detectors in work [109].

It should be noted that the value of γ on reduced samples of ZnO is far greater than on the oxidized samples. Besides with the singlet oxygen treatment, the coefficient of $O_2(^1\Delta_g)$ deactivation on the reduced samples continuously decreases. It is evidence that the centres of singlet oxygen chemisorption on ZnO are superstoichiometric atoms of zinc and oxygen vacancies. inference is also supported by the data on measurements of the electrical conductivity of ZnO films under the action of $O_2(^1\Delta_g)$. If the case is with reduced films, we have a signal of a decrease in the electrical conductivity, a signal traceable to oxygen chemisorption (see 5.7). A similar signal appears when other forms of oxygen $(O_2(^3\Sigma_g^-)$, O-atoms) are adsorbed on ZnO. As shown above, these forms are absorbed on excess atoms of zinc. For strongly oxidized samples of ZnO, the effect of singlet oxygen leads to an increase in the electrical conductivity, i.e. the physical ways of energy dissipation not associated with the chemisorption processes become disabled. A more complete pattern of the interaction between singlet oxygen and a ZnO surface is obtained by Zav'yalov et al. [110], a work in which precision measurements of adsorption were taken by the quartz-crystal microweighing method in parallel with measurements of electrical conductivity of zinc oxide. The microbalance that was used in this work had an oxygen 10^{-3} monolayer sensitivity. The microbalance was furnished with specific protection against thermal gradients and electromagnetic interference [111]. The high sensitivity made it possible to reliably distinguish adsorption of singlet oxygen against the triplet oxygen background. Figure 5.14 shows changes in the mass and in the electrical conductivity of ZnO films as functions of the time taken to treat the films with singlet oxygen. The measurement were taken after the adsorption equilibrium had been set in the system relative to triplet oxygen. In order to obtain $O_2(^1\Delta_g)$ use was made of a photochemical source based on quartz with V_2O_5 complexes applied to. The concentration of the $O_2(^1\Delta_g)$ was equal or approximately equal to $4 \cdot 10^{10}$ cm^{-3}.

Referring to the figure, additional chemisorption takes place under the action of singlet oxygen, a phenomenon that is accompanied by changes in the electron concentration in the ZnO conduction band. The initial rates of adsorption and changes in the electrical conductivity of ZnO are equal to $1.4 \cdot 10^{10}$ molecule/s and 10^4 electron/s, respectively. From these data one can evaluate the probability of $O_2(^1\Delta_g)$ deactivation on a ZnO surface by the chemisorption mechanism. The deactivation is equal or approximately equal to $1.6 \cdot 10^{-4}$ and agrees well with the values of the coefficient dealing with heterogeneous deactivation of singlet oxygen, the values that were obtained by the Smith method and cited above. Another parameter of importance that can be evaluated on the basis of the data given is the coefficient of singlet oxygen ionization. This magnitude makes up 10^{-6}. On a signal of change in the sensor electrical conductivity, this coefficient allows one to estimate the number of oxygen molecules adsorbed during the process of interaction between ZnO and $O_2(^1\Delta_g)$.

Prolonged seasoning of the adsorbent in triplet oxygen leads to loss of the adsorbent's activity relative to the adsorption of $O_2(^1\Delta_g)$. This result suggests an idea about the identity of chemisorption centres of singlet and triplet oxygen on a ZnO surface. This inference is supported by the temperature measurements as well [112]. The energies of activating adsorption of triplet and singlet oxygen practically coincide (33 kJ/mole). The measurements of ZnO electrical conductivity being acted upon by these two forms of oxygen also coincide and have the same values of activation energy. Neither additional adsorption of $O_2(^1\Delta_g)$ changes the shape of the curves showing changes in the electrical conductivity of oxygen-treated ZnO films when the films are subject to linear heating in vacuum. Therefore, one may infer that the singlet oxygen occupies the same centres on a ZnO surface, as the triplet oxygen does, and is adsorbed in the same states.

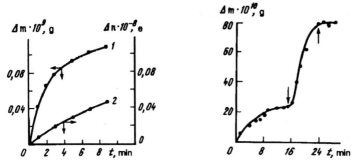

Fig. 5.14. Changes in mass (*1*) and electron concentration (*2*) in ZnO film during adsorption of singlet oxygen [100]
Weight of ZnO batch is $4.5 \cdot 10^{-6}$ g; $T = 296$ K

Fig. 5.15. Changes in mass of resonator with Co_3O_4 applied to its electrodes during adsorption of triplet and singlet oxygen [113]
Arrow-heads ↓ and ↑ designate on/off switching of signlet oxygen source. $P_0 = 3 \cdot 10^{-1}$ Torr, $T_{Co_3O_4} = 296$ K.

The high chemisorption ability of singlet oxygen compared with triplet oxygen is traceable to a significantly larger value of pre-exponential factor. In other words, molecules of singlet oxygen readily occupy adsorption centers that are not easily accessible to triplet oxygen.

The inhomogeneity of the ZnO surface with respect to the oxygen adsorption can be either of structural (inhomogeneous distribution of defects), or charge (nonuniform charging of surface during adsorption) natures. Adsorption of singlet oxygen in centres that disadvantageous from the energy standpoint to the triplet molecules may be due to a few reasons. This can be a distinction between the $O_2(^1\Delta_g)$ electron shell structure and the ground state, a distinction that promotes formation of a direct adsorption coupling with these centres. At the same time there may be processes conditioned by the effects of energy transfer. These processes are a change in the charge state of ZnO surface defects under the action of electron excitation of $O_2(^1\Delta_g)$ molecule, and also a local structural disordering in the course of excitation energy dissipation in the surface layers of the adsorbent. The latter standpoint may be supported as follows. Zav'yalova et al. [113] revealed a correlation between the efficiency of singlet oxygen adsorption by the adsorbents and the heats of adsorbents decompositions reduced to a mole of oxygen and solid phase in the lowest state. For the ZnO this parameter is 698.358 kJ, for Co_3O_4 – 478.132 kJ, and for MgO – 323.286 kJ. In case of singlet oxygen adsorption on a zinc-oxide surface presaturated with triplet oxygen, an additional change of the adsorbent mass makes up a few percent of its change due to adsorption of $O_2(^3\Sigma_g^-)$. A less strong oxide of Co_3O_4 is characterized by far more intensive adsorption of singlet oxygen. Figure 5.15 shows how the mass of a Co_3O_4 weighed sample (batch) changes during interaction with triplet and singlet oxygen. It is seen that enabling the source of singlet oxygen, after a triplet oxygen adsorption equilibrium has set in, leads to a fairly considerable increase in the adsorption. In case of interaction between $O_2(^1\Delta_g)$ and MgO at a temperature above 50°C, no additional adsorption is recorded at all, though the efficiency of singlet oxygen de-excitation on magnesium oxide is close to unity.

Another factor that supports the probability of structural disordering that may occur in the interaction between $O_2(^1\Delta_g)$ and a surface of oxides is detection of an inverse process, i.e. emission of singlet oxygen into the gaseous phase during relaxation of disorder on a surface of SiO_2 [114]. This phenomenon will be considered in detail by the next chapter.

5.6. Singlet Oxygen Interaction with Films of Dyes and Aromatic Hydrocarbons

It is an important role plaid by singlet oxygen in oxidizing organic substances that does determine the interest in the study of interaction between $O_2(^1\Delta_g)$ and the surface of these substances. Application of semiconductor sensors in this field also assists in obtaining unique data. Grigor'ev et al. [115] used the sensor technique to survey deactivation of $O_2(^1\Delta_g)$ on the surface of thick (2 – 3 microns) trypaflavine film sublimated on a glass surface. It is shown that the coefficient γ of heterogeneous deactivation on the surface of trypaflavine varies within the limits of $10^{-3} - 10^{-2}$, depending on the experiment conditions. The efficiency of de-exciting singlet oxygen materially increases with increasing of the surface temperature. The activation energy of γ change was equal or approximately equal to 12 kcal/mole. The magnitude of γ also decreases when the oxygen pressure in the set-up is increased. The general nature of the dependencies obtained is the same as that for the decay of singlet oxygen on a zinc-oxide surface. Based on these data, the authors have arrived at an conclusion about the chemisorption nature of $O_2({}^1\Delta_g)$ deactivation on the trypaflavine surface.

More comprehensive information was obtained [116] about the interaction of singlet oxygen with anthracene films. The quartz microweighing and mass spectrometry were used in this work along with the methods of semiconductor detectors.

Increasing of a weighed sample mass of anthracene caused by periodically enabling the source of $O_2({}^1\Delta_g)$ is shown in Fig. 5.16, the weighed sample temperature being 243 K. It is seen that the mass of weighed sample gains weight being acted upon by singlet oxygen, and remains unchanged when the source of singlet oxygen is disabled. The mass-spectrometric analysis of gaseous phase shows concurrent decay of partial oxygen pressure in the set-up. These facts allow one to arrive at an conclusion about adsorption of singlet oxygen on films of anthracene. Heating the samples to room temperature leads to liberation of oxygen from the surface of adsorbent. The liberated oxygen reduces the electrical conductivity of the ZnO-sensor positioned above the adsorbent, the sensor being prepassivated with respect to triplet oxygen. This fact allows one to suppose that molecules of singlet oxygen desorb from the surface together with $O_2({}^3\Sigma_g^-)$. This situation can be realized, provided the singlet oxygen at low temperatures forms unstable compounds – endoperoxides and does not participate in deep oxidation. The supposition made herein agrees with the data of work [117] that has shown that anthracene solutions chilled to 213 K are able to "stock up" singlet oxygen and to release it readily in the same state when heated.

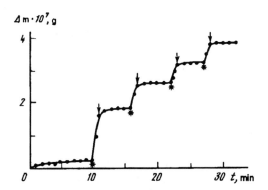

Fig. 5.16. Changes in mass of anthracene weighed sample (batch) against time during its treatment by singlet oxygen [116]

$T_{anthracene}$ = 243 K, P_{O_2} = $4 \cdot 10^{-1}$ Torr. An asterisk designates on- and an arrowhead, off-switching of singlet oxygen source

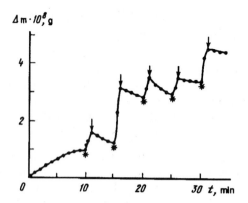

Fig. 5.17. Changes in mass of anthracene weighed sample (batch) against time during its treatment by singlet oxygen at 296 K [116]
$P_{O_2} = 2 \cdot 10^{-1}$ Torr. An asterisk designates on- and an arrow-head, off-switching of $^1O_2$

At room and higher temperatures the pattern of interaction between the singlet oxygen and anthracene changes. Changes in the mass of weighed sample of anthracene being acted upon by $O_2(^1\Delta_g)$ at room temperature are exhibited in Fig. 5.17. At this point the singlet oxygen is also captured by the anthracene, but during the intervals between the periods when the source is one observes a decrease in the mass of weighed sample. At the same time the mass-spectrometry of the gaseous phase records an increase in the partial pressure of CO_2. This suggests that a process of deep oxidation of anthracene with singlet oxygen is taking place in the system.

Therefore, the experiments in the interaction of $O_2(^1\Delta_g)$ with films of anthracene make it possible to distinguish two stages in this process. These are capture of singlet oxygen by molecules of anthracene with formation of endoperoxides and the oxidation of the latter endoperoxides by singlet oxygen, a fact that can be judged of by liberation of carbon dioxide. The first stage is rapid; it proceeds virtually without activation energy. The other stage is somewhat slower. It proceeds at a noticeable rate, at temperatures above 258 K. The result obtained allows one to explain certain principles of partial oxidation of aromatic hydrocarbons on oxide – vanadium catalysts [118], if it is supposed that some roles in these reactions is plaid by singlet oxygen formed in the course of structural-chemical transformations of the catalyst surface.

5.7. Interaction between Metastable Atoms of Rare Gases and Surface of Oxide Semiconductors

The rare gas metastable atoms (RGMAs) are an example of EEPs that feature a minimum chemical activity. While for the singlet oxygen, the major ways of heterogeneous de-excitation are represented by chemisorption and chemical reactions, the RGMAs are not characterized by these processes at all. Neither in the ground state, nor in the excited state, the rare gases are adsorbed on a solid surface at temperatures above room temperature. Therefore, one should expect that energy relaxation on a surface will occur solely by physical ways.

Large excitation energies and long lifetimes of RGMAs (see Table 5.1) make these atoms suitable subjects to study transfer of EEP excitation energy to a solid body.

This process is an important way (channel) of energy exchange between the gaseous phase and a surface in the course of heterogeneous – homogeneous, thermal, plasma, and photochemical reactions. The processes of energy transfer, seemingly, play a role of importance in such phenomena as activation of reactor walls by an explosion [119], increasing catalytic activity of the walls after discharge treatment [26], and the like. Nevertheless, the mechanism of gaseous phase EEP energy transfer to a solid body is studied, but not enough. In many aspects, this is connected with imperfection of experimental methods used for the study of the interaction between EEPs of rare gases, in particular, and the surface of semiconductors and dielectrics. The present section deals with the problems of developing sensors sensitive to metastable atoms and with application of these sensors for the study of heterogeneous de-excitation of RGMAs on a surface of pure material and on a surface promoted with oxide additives.

5.7.1. Heterogeneous Processes with Participation of Rare Gas Atoms

Metastable atoms are an important constituent of rare gas plasma. Depending on the chemical nature, these atoms are characterized by excitation energies ranging from 8 to 20 eV and by radiative lifetimes lasting from hundredth fractions of a second for $Xe(^3P_0)$ to about 2 hours for the He (^3S) state. The above-cited parameters of RGMAs account for the diversity of processes in the gaseous phase with participation of atoms in question. These processes may be excitation and deactivation under two- and three-particle collisions, chemi-ionization, Penning Effect ionization, transfer of electron-excitation energy, dissociation, dissociative ionization, chemiluminescence reactions, etc. A like analysis of RGMA mechanism in the gaseous phase can be found in reviews and monographs [6, 26, 27, 28, 29]. In this section we shall dwell upon the literature data on the heterogeneous de-excitation of metastable atoms in more detail.

Most of the publications dedicated to the interaction between the RGMAs and a solid surface refer to the rare gas – metal system. The secondary electron emission that occurs in the system allows one to judge of the mechanism that deactivates metastable atoms on a metal surface, as well as to evaluate the concentration of metastable atoms in the gaseous phase.

The earlier classical publications in this field belong to Oliphant [120]. In his survey he used a beam of rapid metastable atoms of helium obtained by neutralizing ions on the walls of platinum capillary. Oliphant was the first to observe emission of electrons from a surface of magnesium and molybdenum under the action of metastable atoms, and also rebounding of metastable atoms from a molybdenum surface.

Oliphant and Moon theoretically considered the possibility of electron emission by resonance ionization of metastable atoms near a metal surface. Shekter [122] investigated the Auger-neutralization of ions on a metal surface. Hagstrum [124, 125] carried out an generalized analysis of metastable atoms with a metal surface.

The major process that proceeds under the interaction of RGMAs with a metal surface is deactivation of RGMAs accompanied by electron emission. Two mechanisms of such deactivation are feasible [126]; for their energy diagrams, see Fig. 5.18. If the case is that the work function of metal $q\varphi$ is greater than the potential of metastable atom ionization (a), a resonance tunneling transition of excited electron is possible from the atom electron shell to an unoccupied electron energy level of metal. The formed ion of rare gas undergoes Auger-neutralization: one of the metal valence electrons occupies an unoccupied orbital of the ground state of striking atom, and the other is ejected into the gaseous phase. In this case, the secondary electrons have a wide energy spectrum whose parameters are determined by the ionization potential of the ground state of rare gas atom and by the work function of surface.

If the work function is smaller than the ionization potential of metastable state (see, Fig. 5.18b), then the process of resonance ionization becomes impossible and the major way of de-excitation is a direct Auger-deactivation process similar to the Penning Effect ionization: a valence electron of metal moves to an unoccupied orbital of the atom ground state, and the excited electron from a higher orbital of the atom is ejected into the gaseous phase. The energy spectrum of secondary electrons is characterized by a marked maximum corresponding to the difference in the excitation energy of the metastable atom and the work function of surface.

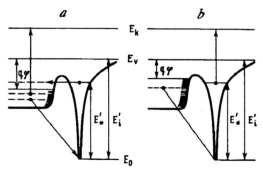

Fig. 5.18. Energy diagrams of possible mechanisms for deactivation of rare gas metastable atoms on a metal surface [126]
a – resonance ionization with subsequent Auger-neutralization; b – Auger-deactivation.
$q\varphi$ – work function of metal; E'_*, E'_i – excitation energy and ionization potential of striking atom, respectively; E_0 – energy of electron in the ground state of atom

This relationship of the metastable atom deactivation mechanisms is valid for atomically pure metal surfaces and is proved true in a series of works [60, 127, 128]. Direct demonstrations of resonance ionization of metastable atoms near a metal surface are given by Roussel [129]. The author observed rebound of metastable atoms of helium in the form of ions from a nickel surface in the presence of an adsorbed layer of potassium. In case of large coverages of the target surface with potassium atoms, when the work of yield becomes less than the ionization potential of metastable atoms of helium, the signal produced by rebounded ions disappears, i.e. the process of resonance ionization becomes impossible and the de-excitation of metastable atoms starts to follow the mechanism of Auger deactivation.

Adsorbed layers, thin films of oxides, or other compounds present on the metal surface aggravate the pattern of deactivation of metastable atoms. The adsorption changes the surface energy structure. Besides, dense layers of adsorbate may hamper the approach of metastable atom sufficiently close to the metal to suppress thus the process of resonance ionization. An example can be work [130], in which a transition from a two- to one-electron mechanism during deactivation of He* atoms is exemplified by the Co – Pd system (111). The experimental material on the interaction of metastable atoms with an adsorption-coated surface of

metals is fairly large being represented by a large number of works
[127, 128, 131 – 135].

The results of work [135] are of specific interest. The work surveyed
the influence of the nature and structure of adsorbed layers upon the
mechanism of deactivation of $He(2^1S)$ atoms. It has been shown that on
a surface of pure Ni(111) coated with absorbed bridge-positioned molecules of CO or NO, the deactivation of metastable atoms proceeds by
the mechanism of resonance ionization with subsequent Auger-
neutralization. With large adsorbent coverages, when the adsorbed
molecules are in a position normal to the surface, deactivation proceeds
by the one-electron Auger-mechanism. The adsorbed layers of C_2H_4 and
H_2O on Ni(111) de-excite atoms of $He(2^1S)$ by the two-electron
mechanism solely. In case of NH_3 adsorption, both mechanisms of deactivation are simultaneously realized. Based on the given data, the
authors infer that the nature of metastable atoms deactivation on an
adsorbate coated metal surface is determined by the distance the electron density of adsorbate valance electrons is removed from the metal
lattice.

The presence of adsorbed layers also affects the other parameters of
the interaction between metastable atoms and a metal surface. Titley et
al. [136] have shown that the presence of an adsorbed layer of oxygen
on a W(110) surface increases the reflection coefficient of helium metastable atoms. The reflection is of irregular nature and grows higher when
the incidence angle of the initial beam increases. A series of publications
[132, 136, 137] indicate that the presence of adsorbed layers causes an
increase in the quantum yield of electron emission from a metal under
the action of rare gas metastable atoms.

In the literature dedicated to the interaction of rare gas metastable
atoms with a metal surface, the most discrepant problem is represented
by the reflection in the excited state. The evaluations of reflection coefficient of metastable atoms given by various authors diverge a few orders of magnitude. Large values of reflection coefficient were obtained
in the earlier works whose authors used intensive beams of metastable
atoms having high kinetic energy. Thus, Oliphant [120] states that the
coefficient of He^* reflection from Mo rises in his conditions from 0.25 to
0.5 when the energy of ions producing metastable atoms changes from
2100 to 120 eV. Green [138] reports that the coefficient of He^* reflection from Mo ranges from 0.4 to 0.75 within the 300 – 1100 eV range of
producing ion energies. Much smaller reflection coefficients are found in
the later publications, whose authors have used metastable atom beams
with thermal energy of relatively low intensity. Thus, Allison et al.
[137] estimate coefficients of $He(2^{1,3}S)$, $Ne(^3P_{0,2})$, and $Ar(^3P_{0,2})$ reflection from a stainless steel surface to be 0.01, 0.02, and 0.03, respectively. Conrad et al. [60] tell us that the reflection of $He(2^1S)$ from

surfaces of pure Pd(220) and Cu(110) does not exceed 10^{-6} of the primary flux of excited particles. The presence of a Co layer on Pd(120) somewhat adds to the reflection coefficient, but even in this case the coefficient does not exceed 10^{-3}. At the same time, Titley et al. [136] report a well seen reflection of metastable atoms of helium from tungsten face (110). According to these authors, the reflection coefficient amounts to 0.5 at large angles of incident beam of metastable atoms. The cause of existing discrepancies lies, probably, in the diversity of surface conditions, and also in different degrees of actions exerted by primary beams having different intensities on the surface involved.

The pattern of interaction between metastable atoms of rare gases and a semiconductor or dielectric surface is not yet clear. the literature data in this field are incomplete and uncoordinated, a fact that is primarily associated with the lack of convenient techniques suitable for studying these systems.

In the course of interaction between metastable atoms and nonmetallic surfaces, some electron emission is generally possible due to the processes of surface ionization caused by the Penning Effect. Such emission was observed during the experiments in molecule crystals of aromatic hydrocarbons [139 through 143], halides of alkali metals [144], and in selenium [145]. However, solid bodies in all these cases were thin (up to 500 Å) films applied to metallic substrata, these films being in some cases [141, 142] dispersed polycrystalline structures. At the same time, no electron emission from bulky samples of Peryxes [137] were observed. There probably exist essential distinctions in terms of electron emission under the action of metastable atoms between the cases of thin films applied to a metal and bulky samples. Rapid relaxation of induced charges takes place in the thin samples due to tunneling of electrons from a metallic substratum. As this happens, the flux of metastable atoms should not strongly affect the electronic properties of the target surface. With the bulky samples, the relaxation of charges proceeds slowly. Should it be that at first some electron emission takes place under the action of metastable atoms, the process must soon cease due to changes in the electronic properties of the surface. Therefore, the emission methods are not suitable for studying deactivation of metastable atoms of rare gases on bulky nonmetallic surfaces.

Because of these factors, the fundamental experimental information about the interaction of metastable atoms with semiconductors and dielectrics is meant for the reflection coefficients that are determined with the aid of beam methods and for the coefficients of heterogeneous deactivation which are evaluated under diffusion conditions. However, the data in this event are fairly scarce and conflicting. The results obtained by the methods of electronic beams do not agree with diffusion experiments. Thus, Allison et al [137] report that the coefficients dealing with reflection of He, Ne, and Ar metastable atoms from a Peryx surface are of the order of 10^{-2}. According to Conrad et al. [60] the reflection does not exceed 10^{-4} in dispersion of metastable atoms of He and Ar over a surface of LiF and NaCl monocrystals. In the event of Ne, the reflection is 10^{-2}. Craig and Dickinson [146] have established that excited atoms of He, Ne, and Ar are not practically reflected from a pure germanium face (100), but the reflection factor rapidly rises when oxygen is absorbed at the face. At the same time, Lisitsin et al. [147] have found out (by the adsorption spectroscopy) that the coefficient of heterogeneous deactivation of helium 2^3S and 2^1S on glass under the glow discharge conditions was 0.11 and 0.35, respectively, figures that correspond to 89% and 65% reflection of bombarding atoms. Using the same method, Sadeghi and Pebay-Peyroula [148] gave an estimation of Pyrex reflection coefficients under the gas pulse-excitation conditions. The estimation is $R_{ref} < 0.75$ for He (2^3S) and $0.15 < R_{ref} < 0.5$ for Ar($^3P_2$). An indirect estimation of the reflection coefficient of Ar($^3P_2$) atoms from a quartz surface in discharge afterglow was given by Tabachnik et al. [46]. The estimation value reads $R_{ref} < 0.5$. A group of Canadian authors [44, 149] have obtained radial profiles of the density distribution underwent by metastable atoms of Ar($^3P_2$) in case of gas excitation by electrodless high-frequency discharge. The profiles are nearly uniform, and although the authors have made no attempts to evaluate the coefficients of heterogeneous deactivation in this case, it is of reason to suppose that they are fairly small.

As evident from the above-given data the two above-mentioned methods of evaluating the coefficients of metastable atom reflection from nonmetallic surfaces yield essentially different results. The information found in the literature is yet insufficient to understand the cause of this discrepancy.

Still more impressive discrepancies result when one works under the diffusion conditions with low and very low concentrations of excited atoms. Thus, the authors of work [150] have repeatedly found metastable atoms in the gas far later after the excited discharge was disabled by using a very sensitive method of detecting metastable atoms based on the principle of initiating gas discharge. For helium periods of after time amount to thousands of seconds, and for other rare gases, to hundreds of seconds, i.e. the periods of after time are comparable with the radiative lifetimes. This result is relevant to negligibly small coefficients of heterogeneous deactivation. Bosan and Lukic [151] have recorded the presence of N_2 molecules and Ar atoms in the long-lived metastable states of inherent gas flow, 450 cm from the source at a flow rate of 10 cm/s. This result also indicates low efficiency of heterogeneous de-excitation of metastable states in the environment of the described experiment.

Note that a similar situation arises in the study of heterogeneous deactivation of electron-excited molecules of N_2. Thus, an opinion expressed by Clark et al. [152] states that the coefficients of heterogeneous deactivation of $N_2(A^3\Sigma_u^+, v = 0.1)$ for all surfaces are close to unity. On the other hand, Vidaud with his coworkers [59, 153] have obtained $3\cdot10^{-2}$ and $(1.8 \div 1.2)\cdot10^{-5}$ values for these coefficients shown by platinum and Pyrex, respectively. Tabachnik and Shub [154] investigated heterogeneous decay of $N_2(A^3\Sigma_u^+)$ molecules on a quartz surface by the method of bulk-luminescence spectroscopy. The authors carried out a series of experiments within a broad (about four orders of magnitude) range of active particle concentrations and arrived at a conclusion that at a concentration of $N_2(A^3\Sigma_u^+)$ in excess of 10^8 mole/cm^3, the heterogeneous deactivation proceeds at high efficiency ($\gamma > 0.5$). At lower concentrations of excited molecules, the heterogeneous deactivation coefficient is small and equals $7\cdot10^{-4}$. The causes of this concentration dependence are unknown. However, one may suppose that the same dependence takes place in case of heterogeneous de-excitation of metastable atoms of rare gases as well.

The absence of sufficient amount of reliable experimental data accounts for the complete vagueness in the field of understanding the mechanism of energy transfer during deactivation of metastable atoms of rare gases on a surface semiconductors and dielectrics. Probably, there exist a few probable ways of energy transfer during processes like this. Some prevalence of one or other way is most likely dictated by the particular properties of the excited particle – surface pair. For example, Svejda et al. [155] have observed formation of radicals on a hydrogenated surface of MgO being acted upon by excited atoms of helium and argon. The formation of radicals was observed in a flow-type system by the EPR (electron paramagnetic resonance) technique. At the same time, the authors of work [156] propose an electron excitation method of deactivating $Xe(^3P_2)$ atoms on the surface of a solid xenon matrix. The possibility of an electron excitation mechanism of deactivating $Ar(^3P_2)$ on a quartz surface is mentioned by Tabachnik at al. [46]. Radiative deactivation of metastable atoms near the surface is also probable due to some broadening of levels in the field of surface potential and radiationless transition of an excited electron to an allowed level [46, 126]. Discovery of a change in the electrical conductivity of thin semiconductor films being acted upon by metastable atoms in work [157] speaks of the possibility that energy can be transferred to the electron subsystem of a solid body. All possibilities mentioned, however, are insufficiently studied and remain far from the stage of theoretical discussion.

Doyen [158] was one who theoretically examined the reflection of metastable atoms from a solid surface within the framework of a quantum – mechanical model based on the general properties of the solid body symmetry. From the author's viewpoint the probability of metastable atom reflection should be negligibly small, regardless of the chemical nature of the surface involved. However, presence of defects and inhomogeneities of a surface formed by adsorbed layers should lead to an abrupt increase in the reflection coefficient, so that its value can approach the relevant gaseous phase parameter on a very inhomogeneous surface.

From the above-made review of literature, one may infer that the interaction of metastable atoms of rare gases with a surface of semiconductors and dielectrics is studied, but little. The study of the mechanism of transferring energy of electron-excited particles to a solid body during the processes under discussion is urgent. The method of sensor detection of rare gas metastable atoms makes it possible to obtain new information about the heterogeneous de-excitation of metastable atoms inasmuch as it combines high sensitivity with the possibility to conduct measurements under different conditions.

5.7.2. Rare Gas Metastable Atom Sensors Based on the Au/ZnO Structure

Panesh et al. [157] were the first to make an attempt to detect rare gas metastable atoms (RGMAs) with the aid of semiconductor sensors. The sensing element (a sensor) was represented by a sintered polycrystalline film of ZnO; metastable atoms were obtained in a neon ambient by electron impact. It was shown that electrical conductivity of ZnO film irreversibly increases under the action of RGMAs. However, the signals obtained were too small and that did not allow one to utilize the sensing technique to survey the processes with participation of metastable atoms.

The most obvious way to raise the sensitivity of sensors to RGMAs is by activating their surface with additives that actively interact with metastable atoms and have some electron coupling with semiconductor. These additives can be microcrystals of metals. As previously shown, the de-excitation of RGMAs on a metallic surface truly proceeds at high efficiency and is accompanied by electron emission. Microcrystals of the metal being applied to a semiconductor surface have some electron coupling with the carrier [159]. These two circumstances allow one to suppose that the activation of metals by microcrystals adds to the sensitivity of semiconductor films to metastable atoms.

This supposition is experimentally substantiated by Kupriyanov et al. [160]. In this work they investigated the influence of RGMAs upon the electrical conductivity of pure zinc oxide films and films activated by microcrystals of gold. The gold was chosen as the activator because of its chemical inactivity and high lateral mobility. This makes it possible to obtain "islet" films on a ZnO surface at room temperature, thus avoiding probable metallurgical processes.

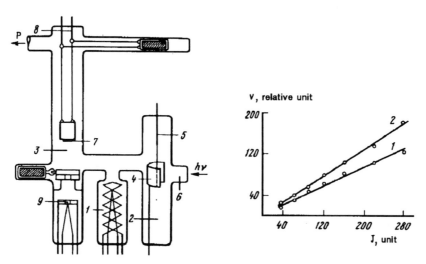

Fig. 5.19. Experimental set-up to study the influence of metastable atoms of rare gases on electrical conductivity of Au/ZnO films [160]
1 – source of excited particles; 2, 3 – measurement chambers; 4 – magnesium coated nickel cathode; 5 – Collector of secondary electrons; 6 – quartz port; 7 – ZnO film; 8 – sliding contacts; 9 – Au atomizer

Fig. 5.20. relationship between the starting velocity v of changes in electrical conductivity of ZnO films with different surface concentration of gold and the current of secondary electron emission J from Mg surface under the action of He* [160]. The Au concentration grows in series $2 > 1$

The experiments were carried out with the aid of a set-up exhibited in Fig. 5.19. The source of RGMAs was a d.c. discharge with a glowing cathode arranged in gas diode 1. The operating pressure of gases (helium, neon, krypton, xenon) was 10^{-2} Torr. The voltage across the electrodes was varied within the range from 10 to 80 V. The discharge current was 4 – 5 mA. The metastable atoms were diffusing from the discharge zone to measurement chambers 2 and 3. Chamber 2 is an electrometer cell used to take measurements of RGMAs concentration by the secondary electron emission. Electrons were knocked out of nickel cathode *4* coated by vacuum-atomized magnesium and collected at collector *5*. Chamber *3* is designed for developing the sensor methods of detecting RGMAs. This chamber includes a cylinder housing a carriage with zinc oxide film that can mover axially and gold atomizer *9* separated out by a glass slide. The electrical conductivity of the film was measured by means of a d.c. bridge circuit via platinum sliding contacts *8*.

Some preliminary experiments show that a very weak signal indicating an irreversible increase in the ZnO film electrical conductivity similar to the signal obtained in work [157] was observed under the action of RGMAs. The pattern changes after application of Au microparticles to the film. One observes a decrease in the film electrical conductivity when the metal-activated ZnO film is acted upon by RGMAs and relaxation of the conductivity back to the starting value after disabling the RGMA source. The change in the electrical conductivity of gold-activated ZnO film being acted upon by RGMAs exceeds about two orders of magnitude the similar value for a nonactivated film and amounts to a few percent of the starting value.

Simultaneously with the signal of changes in the electrical conductivity of the Au/ZnO film, a current was recorded of electrons being picked by metastable atoms out of the magnesium cathode. Inasmuch as the method of recording RGMAs by the current of secondary electrons is not absolute owing to the vagueness of quantum yield, relative measurements were taken. The concentration of RGMAs was being varied by changing the anode voltage across the source of active particles. It turns out that with helium, the current of secondary electron emission appears at some anode voltage of 20 V. This approximates the excitation threshold of metastable atoms of helium. With increasing of the anode voltage the emission current rises. In this case the general shape of the curve is similar to the plot exhibiting the fraction of inelastic collisions as a function of the electron energy obtained during Maier – Leibnits experiments [161] dealing with helium excitation by electron impact. Allowing for the fact that metastable atoms arise in the discharge source not only due to direct excitation by electron impact, but also in transitions from the above found excited states and ion neutralization, one can infer that the secondary electron current observed in the course of these experiments is a current traceable to the interaction between metastable atoms of helium and the magnesium surface.

The signal of changes in the electrical conductivity of the Au/ZnO due to changes in the anode voltage across the source also has a threshold of appearance at about 20 volts. With an increase in the anode voltage, the starting velocity of change in Au/ZnO electrical conductivity rises in the agreement with (has similar curve slopes) the emission current of electrons knocked by helium metastable atoms out of the magnesium cathode. The values of the emission current and starting velocity of changes in the electrical conductivity of Au/ZnO films are compared in Fig. 5.20 for different concentrations of He metastable atoms. Referring to the figure, the proportionality between these values remain unchanged within the whole of the studied concentration range, the concentration lying in this case within $10^5 - 10^6$ atoms/cm^2. From the comparison of straight lines *1* and *2* it follows that the proportionality is not disturbed when the concentration of applied gold varies.

Allowing for the fact that the secondary electron current is in proportion to the concentration of metastable atoms in gaseous phase [62], one may infer that the Au/ZnO structures are highly sensitive proportional sensors of helium metastable atoms. Similar results were obtained for other rare gases.

The sensor detection of RGMAs involves a series of mechanical problems associated with the nonselectivity of gaseous- phase excitation. On enabling the gas excitation source, the origination of RGMAs in the source is accompanied by the birth of other active particles that can reach the source and affect the sensor electrical conductivity. The problems of assessing such stray phenomena and selecting an optimum source of RGMAs are considered in [39]. Different types of discharge (high frequency-frequency, ultra-high frequency, and d.c. discharge with glowing anode) were tried to evaluate the contribution of the following factors to the changes in the Au/ZnO electrical conductivity:

(1) getting of charged particles onto the film;

(2) optical radiation permeating from the gas excitation zone;

(3) desorption of chemically-active impurities from the walls and the impurities' activation during discharge.

A conclusion has been drawn based on the results of the studies that a most convenient source of RGMAs suitable for sensor measurements is represented by discharge sources of the diode type, sources that operate in the d.c. discharge, glowing cathode mode.

Another methodical trait of the Au/ZnO sensor application to detect metastable atoms of rare gases is the limitation of the range of operating temperatures. When heated to above 500 K, these sensors irreversibly loose their sensitivity to RGMAs. The loss of sensitivity is associated with the coalescence of Au microcrystals applied to a ZnO surface. The causes of this will be discussed later.

5.7.3. Mechanism of Changes in the Au/ZnO-Sensor Electrical Conductivity under the Action of Rare Gas Metastable Atoms

Variation of the Au/ZnO – sensor electrical conductivity under the action of RGMAs is a complicated process including a stage of restructuring the semiconductor electron subsystem due to the excitation energy of RGMAs deactivated on the sensor surface. Metastable atoms truly exert only minute effects on the electrical conductivity of nonactivated films of ZnO. At the same time, the maximum sensitivity of Au/ZnO-sensors to RGMAs is realized in the region of Au surface concentrations within $10^{15} - 10^{16}$ atoms/cm^2. when the electrical conductivity of applied islet-like films of Au is shunted by a comparatively low-resistance semiconductor substratum. Therefore, the change of the Au/ZnO electrical conductivity under the action of RGMAs occurs owing to changes in the electron properties of the semiconductor. involved. The acceptor nature of the signal is evidence that in this case the nature of changes in the electron properties of ZnO drastically differs from the sensitizing of semiconductors by dyes [63, 64] and the nature must be described by another mechanism of deactivation.

An important role in the mechanism is plaid by the applied particles of metal. One can conclude so from the experiments in studying heterogeneous deactivation of RGMAs on a pure and Au microcrystal-activated surfaces of glass and zinc oxide [162]. The experiments were conducted by the Smith method, Au/ZnO sensors being used as RGMA detectors. The results of these investigations are tabulated in Tables 5.3.

TABLE 5.3

Surface	Weight thickness of Au film, Nm	Average size of particle in film, Nm	Metal-coated fraction of surface	Coefficient of heterogeneous deactivation, γ		
				He$^\bullet$	Kr$^\bullet$	Xe$^\bullet$
Glass	0	0	0	$7.5 \cdot 10^{-4}$ $2.2 \cdot 10^{-3}$	$7.3 \cdot 10^{-3}$	$3.7 \cdot 10^{-3}$
Au/Glass	0.2	4.0	0.13	0.20	–	–
Au/Glass	3.3	12.0	0.45	0.45	–	–
ZnO	0	0	0	$4.2 \cdot 10^{-3}$	$4.2 \cdot 10^{-3}$	$4.6 \cdot 10^{-3}$
Au/ZnO	0.2	4.0	0.13	0.13	0.10	0.25
Au/ZnO	3.3	12.0	0.45	0.50	–	–

Referring to the above-tabulated data, one can see that the coefficient of heterogeneous deactivation of RGMAs on gold-activated samples is two orders of magnitude greater than that on pure samples, the coefficient of heterogeneous deactivation, γ being weakly dependent on the substratum nature and metastable particle. Note the correlation between the value of γ and the sample surface fraction plated by the metal of islet film. It is of reason to suppose that the Au microcrystal involved by the described experiment interact with RGMAs in a manner similar to bulk gold, i.e. essentially each collision of a metastable atom with an Au particle leads to deactivation. At the same time, the interaction of RGMAs with the substratum surface mostly leads to the reflection in the excited state, a factor that is proved true by the independence of the results of the chemical nature of the substratum.

It should be noted that the values of γ obtained in this work for pure substrata are fairly small and do not agree either with the coefficients of metastable atom refection from a glass surface [137], or with the values of γ obtained in [147]. However, the above-mentioned values of γ do agree with the data of work [150, 151]. The cause of discrepancies may lie in the state of sample surfaces. Thus, Kupriyanov et al. [162] show that strongly adsorbed water abruptly increases the glass activity with regard to de-exciting He*. Since water is one of the main components of a vacuum system, it comes as no surprise that distinct experiments produce distinct results. The more so as the coefficient of RGMA heterogeneous deactivation is dependent upon the concentration.

The next step in surveying the mechanism of de-exciting RGMAs on an Au/ZnO surface is the comparative study of the secondary emission of electrons and changes in the electrical conductivity of gold-activated films of zinc oxide [164]. The ZnO film under survey was applied to the internal surface of quartz substratum in the form of a hollow open cylinder. An electron collector was axially placed in the cylinder. Sintered-in platinum contacts allowed the U_c potential to be applied to the film in order to collect or retain emitted electrons, and to measure the electrical conductivity of the film. The source of metastable atoms was represented by a glowing cathode d.c. discharge. First measurements were taken at the pure ZnO film. The gold was applied to its surface with the aid of a calibrated thermal source.

No electron emission from the pure ZnO surface were detected under the action of RGMAs. After the ZnO had been activated with gold, an electron current at the collector was recorded. The volt – ampere characteristics of the electron collector with the Au/ZnO film being acted upon by helium metastable atoms of different concentrations is shown by Fig. 5.21,a. The U_c voltage is equal to –30 V and corresponds to the complete collection of emitted electrons. The U_c = +30 V means their complete retention. The concentration of He* atoms was set by the anode voltage of the source and increase in a series $1 < 2 < 3$. Referring to the figure , the current being measured increases with the concentration of He* in the gaseous phase, the derivative maximum on the retaining curves lies in the region of U_c = +15 V, a value that approximately corresponds to the difference between the excitation energy of He(^{3}S) (19.82 eV) and the gold yield work (4.7 – 5.3 eV) [165]. There is no reason to speak of complete coincidence in view of value uncertainty of the potentials in the set-up. The curves of electron collector of retention current with the Au/ZnO film being acted upon by metastable atoms of He and Ne are shown in Fig. 5.21,*b*. Curve *1* (Au/ZnO−Ne*) has the derivative maximum in the region of U_c = +11V, a value that corresponds to the difference between the excitation energy of Ne (16.6 eV) and the work of Au yield. The above-given experimental data clearly show that the collector current is traceable to the emission of electrons from the surface of gold particles applied to ZnO under the action of RGMAs. Of utmost interest are the dependence of the secondary emission current and of the signal indicating changes in the Au/ZnO electrical conductivity being acted upon by RGMAs on the amount of gold.

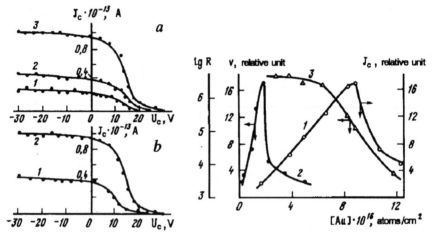

Fig. 5.21. Volt – ampere characteristic of the secondary electron collector with the Au/ZnO film being acted upon by helium metastable atoms of different concentration (*a*) and neon (*1*) metastable atoms (*2*) (*b*) [164]

Fig. 5.22. Changes in parameters of interaction between Au/ZnO and He* as the concentration of applied metal increases [164]
1 – current of electron external emission; *2* – starting velocity of Au/ZnO electrical conductivity signal; *3* – impedance of Au/ZnO film

For the results of measurements, see Fig. 5.22. Curve *1* describes the electron emission current as a function of the Au concentration on the ZnO surface at a constant concentration of He* atoms in the gaseous phase. This dependence is of extremal nature with a linear rise and a maximum peak at the Au concentration of $0.9 \cdot 10^{17}$ atoms/cm^2. A similar curve having its maximum at the same point resulted for Ne* atoms. It can be seen from curve *3* that near the emission current maximum, an abrupt (by more than three orders of magnitude) drop of the Au/ZnO film resistance commences, a drop that is associated with the coalescence of Au islet and formation of continuous metallic film on the zinc oxide surface.

The dependence of the initial velocity of changes in the electrical conductivity of Au/ZnO being acted upon by He* on the surface concentration of Au (curve 2) also has its clear maximum at some point of $2\cdot10^{16}$ atoms/cm^2. The fact that peaks of curves 1 and 2 are abscissa-spaced apart almost an order of magnitude engages our attention, i.e. we see that there is no direct correlation between the secondary emission currents and the sensitivity of Au/ZnO film to metastable atoms.

Comparing the above-mentioned results with the data of morphological survey of islet films of gold on a ZnO surface [116 – 168] leads to an inference that the maximum of curve 1 is associated with changes in the geometric surface of the Au islet film as it grows, while the maximum at curve 2 is connected with changes in the mean size of microcrystals in the islet film.

The information obtained has made it possible to propose a mechanism of interaction between RGMAs and Au/ZnO structures based on the concept of some change in the nature of the interaction between the metal microcrystal applied and the substratum as its size increases.

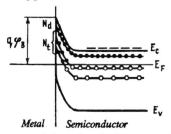

Fig. 5.23. Band diagram of the Schottky barrier at the gold – zinc oxide interface

As will be recalled [169, 170], the gold-to-zinc oxide contacts are formed in a manner similar to the Schottky barrier type. A diagram exhibiting such a contact appears in Fig. 5.23. The zinc oxide is a wide-band semiconductor with admixture of superstoichiometric zinc, an admixture that provides n-type conductivity. The polycrystalline sintered films of ZnO are characterized by a quasi-continuous spectrum of surface states and deep traps of the donor-type in the bandgap, factors that influence the formation of a metal-semiconductor contact, cause, in particular, fixing of the Fermi level on the surface. Thus, Rabadanov et al. [171] show that various metals being thermally deposited on a ZnO surface form nonohmic contacts having a barrier of 0.3 – 0.5 eV, the barrier height being dependent, but little, on the work function of deposited metal, a fact that is indicative of high density of states in the semiconductor bandgap, near the interface. In the depletion zone of the Schottky barrier part of the deep traps reach the Fermi level due to band bending and gives off electrons to the conduction band, gaining in that a positive charge. Emission of electrons under the action of metastable atom from the metallic particle in contact with the semiconductor is equivalent to applying a positive bias to the barrier. The band bending decreases, part of charged traps lower beneath the Fermi level and can capture an electron from the conduction band in the time the barrier remains in the nonequilibrium state. After compensating for the induced charge, the barrier goes over to the starting state, and the traps capturing electrons must give them back to the conduction band. However, this inverse process proceeds far more slowly than the direct process with relaxation times of tens and hundreds of seconds owing to the slowed relaxation phenomenon [172] characteristic of wide-band spatially-inhomogeneous semiconductors. Therefore, electrons are accumulated at the traps during the interaction between the Au/ZnO structure and RGMAs. As this happens, the semiconductor electrical conductivity must perceptibly decrease. Upon switching off the RGMAs source, the sensor electrical conductivity must, obviously, relax to get its starting value.

From the mathematical standpoint, this model can be formulated as follows. Neglecting the intrinsic conductance, the condition of electrical neutrality of the semiconductor may be

$$N_d + N_t - n_t = n_q,$$
(5.27)

where n_q is the concentration of electrons in the conduction band; N_d is the number of ionized donors; N_t is the number of ionized deep traps in the equilibrium state; n_t is the number of deep traps that have captured electrons in the course of interaction with RGMAs.

The rate of charges accumulation at the traps is in direct proportion to the concentration of vacant traps $(N_t - n_t)$ and to the frequency of interactions between the metallic particles on the ZnO surface and RGMAs, which in turn is proportional to their concentration in the gaseous phase (N_m). The relaxation of charge at the traps can be described by instant time of relaxation $\tau(t,T)$ that are generally a time and a temperature functions. The equation that describes the process of trap recharging is as follows

$$\frac{dn_t}{dt} = -\frac{n_t}{\tau(t,T)} + AN_m(N_t - n_t),$$
(5.28)

where A is the coefficient of proportionality.

This equation can be readily integrated by quadrature, and the initial velocity of changes in the electrical conductivity of the Au/ZnO structure can be obtained in the form

$$\left.\frac{d\sigma}{dt}\right|_{t\to0} = -\mu q N_t A N_m \, , \qquad (5.29)$$

where q is the electron charge; μ is the electron mobility in the conduction band of the semiconductor.

By this means the experimental fact of the proportionality of the starting velocity of change in the electrical conductivity of the Au/ZnO-sensor and in the concentration of RGMAs in the gaseous phase. The above-described approach also allows one to explain the dependence of sensor sensitivity to metastable atoms on the concentration of activator metal on the semiconductor surface (see Fig. 5.22, curve 2). In the course of thermal deposition of islet gold films on zinc oxide at room temperature, the saturation concentration of microcrystal nuclei commences at the Au concentration of 10^{15} atoms/cm^2 [167], and the further growth of the film proceeds due to the growth of the nuclei formed without perceptible changes in their density [173]. At this stage, the number of microcrystals on the ZnO surface is large, while their sizes are fairly small, so that changes in the charge of Au particles being acted upon by RGMAs have a strong effect on the state of the near-contact regions of the semiconductor. For example, knocking an electron out of a particle, a few nanometers in size, is equivalent to applying a few-volt bias to the barrier. Growth of particular stable islets adds to the Au/ZnO sensitivity to RGMAs, inasmuch as the dimensions of the near-contact regions increase, and the probability of metastable atom striking a metallic particle increases.

The coalescence of Au microcrystal in the islet film occurs during thermal annealing, as well as when the metal concentration is increased. It is probably the coalescence that does determine the above-mentioned loss of Au/ZnO-sensor sensitivity during high-temperature heating operations.

To further substantiate the proposed model, they have carried out some investigations connected with modification of semiconductor electron subsystem [174, 175]. Temperature is one of the important factors. Having no effect on the electron emission from the metal under the action of RGMAs, temperature strongly affects the current-transfer processes at the metal – semiconductor contacts. The impact of temperature on the interaction of RGMAs with Au/ZnO structures can be evaluated as follows.

The rate of electron accumulation at ionized traps in the depletion zone of the Schottky barrier in the Au/ZnO contact is in proportion to the concentration of unoccupied traps, frequency of metal particle/metastable atom interaction events, and to the probability of electron capture per a trap in a single event of interaction between metastable atoms and metal particle.

The frequency of interaction events can be evaluated as

$$\eta = \frac{1}{4} N_m \upsilon_m \psi S_M N_M ,$$ (5.30)

where N_m is the concentration of RGMAs in the gaseous phase; υ_m is the heat velocity of RGMAs; ψ is the quantum yield of emission of electrons from the metal particles being acted upon by RGMAs; S_M is the mean surface area of metal particle; N_M is the density of metal particles on the semiconductor surface.

The probability of charging a trap when a metal particle is ionized by a metastable atom can be characterized by the ratio of the time t_n the barrier stays in nonequilibrium state to the time constant τ_t of trap recharging.

The value of t_n is the time taken to compensate for a charge arising at a metal particle as a result of its interaction with a metastable atom. This time can be evaluated within the scope of the theory of current transfer over the barrier [176] and in a first approximation it takes the form

$$t_n = \frac{ckg}{A^* qT} \exp\left(\frac{q\varphi_B}{kT}\right) ,$$ (5.31)

where c is the capacitance of the metal-semiconductor contact; k is the Boltzmann's constant; A^* is the Richardson efficient constant; g is the coefficient of barrier ideality; φ_B is the barrier height; q is the electron charge.

The value of τ_t is evaluated [177] as $\tau_t = (2C_n n_{q1})^{-1}$, where C_n is the coefficient of electron capture by a trap; n_{q1} is the local density of electrons at the instant the Fermi level intersects the trap level when a direct bias signal is applied to the barrier.

The values of t_n and τ_t in fact must vary from contact to contact, as dictated by the conditions on the surface, depth position, and physical nature of traps, i.e. averaging is required for all metal-semiconductor microcontacts and all types of traps. We shall assume that all barriers and all traps are similar. If that is the case, one can write coefficient A in the second term of the right-hand side of equation (5.28) that describes the rate of charges accumulation at the traps as follows

$$A = b \exp\left(\frac{q\varphi_B}{kT}\right),$$

(5.32)

where $b = (\alpha ckqv_m\psi S_M N_M C_n n_{q1}) / (2A^*Tq)$ is the coefficient of proportionality that weakly depends on temperature (α is a constant factor). Substituting (5.32) into (5.31) explicitly gives us the temperature dependence of the signal caused by changes in the electrical conductivity of Au/ZnO sensor being acted upon by RGMAs:

$$\left.\frac{d\sigma}{dt}\right|_{t\to 0} = -\mu q N_t b N_m \exp\left(\frac{q\varphi_B}{kT}\right).$$

(5.33)

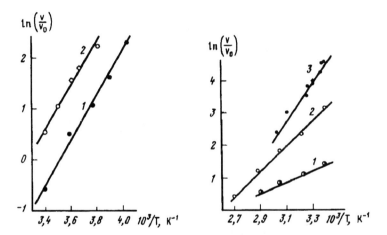

Fig. 5.24. Temperature dependence of the starting velocity of change in Au/ZnO film electrical conductivity being acted upon by helium (1) and xenon (2) metastable atoms [174]

Fig. 5.25. Temperature dependence of the starting velocity of change in Au/ZnO film being acted upon by helium metastable atoms after different preliminary treatment [174] (1 – film treated by active hydrogen; 2 – starting film; 3 – film treated by active oxygen)

The experiments carried out by Kupriyanov et al. [178] support the validity of equation (5.33). Figure 5.24 shows the temperature dependencies of the starting velocity of changes in the electrical conductivity of Au/ZnO film being acted upon by metastable atoms of helium and xenon. From the figure we notice that these dependencies (curves) yield straight lines when plotted on the Arrhenius coordinates, and that the active energy of activation is negative. For helium the energy is 0.38 ± 0.03 eV, and for xenon, 0.36 ± 0.05 eV. These values coincide within the experiment accuracy, in spite of a large distinction between the excitation energies of $He(^3S)$- and $Xe(^3P_2)$-atoms. It is of reason to infer that the value of efficient energy of activation is in fact characteristic of the averaged height of Schottky barrier at the interface between the Au microcrystal and zinc oxide.

This inference is also supported by the results of the experiments associated with the preliminary adsorption of chemically active gases on the Au/ZnO surface (Fig. 5.25). The adsorption of atomic hydrogen (straight line *1*), an adsorption that reduces the barrier leads to a decrease in the absolute value of the signal indicating changes in the Au/ZnO film electrical conductivity. The oxygen treatment that increases the barrier adds to the same values (straight line *3*). Treatment with active nitrogen produces an effect similar to the oxygen treatment.

The results described are of particular interest as well in terms of the Au/ZnO-sensor selectivity of RGMAs. Inasmuch as the signal indicating changes in the electrical conductivity of semiconductor films, a signal that is associated with the adsorption of chemically active particles, decreases as the temperature decreases, while the signal caused by RGMAs rises in that, one can select a temperature range that permits reliable detection of metastable atoms against the background of other particles.

Experiments with joint effects of RGMAs and optical radiation [175] upon a sensor are direct evidence that deep traps do participate in the formation of an Au/ZnO-sensor signal. If the Au/ZnO structure is exposed to light in the red- and near infrared-region, the result may be some change in the concentration of ionized traps in the surface region of semiconductor, having avoided in doing so the effects associated with band-to-band transitions and injection of hot electrons into the semiconductor from the metal [178, 179].

Let n_t^1 be the concentration of uncharged traps in the depletion band of the Au-ZnO microparticle contacts in the equilibrium conditions. Being exposed to light, the traps will be ionized giving electrons to the conduction band of the semiconductor. If we designate the concentration of nonequilibrium ionized traps N_t^1, then the rate of traps accumulation under illumination will be

$$\left(\frac{dN_t^1}{dt}\right)_{accumulation} = I\tilde{\sigma}(n_t^1 - N_t^1),$$

(5.34)

where I is the intensity of incident light; $\tilde{\sigma}$ is the cross-section of optical excitation of electrons from the traps.

Simultaneously a process of recombination proceeds. In our case, when the level of optical excitation is low, the rate of recombination can be evaluated [180] as follows

$$\left(\frac{dN_t^1}{dt}\right)_{\text{recombination}} = C_n N_t^1 n_q \,, \tag{5.35}$$

where C_n is the coefficient of electron capture at the traps; n_q is the concentration of electrons in the conduction band. The stationary concentration of nonequilibrium ionized traps under continuous illumination from (5.34) and (5.35) will be

$$(N_t^1)_{\text{st}} = n_t^1 \left(1 + \frac{C_n n_q}{I\tilde{\sigma}}\right)^{-1}. \tag{5.36}$$

Note that all parameters mentioned in the formulas should be considered as certain averaged magnitudes characteristic of the entire aggregate of the Au-ZnO microcontacts on the sensor surface.

When the illuminated part of Au-ZnO film is acted upon by RGMAs, the light-ionized traps will take part in recharging, as well as the equilibrium ionized traps N_t. Substituting the sum $(N_t + N_t^1)$ into equation (5.28) in place of N_t, we shall obtain an expression for the starting velocity of change in the electrical conductivity of the illuminated Au/ZnO structure under the action of RGMAs:

$$v = \frac{d\tau}{dt}\Big|_{t\to 0} = -\mu q A N_m \left[N_t + \frac{n_t^1}{1 + C_n n_q / I\tilde{\sigma}}\right]. \tag{5.37}$$

The designations used herein are taken from (5.28).

Substituting the value $v_0 = -\mu q A N_m N_t$ corresponding to the dark conditions into (5.37) will give us

$$\frac{v}{v_0} = 1 + \frac{n_t^1}{N_t}\left(1 + \frac{C_n n_q}{I\tilde{\sigma}}\right)^{-1}. \tag{5.38}$$

This expression describes well the illumination intensity dependence of the signal indicating changes in the electrical; conductivity of Au/ZnO being acted upon by RGMAs. Figure 5.26 illustrates the ex-

340

perimentally obtained dependence of v/v_0 upon the intensity of incident light, dependence that relates to the interaction between the Au/ZnO film and the metastable atoms of helium. From the figure we notice that the mentioned dependence plotted on coordinates

$$1/I - \left(\frac{v_1}{v_0} - 1\right)^{-1}$$ is linear, i.e. expression (5.38) consistently describes

the phenomena under experiment.

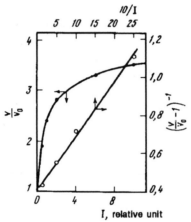

Fig. 5.26. Dependence of the starting velocity of change in the electrical conductivity of illuminated Au/ZnO film being acted upon by He* upon light intensity [175]

Summing up the results of the investigation described, one can fancy the following mechanism of the interaction between RGMAs and a metal-activated surface of semiconductor. The metastable atoms are mainly de-excited at metal microcrystals. The de-excitation is accompanied by emission of electrons into the gas-phase. Most excitation energy of incident metastable atoms is carried away in the form of kinetic energy of secondary electrons. A smaller part of energy is transferred to the electron subsystem of semiconductor due to strong metal – carrier coupling. This energy accounts for restructuring of spatial charge in the depletion bands of the metal – semiconductor contacts. In the course of this restructuring, electrons from the conduction band are deep-trap captured in the conduction band gap. Upon disabling the RGMAs flux, the electron subsystem of semiconductor relaxes back to the starting state. Since the role of deep traps in a semiconductor is plaid by structure defects, admixture and adsorbed atoms and molecules, one should expect that a RGMA-induced change in their charge state will affect the interaction between the sample surface and the other components of gaseous phase, in particular, the adsorption and catalytic properties. The probability of these phenomena should be taken into account in the analysis of heterogeneous plasmochemical reactions and other processes with participation of high concentrations of RGMAs.

Note that the above-described pattern of the RGMA interaction with metal-activated oxides applies not only to the Au/ZnO structures. Similar phenomena may be expected for other metal – semiconductor pairs, in which metal microcrystals form barrier contacts with the carrier, and processes of prolonged relaxation occur. Work [181] may be an example to illustrate this, a work in which the interaction between the Au/TiO$_2$ films and metastable atoms of helium is studied. It is shown in this work that the signals of changes in the electrical conductivity under the action of RGMAs, signals that are characteristic of the Au/TiO$_2$ films, are like the signals for the Au/ZnO films, and that there is proportionality between the starting velocity of changes in the electrical conductivity of Au/TiO$_2$ and the concentration of RGMAs in the gaseous phase. Further investigations into the interaction of RGMAs with semiconductors are to show what a wide class of materials is available for the development of sensors for metastable atoms and what degree of generality is featured by the mechanism of de-exciting RGMAs on gold-activated zinc oxide.

5.8. Prospects of Developing Sensor Methods of Detecting Electron-Excited Particles

The previous section of this chapter are dealing with examples of employing semiconductor sensors with a view to studying the mechanism of heterogeneous de-excitation of electron-excited particles (EEP). The examples in question well illustrate the advantages and disadvantages of the sensors used as devices for scientific survey. Their positive advantages are high sensitivity in conjunction with ability to operate under diffusion conditions, small size, and the simplicity of the hardware support. The major disadvantages of EEP sensors are low selectivity, impossibility of identifying an active particle against a signal of changes in the electrical conductivity, and difficulties in taking measurements of absolute concentrations of EEPs.

The combination of advantages and disadvantages is a factor that determines the role that may be plaid by the semiconductor sensor techniques among the other experimental methods of physical chemistry, and the prospects of the technique's evolution.

Vigorous development in the recent years of highly-sensitive methods of surface and gaseous-phase analysis (the electron spectroscopy of surface, time of flight mass-spectrometry, thermodesorption analysis, etc.) much facilitate the possibility of accomplishing sensor studies. The high-sensitivity of present-day equipment is comparable with the sensitivity of sensors, and the parallel application of the equipment for sensor calibration and identification of active particles is fairly fruitful. For example, there appears an opportunity first to study heterogeneous processes under vacuum conditions with the aid of present-day techniques, then to calibrate the sensors involved, and as the next step to continue the study of the process in question with the calibrated sensor doing so under high-pressure conditions under which no other methods can be used.

The high sensitivity of the sensors makes it possible to use low-capacity sensors of active particles that feature selective generation in the study of heterogeneous processes. Thus, we are in a position to eliminate the influence of gaseous phase on the surface properties and to study in succession the interaction between the surface and certain constituents of an excited gaseous phase.

Expansion of the number of detectable active particles is a prerequisite for a wide application of sensors in physicochemical studies. Of the promising guidelines in this field one can distinguish the development of semiconductor detectors of electron-excited molecules of aromatic compounds. The class of aromatic substances is fairly broad, and these substances play an important role in may photochemical processes [3], herein there are many problems in which new unique results can be obtained through the application of sensors. The first steps in this direction are undertaken in works [72, 182]. These works show that the sensitization of a zinc oxide surface by phtalocyanines of metals allow one to proportionally detect triplet molecules of benzene and naphthalene, and have studied the principles of excitation propagation in the vapour of parent substance. The works also estimate the contribution of heterogeneous constituent of diffusion decay suffered by the electron-excited molecules of benzene. In addition to the problem of sensor detection of EEPs, there arises a problem of developing sensors for vibration-excited molecules (VEM). Compared to EEPs, these particles possess a less resource of energy, but their role in the physical chemistry of heterogeneous processes is important, since these particles arise in large quantities in the course of catalytic-thermal reactions as a result of EEP energy relaxation when gas is accumulated on a hot surface. The first works [183 through 185] aimed at the study of the interaction with the surface of semiconductor oxide have yielded promising results. It turns out that the vibration-excited molecules of oxygen and nitrogen markedly change the electrical conductivity of ZnO and NiO films. As the case is with EEPs, the polarity of the signal caused by a change in the electrical conductivity of sensors being acted upon by RGMAs may vary with temperature, a fact that testifies to the competition of a few ways of

VEM excitation energy relaxation. Such ways (channels) be probably chemisorption with charge transfer, production of phonons, ejection of electrons from surface states and traps, and the like. The further studies in this field will, obviously, make it possible to give a more complete characteristic of the VEM interaction with the surface of solid bodies and the possibilities of VEM detecting with the aid of semiconductor sensors.

The main problem of semiconductor EEP sensors consists of course in improving their selectivity. A method to raise the selectivity of sensors by applying surface additives have been discussed. There is one more way of raising the selectivity. It consists in utilizing composite materials and more intricately arranged structures for making sensors, rather than pure elements and compounds. This way of developing the sensor technology is rather time-consuming, very promising, regretfully still insufficiently studied. As of now, the method of selective, sensor detection of EEP based on the use of complex transducers is probably more promising. The mentioned transducer consists of several sensors that are sensitive to diverse constituents of the gaseous phase. Simultaneously the measurements from all sensors and computer-aided processing of information in principle allow one to estimates at once concentrations of a few active particles in the gaseous phase. The rapid development of microelectronics and computer technology allows us to hope for the practical use of such devices in the physicochemical experiment.

References

1 J.G. Calvert and J.N. Pitts, Photochemistry, New York, Wiley, 1966.
2 H. Okabe, Photochemistry of Small Molecules, N.Y., Wiley, 1978.
3 L. Barltrop and J. Coule, Excited States in Organic Chemistry, London, Wiley, 1975.
4 A.N. Terenin, Fotonika Molekul Krasitelei i Rodstevennykh Orgnicheskikh Soedinenii (Photonics of Dye Molecules and Related Organic Compounds), Leningrad: Nauka, 1977.
5 G. Breit, E. Teller, J. Astrophys, 1940, Vol. 91, p. 215-223.
6 M.F. Gold, Gas Kinet. and Energy Transfer. 1977, Vol. 2, p. 123 – 174.
7 G.W.F. Drake, G.A. Victor, A. Dalgarno, Phys. Rev. 1969, Vol. 180, p. 25 – 31.
8 G.W.F. Drake, Phys. Rev.A. 1971, Vol. 3, p. 908 – 923.
9 N.E. Small-Warren, L.-Y. Chin, C., Ibid. 1975, Vol. 11, p. 1777 – 1785.

10   K. Schofield, J. Phys. and Chem. Rev. Data, 1979, Vol. 8, p. 723 – 834.

11   E.C. Zipf, Canad. J. Chem., 1969, Vol. 47, p. 1863 – 1878.

12   G. Nowak, W.L. Borst, J. Fricke, Ibid. A., 1971, Vol. 3, p. 908 – 916.

13   R.M. Badger, A.C. Wright, R.F. Whitlock, J. Chem. Phys., 1965, Vol. 43, p. 4191-4197.

14   R.W. Nicholls, Amer. Geophys., 1964, Vol. 20, p. 144-156.

15   N.A. Borisevich and V.V. Gruzinskii, Dokl. Akad. Nauk SSSR, 1967, Vol. 175, p. 852-857.

16   C.S. Parmenter, B.L. Ring, J. Chem. Phys., 1967, Vol. 46, p. 1998-2010.

17   V.L. Ermolaev, E.N. Bodunov, E.B. Sveshnikova and T.A. Shkhoverdov, Bezizluchtelnyi Perenos Energii Elektronnogo Vozbuzhdeniya (Nonradiative Transfer of Electron Excitation Energy), Laningrad: Nauka, 1977, p. 311.

18   I.Ya. Fugol and P.L. Pakhomov, Zh. Exp. Teor. Fiz, 1967, Vol. 53, p. 866-875.

19   P.L. Pakhomov, G.P. Reznikov and I.Ya. Fugol, Optika i Spektroskopiya, 1966, Vol. 20, p. 10-21.

20   O. Svelto, Principles of Lasers, N.Y. Plenum, 1976.

21   S.P. Vetchinin and A.A. Rakovets, Vliyanie Elementarnykh Protsessov, Protekayushchikh s Uchastiem Metstabilnykh Chastits, Na Kharakteristiki Tleyushchego Razryada (Influence of Elementary Processes Proceeding with Participation of Metastable Particles on Characteristics of Glow-Discharge), Preprint of Inst. Vys. Temp. Akad. Nauk SSSR, 1982.

22   Singlet Oxygen, N.Y.: Acad. press, 1979, 573 p.

23   S.D. Razumovskii, Kislorod – Elementarnye Formy i Svoistva (Oxygen – Elementary Forms and Properties), Moscow: Khimya, 1979, 314 p.

24   A.N. Wrigt and C.A. Winkler, Active Nitrogen, N.Y., L.: Acad. press, 1968, 542 p.

25   L.A. Palkina, B.M. Smirnov and M.I. Chibisov, Zh. Exp. Teor. Fiz., 1969, Vol. 56, p. 340 –348.

26   L.I. Slovetskii, Mekhanizmy Khimicheskikh Reaktsii v Neravnovesnoi Plazme (Mechanisms of Chemical Reactions in Nonequilibrium Plasma), Moscow: Nauka, 1980, 310p.

27   B.M. Smirnov, Vozbuzhdennye Atomy (Excited Atoms), Moscow: Energoizdat, 1982, 232 p.

28   R.J. Donovan, Prog. React. Kinet., 1979, Vol. 10, p. 254 – 277.

29   I.Ya. Fugol, UFN, 1969, Vol.97, p. 429-463.

30   D.R. Snelling, Chem. Phys. Lett. 1968. Vol. 2, p. 346-349.

31 R.A. Young, Canad. J. Chem., 1966, Vol. 44, p. 1171-1178.
32 Meyer J.A., D.H. Klasterboer, D.W. Sester, J. Chem. Phys. 1971, Vol. 67, p. 2931-2935.
34 I.A. Myasnikov, E.I. Grigorev and V.I. Tsivenko, Usp. Khim., 1986, Vol. 55, p. 161-205.
35 W.V. Smith, J. Chem. Phys., 1943, Vol. 11, p. 110-116.
36 N.N. Semonov, Acta Physicochim. URSS, 1943, Vol. 18, p. 93-97.
37 H. Wise, C. Ablow, J. Chem. Phys. 1958, Vol. 29, p. 634-639.
38 H. Mots, H. Wise, Ibid. 1960, Vol. 32, p. 1893-1896.
39 L.Yu. Kupriyanov, Vzaimodeistvie Metastabilnykh Atomov Inertnykh Gazov s Chstoy i Aktivirovannoy Metallami Povrkhostyu Okislov (Interaction of Rare Gas Metastable Atoms with Pure and Metal-Activated Surface of Oxides): Dissertation, Cand. Sci. (Phys.-Math.)Moscow, 1985, 140 p.
40 M.E. Ryskin and B.R. Shub, Khim. Fiz., 1982, Vol. 1, p. 212-219.
41 P.G. Dickens, D. Schofieldsd and J. Walsh, Trans. Faraday Soc. Pt. II, 1960, Vol. 56, p. 225-234.
42 J.M.W. Milats and L.S. Ornstein, Physika, 1935, Vol. 2, p. 355-362.
43 J.H. Kolts and D.W. Setser, J. Chem. Phys. 1978, Vol. 68, p. 4848.
44 M. Moison, R. Pantel and A. Richard, Canad, J. Phys., 1982, Vol. 60, p. 379-392.
45 J.M. Cook, T.A. Miller and V.E. Bondybey, J. Chem. Phys., 1978, Vol. 68, p. 2001-2004.
46 A.A. Tabachnik, S.Yu. Umanskii and B.R. Shub, Khim. Fiz., 1983, Vol. 2, p. 938-945.
47 T.A. Delchar, D.A. McLennon and A.M. Landers, J. Chem. Phys. 1969, Vol. 50, p. 1779-1787.
48 R.S. Van Dyck, C.E. Jonson and H.A. Shugar, Ibid. Lett. 1970, Vol. 25, p. 1403-1405.
49 J.R. Woodworth and H.W. Moos, Ibid. A. 1975, Vol. 12, p. 2455-2470.
50 W. Wieme and J. Wieme-Lenaerts, Phys. Lett.A., 1974, Vol. A74, p. 37-43.
51 W. Wieme, Physica B + C, 1980, Vol. 98, p. 229-243.
52 W.H. Brechenridge and T.A. Miller, Chem, Phys. Lett., 1972, Vol. 12, p. 437-441.
53 A.M. Fallick, B.H. Mahan and R.J. Meyer, J. Chem Phys., 1965, Vol. 42, p. 1837-1842
54 V.L. Orkin and A.M. Chaikin, Kinet. Katal., 1979, Vol. 20, p. 1367-1459.
55 A. Holin, B. Vidal and P. Goudman, J. Chem Soc. Commun., 1976, N 23, p. 960-966.

56   E.J. Corey and W.C. Taylor, J. Amer. Chem. Soc., 1964, Vol. 86, p. 3880-3896.

57   J.R. Scheffter and D.M. Ouch, Tetrahedron Lett., 1970, N. 3, p. 223-227.

58   R.B. Cairas and I.A.R. Samson, Ibid. A., 1965, Vol. 139, p. 1403-1411.

59   P.H. Vidaud and A. von Engel, Proc. Roy. Soc., 1969, Vol. 313, p. 531-547.

60   H. Conrad, G. Ertl, and J. Kuppers et al., Surface Sci., 1982, Vol. 117, p. 98-105.

61   D.A. McLennon and T.A. Delchar, J. Chem Phys., 1969, Vol. 50, p. 1772-1777.

62   F.B. Dunning, J. Phys. E: Sci. Instrum., 1972, Vol. 5. p. 263-267.

63   A.N. Terenin, Selected Works, Leningrad: Nauka, 1974, Vol. 2, p. 473.

64   S.V. Vintsenets, P.K. Kashkarov, V.F. Kisilev and G.S. Plotnikov, Dokl. Akad. Nauk SSSR, 1983, Vol. 268, p. 373-378.

65   S.V. Vintsenets, V.F. Kisilev, L.V. Levshin et al., Dokl. Akad. Nauk SSSR, 1984, Vol. 274, p. 96.

66.   I.A. Myasnikov, S.A. Zav'yalov, E.I. Grigor'ev et al., Zh. Fiz. Chim., 1981, Vol. 55, p. 1840-1842.

67   I.A. Akimov, Spektroskopiya Fotoprevrashchenii v Molekulakh (Spectroscopy of Phototransformations in Molecules), Leningrad: Nauka, 1977, p. 239-273.

68   G. Scheibe and F. Dorr, Bayer. Acad. Wiss. Math. Naturwiss, Kl., 1959, N. 8, p. 183-198.

69   H. Kokado, I. Nakayame and E. Inone, J. Phys. and Chem. Soc. 1971, Vol. 32, p. 2785-2793.

70   I.A. Myasnikov and E.V. Bol'shun, Dokl. Akad. Nauk SSSR, 1978, Vol. 240, p. 906-913.

71   S.A. Zavyalov and I.A. Myasnikov, Dokl. Akad. Nauk SSSR, 1981, Vol. 257, p. 392-398.

72   V.I. Tsivenko, I.A. Myasnikov, E.V. Bol'shun and O.Yu. Nikola-eva, Zh. Fiz. Chim., 1986, Vol. 60, p. 1314-1321.

73   G.A. Melin and R.J. Madix, Trans. Faraday Soc. Pt. I, 1974, Vol. 76, p. 198-207.

74   M.P. Weinreb and G.G. Manella, J. Chem. Phys., 1969, Vol. 51, p. 4973-4984.

75   N.V. Shinkarenko and V.B. Aleskovskii, Usp. Khim., 1981, Vol. 50, p. 406-443.

76   E.A. Ogryzlo, Photophysiology, 1975, Vol. 5, p. 35-42.

77   V.Ya. Shlyapintokh and V.B. Ivanov, Usp. Khim., 1976, Vol. 45, p. 202-237.

78 S.J. Arnold, N. Finlayson and E.A. Ogryzlo, J. Chem. Phys., 1966, Vol. 44, p. 2529-2543.

79 A.M. Winer, K.D. Bayes, Ibid. Vol. 70, p. 302-313.

80 R.P. Steer, R.A. Ackerman and J.N. Pitts, Ibid. 1969, Vol. 51, p. 843-853.

81 I.D. Clark and R.P. Wayne, Chem. Phys. Lett., 1969, Vol. 3, p. 93-97.

82. R.C. Dervent and V.A. Trush, Trans. Faraday Soc. Pt. I. 1971, Vol. 67, p. 2036-2053.

83 S.J. Arnold and E.A. Ogryzlo, Canad. J. Phys, 1967, Vol. 45, p. 2053-2065.

84 T.P.J. Irod and R.P. Wayne, Proc. Roy. Soc., London A., 1968, Vol. 308, p. 81-96.

85 T.C. Frankievicz and R.S. Berry, J. Chem. Phys., 1973, Vol. 58, p. 1787-1799.

86 O.J. Giachard, G.W. Marris and R.P. Wayne, J. Chem. Soc. Faraday Trans. Pt. II, 1976, Vol. 72, p. 619-630.

87 R.P. Wayne, Adv. Photochem., 1969, Vol. 7, p. 311-337.

88 K. Furukawa, E.W. Gray and E.A. Ogryzlo, Ann. N.Y. Acad. Sci., 1970, Vol. 171, p. 175-187.

89 M.E. Ryskin and B.R. Shub, React. Kinet. and Catal. Lett., 1981, Vol. 17, p. 41-45.

90 M.E. Ryskin and V.I. Chernysh, Abstracts of Papers, On Mechnism of Catalytic Reactions, 4th All-Union Conference, Moscow, 1986, Part 2, p. 358.

91 M.E. Ryskin, Geterogennaya Dezaktivatsiya Singletnogo Kisloroda na Poverkhnosti Metallov i Dielektrikov (Heterogeneous Deactivation of Singlet Oxygen on Surface of Metals and Dielectrics): Dissertation Cand. Sci.(Phys.-Math.), Moscow, 1983, 113 p.

92 M.E. Ryskin, B.R. Shub, J. Pavlicek and S. Enor, Chem. Phys. Lett., 1983, Vol. 99, p. 140-143.

93 I.A. Myasnikov, V.I. Tsivenko and M.V. Yakunichev, Dokl. Akad. Nauk SSSR, 1982, Vol. 267, p. 873-881.

94 M.V. Yakunichev, I.A. Myasnikov and V.I. Tsivenko, Zh. Fiz. Khim., 1985, Vol. 59, p. 873-875.

95 E.I. Grigor'ev, I.A. Myasnikov and V.I. Tsivenko, Zh. Fiz. Khim., 1981, Vol. 55, p. 1840-1847.

96 E.I. Grigor'ev, I.A. Myasnikov and V.I. Tsivenko, Zh. Fiz. Khim., 1982, Vol. 56, p. 1748-1756.

97 L.W. Bader and E.A. Ogryzlo, Discuss Faraday Soc., 1964, N. 37, p.46-59.

98 S.N. Whitlow and F.D. Findlay, Canad. J. Chem., 1967, Vol. 45, p. 2087-2099.

99   R.M. Young, K. Werly and R.L. Martin, J. Amer. Chem. Soc., 1974, Vol. 93, p. 5774-5790.

100. E.I. Grigor'ev, Fotosensibilizirovannoe Obrazovanie i Gibel' Singletnogo Kisloroda v Geterogennykh Sistemakh Tverdoe Telo-Gaz (Photosensitized Formation and Decay of Singlet Oxygen in Heterogeneous Systems Solid Body – Gas): Dissertation, Cand. Sci. (Phys. – Math.), Moscow, 1982, 112 p.

101 S.A. Zav'yalov and I.A. Myasnikov, Abstract of Papers, "Gas Analysis and Environment Protection", All-Union Conference, November 23-25, 1981, Laningrad, p. 18.

102 E.I. Grigor'ev, I.A. Myasnikov and V.I. Tsivenko, Zh. Fiz. Khim., 1983, Vol. 57, p. 505-513.

103 V.I. Egorov, Yu.M. Gershenson and V.B. Rosenshtein, Chem. Phys. Lett., 1973, Vol. 20, p. 77-79.

104 S.A. Kovalevskii and B.R. Shub, Probl. Kinet. and Katal., Moscow: Nauka, 1977, Vol. 17, p. 29-36.

105 S. Greg and K. Sing, Adsorption, Specific Surface, Porosity, (in Russian) Moscow: Mir, 1970, 408 p.

106 Ya. De Bur, Dynamic Nature of Adsorption, Moscow: Izd. Inostr. Lit., 1962, 290 p.

107. Yu.M. Gershenzon, V.B. Rozenshtein and S.Ya. Umanskii, Probl. Kinet. and Katal., Moscow: Nauka, 1977, Vol. 17, p. 36-45.

108 M.V. Yakunichev, V.I. Tsivenko and I.A. Myasnikov, Zh. Fiz. Khim., 1986, Vol. 60, p. 1017-1021.

109 I.A. Myasnikov, Mendeleev ZhVKhO, 1975, Vol. 20, p. 19-32.

110 S.A. Zav'yalov, I.A. Myasnikov and L.M. Zav'yalova, Zh. Fiz. Khim., 1986, Vol.60, p. 2490-2495.

111 L.M. Zav'yalova, Osobennosti Adsorbtsii Singletnogo Kisloroda na Poverkhnosti Tel Razlichnoi Prirody (Features of Singlet Oxygen Adsorption on Surfaces of Various Natures): Dissertation, Cand. Sci. (chem.), Moscow, 1988, 108 p.

112 L.M. Zav'yalova, I.A Myasnikov and S.A. Zav'yalov, Zh. Fiz. Khim., 1987, Vol. 61, p. 473-479.

113 L.M. Zav'yalova, I.A. Myasnikov and S.A. Zav'yalov, Chem. Phys., 1987, Vol. 6, p. 473-476.

114 S.A. Zav'yalov and I.A. Myasnikov, Zh. Fiz. Khim., 1982, Vol. 56, p. 2616-2619.

115 E.I. Grigor'ev, I.A. Myasnikov and V.I. Tsivenko, Zh. Fiz. Khim., 1982, Vol. 56, p. 1558-1563.

116 S.A. Zav'yalov, I.A. Myasnikov and L.M. Zav'yalova, Kinet. Katal. , 1988, Vol. 29, p. 1245-1253.

117 J.I. Canva, C. Bandly, P. Douzan and J.C.R. Bourlon, Acad. Sci. C., 1969, Vol. 268, p. 1027-1032.

118 V.P. Ushakova, T.P. Korneichuk, V.A. Roiter and Ya.V. Zhigailo, Ukr. Khim. Zh., 1975, Vol. 23, p. 191-195.

119 V.V. Azat'yan, Khim. Fiz., 1982, Vol. 1, p. 491-498.

120 M.L.E. Oliphant, Proc. Roy. Soc. London A. 1929, Vol 124, p. 228-242.

121 M.L.E. Oliphant and P.B. Moon, Ibid., 1930, Vol. 127, p. 388-401.

122 S.S. Shekter, J. Exp. and Teor. Phys. (USSR), 1937, Vol. 7, p. 750-757.

123 A. Cobas and V.E. Lamb, Phys. Rev., 1944, Vol. 65, p. 327-338.

124 H.D. Hagstrum, Ibid., 1954, Vol. 96, p. 336-365.

125 H.D. Hagstrum, Phys. Rev. Lett., 1979, Vol. 43, p. 1050-1054.

126 C. Boizian, Springer Ser. Chem. Phys., 1982, Vol. 17, p. 48-72.

127 J. Johnson and T.A. Delchar, Surface Sci., 1978, Vol. 77, p. 400-407.

128 J. Roussel, C. Bolzian, R. Nowlone and C. Reynand, Ibid., 1981, Vol. 110, p. 634-637.

129 J. Roussel, Phys. Scr., 1983, Vol. 4, p. 96-99.

130 S.-W. Wang and G. Ertl, Surface Sci., 1980, Vol. 93, p. 75-79.

131 H. Conrad, G. Ertl, J. Kuppers and S.-W. Wang, Ibid., 1979, Vol. 42, p. 1082-1085.

132 C. Bolzian, C. Carot, R. Nowlone and J. Roussel, Surface Sci., 1982, Vol. 91, p. 313-318.

133 H. Conrad, G. Ertl, J. Kupper et al., Ibid., 1982, Vol. 121, p. 161-166.

134 F. Bozso, J.T. Yates, J. Arias and Metiu, J. Chem. Phys., 1983, Vol. 78, Pt. 2, p. 4256-4263.

135 F. Bozso, J. Arias, G. Hanrahan et al., Surface Sci., 1984, Vol. 136, p. 257-261.

136 D.J. Titley and T.A. Delchar, Ibid., 1983, Vol. 103, p. 438-441.

137 W. Allisson, F.B. Dunning and A.C.N. Smith, J. Phys. B. Atom and Mol. Phys., 1972, Vol. 5, p. 1175-1183.

138 D. Greene, Proc. Roy. Soc. London B, 1950, Vol. 63, p. 876-882.

139 T. Shibata, T. Hirooka and K. Kuchitsu, Chem. Phys. Lett, 1975, Vol. 30, p. 241-243.

140 J. Munakata, T. Hirooka and K. Kuchitsu, J. Electron. Spectrosc. Relat. Phenom., 1978, Vol. 13, p. 219-222.

141 J. Munakata, K. Ohno and Y. Harada, J. Chem. Phys., 1980, Vol. 72, p. 2880-2894.

142 H. Kubata, J. Munakata, T. Hirooka and K. Kuchitsu, Chem. Phys. Lett. 1980, Vol. 74, p. 409-412.

143 K. Ohno, N. Muton and Y. Harada, Surface Sci., 1982, Vol. 115, p. 128-133.

144 J. Munakata, T. Hirooka and K. Kutchitsu, J. Electron. Spectrosc. Relat. Phenom., 1980, Vol. 10, p. 51-54.

145 Y. Harada, H. Ozaki and K. Ohno, Solid State Commun., 1984, Vol. 49, p. 71-78.

146 J.H. Craig and J.T. Dickinson, J. Vac. Sci. and Technol., 1973, Vol. 10, p. 319-326.

147 V.N. Lisitsin, A.S. Provorov and V.P. Chebotaev, Optic. Spektrosk., 1970, Vol. 29, p. 226-328.

148 N. Sadeghi and J.C. Pebay-Peyroula, J. Phys. (France), 1974, Vol. 35, p. 353-359.

149 R. Pantel, A. Ricard and M. Moisan, Beitr. Plasmaphys., 1983, Vol. 23, p. 561-567.

150 M.M. Pejovic, B.J. Mijovic and Dj. Bosan, J. Phys. D: Appl. Phys., 1983, Vol. 16, p. 149-160.

151 Dj. Bosan and M.M. Lukic, VI Intern. Conf. Gas Discharges and their Appl., Edinburgh, 1980, L.; N.Y., 1980. Pt. 2, p. 46.

152 W.G. Clark, D.W. Setser, J. Phys. Chem., 1980, Vol. 84, p. 2225-2232.

153 P.H. Vidaud, R.P. Wayne, M. Yaron and A. Engel, J. Chem Soc. Faraday Trans. Pt. II, 1976, Vol. 72, p. 1185-1193.

154 A.A. Tabachnik and B.R. Shub, Khim. Fiz., 1983, N. 2, p. 1242-1247.

155 P. Svejda, R. Haul, D. Mihelic and R.N. Schindler, Ber. Bunsenges Phys. Chem., 1975, Vol. 79, p. 71-77.

156 E.I. Grigor'ev, N.A. Slavinskaya and L.I. Trkhtenberg, Khim. Vys. Energ., 1987, Vol. 21, p. 519-523.

157 A.M. Panesh and I.A. Myasnikov, Zh. Fiz. Khim., 1972, Vol. 46, p. 1902-1903.

158 G. Doyen, Surface Sci., 1982, Vol. 117, p. 85-97.

159 I.V. Miloserdov and I.A. Myasnikov, Zh. Fiz. Khim., 1979, Vol. 53, p. 2330-2342.

160 L.Yu. Kupriyanov, V.I. Tsivenko and I.A. Myasnikov, Zh. Fiz. Khim., 1984, Vol. 58, p. 1156-1162.

161 H. Maier-Leibnitz, Zs. F. Phys., 1935, Vol. 95, p. 499-506.

162 L.Yu. Kupriyanov, V.I. Tsivenko and I.A. Myasnikov, Poverkhnost, 1983, N 9, p. 50-57.

163 L.Yu. Kupriyanov and I.A. Myasnikov, Abstracts of Papers, 4th All-Union Conference on Low Temperature Chemistry (Moscow, December 21-23, 1988), Moscow: MGU Publishers, 1988, 204 p.

164 L.Yu. Kupriyanov, V.I. Tsivenko, and Myasnikov I.A., Poverkhnost, 1985, N 10, p. 78-84.

165 Solid State Surface Science / Ad. by M. Green, V. 1, Marcel Dekker New York 1969.

166 L. Sodomka and T. Chudoba, Czechosl. J. Phys. B., 1981, Vol. 31, p. 895-899.

167 K. Albert-Polacek and E.R. Wasserman, Thin Solid Films, 1976, Vol. 37, p. 65-71.

168 L.L. Kazemerski and D.N. Racine, J. Appl. Phys., 1975, Vol. 46, p. 791-795.

169 R.C. Nevill and C.A. Mead, Tbid., 1970, Vol. 41, p. 3795-3799.

170 S.M. Zi, Fizika Poluprovodnikovykh Priborov (Physics of Semiconductor Devices), Moscow: Energiya, 1973, 386 p.

171 R.A. Rabadanov, M.K. Guseikhanov, I.Sh. Aliev and S.A. Semiletov, Izv. Vysh. Uchebn. Zaved., Fiz., 1981, Vol. 24, p. 72-75.

172 M.K. Sheikman and A.Ya. Shik, Fiz. Tech. Poluprovodn., 1976, Vol. 10, p. 209-233.

173 L.I. Trusov and V.A. Kholmyanskii, Ostrovkovye Metallicheskie Plenki (Islet Metal Films), Moscow: Metallurgiya, 1973, 310 p.

174 L.Yu. Kupriyanov, I.A. Myasnikov and V.I. Tsivenko, Poverkhnost, 1985, N 11, p. 39-46.

175 L.Yu. Kupriyanov, I.A. Myasnikov and V.I. Tsivenko, Poverkhost, 1985, N 12, p. 122.

176 E.Kh. Roderik, Kontakty Metall – Poluprovodnik (Metal-Semiconductor Contacts), Moscow: Radio i Svyaz', 1982, 206 p.

177 G.I. Roberts and C.R. Gowell, J. Appl. Phys., 1970, Vol. 41, p. 1767-1773.

178 S. Norrman, T. Anderson, C.G. Granquist and O. Hundery, Phys. Rev. B., 1978, Vol. 18, p. 674-687.

179 G. Pankove, Optical Processes in Semiconductors, Englewood Cliffs (USA) Prentice-Hall, 1971.

180 V.L. Bonch-Bruevich and S.G. Kalashnikov, Fizika Poluprovodnikov (Physics of Semiconductors), Moscow: Nauka, 1977, 673 p.

181 L.Yu. Kupriyanov and V.I. Tsivenko, Zh. Fiz. Khim., 1987, Vol. 61, p. 828-837.

182 V.I. Tsivenko, I.A. Myasnikov, O.Yu. Nikolaeva and E.V. Bol'shun, Zh. Fiz. Khim., 1987, Vol. 61, p. 2729-2738.

183 E.E. Gutman, N.V. Ryl'tsev, S.A. Kazakov and I.A. Myasnikov, Khim.Fiz., 1984, Vol. 3, p. 1625-1638.

184 S.A. Kazakov, I.A. Myasnikov, N.V. Ryl'tsev and E.E. Gutman, Khim. Fiz., 1986, Vol. 5, p. 31-44.

185 E.E. Gutman, I.A. Myasnikov, S.A. Kazakov and N.V. Ryl'tsev, Khim. Fiz., 1986, Vol. 5, p. 386-397.

186 A.M. Panesh and I.A. Myasnikov, Zh. Fiz. Khim., 1971, Vol. 45, p. 261-264.

187 A.M. Panesh and I.A. Myasnikov, Dokl. Akad. Nauk SSSR, 1971, Vol. 200, p. 1136-1141.

188 S.N. Witlow and F.D. Findlay, Canad. J. Chem., 1967, Vol. 45, p. 2087-2099.

189 I.D. Belova, S.A. Zavyalov and Yu.E. Roginskaya, Zh. Fiz. Khim., 1986. Vol. 60, p. 2338-2342.

190 L. Elias and E.A. Ogryzlo, Canad. J. Chem., 1959, Vol. 37, p. 1680-1695.

191 S.S. Arnold, M. Cubo and E.A. Ogryzlo, Adv. Chem. Ser., 1968, Vol. 77, p. 133-156.

192 B. Sterens and S.R. Perez, Mol. Photochem., 1974, Vol. 136-143.

Chapter 6

Application of semiconductor sensors to study emission of active particles from the surface of solid state

The studies of emission of adsorbed active particles from the surface of solids due to effects of various factors characterizes the broad capacity of the method of SCS to study the accompanying physical and chemical phenomena. On the other hand this process encounters several problems which should be overcome during identification of these particles. Moreover, in above examples we also illustrate the techniques of application of SCS to solve more complex analytical problems.

6.1. Emission of initially adsorbed active particles from disordered surface of solids

The scope of phenomena accompanied by emission of active particles into gaseous phase is broad enough. Initially these are heterogeneous-homogeneous reactions when emission of active particles into the gaseous phase occurs due to the energy of chemical reaction [1], thermodesorption of atoms and radicals from chemisorbed layer [2, 3], photostimulated desorption of active particles [4], exoemission of ions and electrons [5], various emission phenomena accompanied by destruction of solids, etc. It is interesting to single out emission of initially adsorbed active particles from these phenomena.

One can conclude from general consideration that from the energy stand-point an adsorbed active particle should first recombine and then gets desorbed as a neutral molecule. Owing to that the probability of emission of atoms and radicals due to surface recombination during adsorbent heating is small. In [6, 7] the probability of detecting short-living highly surface bound intermediate products of heterogeneous-catalytic reaction was assessed by the flash-desorbtion method. Authors considered two competing processes developing during heating of catalyst: desorption of intermediate products and their transformation into final products. The estimates conducted in these papers enable one to conclude that for real catalytic systems the desorption rates and rates of transformation of intermediate products into final ones will be of the same order when the catalyst heating rate is $10^{10} - 10^{11}$ deg/s, although in usual experiments on flash-desorbtion the heating rates are about $10^3 - 10^4$ deg/s ($10^{10} - 10^{11}$ deg/s are only accessible using lasers). Therefore, the emission of active particles from solids is more probable when one expects low deposition of adparticles on the surface, the particles are weakly bound with the adsorbent surface as well as their migration is restricted due to various reasons. The energy necessary to emit an active particle can be provided on the local scale, for instance, by photons, ions, hot atoms or by a unit chemical action as well as due to relaxation of disordered states in the surface and adjacent layers of solid. These processes result in release of substantial (up to several electron-volt) energy sufficient to brake the bond of an active particle with the surface of a solid state leading to the particle's release.

The order of magnitude of this energy can be assessed by a thermal effect of formation and annihilation of point defects in solids. Up-to-date, there are several experimental techniques to determine the energy of defects [8]. Dilatomic, calorimetric techniques, the method of positron annihilation, and others are among them. Additionally, there are several theoretical methods to calculate the energy of defects which agree fairly well with experimental data [8]. These data indicate that the typical value of the energy of vacancies is about 1 eV and the energy of formation of an interstitial atom is about $2 - 5$ eV. Therefore, the energy released during annihilation of the Frenkel pair can attain $5 - 6$ eV.

Even higher energy can be released during surface restructuring when several atoms quasi-isolated from the surface take part in this process. The order of magnitude of energy release in processes of this kind can be evaluated due to the difference of the energy of the ground state of clusters composed of several atoms with various configuration. Such estimates have been conducted by quantum mechanical methods. For instance, in [9] the energy of the ground state of clusters of lithium Li_3 (equal side triangle), Li_4 (cube), Li_9 (volume centered cube), Li_{13} (octahedron), Li_{13} (icosahedron) was calculated depending on the value of internuclear distance. These data indicate that for sufficiently large size of a cluster the binding energy calculated per atom approaches the cohesion energy. The important result of this study is provided by the strong dependence of the energy of ground state on its configuration and the value of internuclear distance. Thus, for Li_{13} (icosahedron) the energy of the ground state is 2 eV lower than for Li_{13} (octahedron). Increasing the internuclear distance in Li_8 cluster by one Å the energy of the cluster increases by 7 eV if contrasted to its ground state. When two four-atom clusters of lithium get merged into one composed of eight-atoms (such situation is probable at initial stage of deposition of a thin film) there will be a release of 2.92 eV. It should be noted that here we consider only those symmetrical cluster configurations which in general possess the lowest energy of the ground state. The restructuring of non-symmetrical atom configurations results in higher heat effects.

Very thin freshly deposited films of various substances are characterized by high degree of defects. Concentrations of vacancies in them can attain 10^{-3} [10]. Obviously, the energy released during ordering of such films would exert high effects on properties of films and on the character of interaction of the film with the substrate surface. It is highly probable that it is due to release of this energy that several specific phenomena develop. They are called condensation-stimulated effects [11]. These effects result, for instance, in anomalous high mutual condensate – substrate solubility.

6.1.1. Emission of atom hydrogen from the surface of amorphous antimony

It has been already mentioned in preceding section that in process of ordering of disordered adsorbents the energy get released which is sufficient to brake the bonds in the surface compounds. Therefore, the emission of initially adsorbed active particles due to disorder relaxation should be studied in disorder surfaces. It is very convenient to use for such studies the amorphous antimony with adsorbed hydrogen atoms. The properties of thin antimony films have been studied in substantial detail due to their use in manufacturing of photocathodes [12].

The amorphous black antimony is obtained by vacuum deposition on a glass or quartz substrate at room or lower temperatures. The black antimony spontaneously transforms into a gray crystalline antimony. The kinetics of crystallization of gray antimony is determined by the temperature as well as by availability of admixtures. Adsorbed gases and mainly oxygen dramatically slow down the crystallization process so that this process gets shut down in atmosphere [13]. Depending on the thickness of amorphous film the crystallization of antimony can develop from several hours up to several days. Very thin films (up to 20 Å) are almost amorphous whereas thick films (above 20 μm) are always crystalline ones. If the temperature of substrate on which antimony gets deposited is 77 K then the antimony film, as a rule, is amorphous. It gets crystallized during heating, the temperature of intensive crystallization being determined by the thickness of the film.

It is important to mention that antimony is absolutely passive to molecular hydrogen but highly responsive to adsorption of atomic hydrogen [13]. This properties of amorphous films of antimony with adsorbed atoms of hydrogen make them very convenient to study emission of atom hydrogen due to ordering in antimony films.

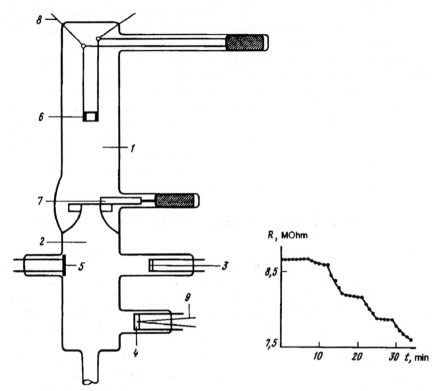

Fig. 6.1. Experimental set-up to study emission of active particles: *1, 2* – chambers; *3* – evaporating tray; *4* – platinum filament; *5* – target; *6* – mobile sensor; *7* – shatter; *8* – contacts; *9* – thermocouple; *10* – weight to move sensor.

Fig. 6.2. Response of the sensor in the vial containing the antimony film with deposited H-atoms: At moments of time $t = 10, 20, 30$ minutes the sensor gets closer to the surface of antimony film. For moments $t = 15, 26$ minutes the sensor gets back to its initial position

The experiment was conducted in a vial [14, 15] (Fig. 6.1) consisting of two chambers *1* and *2* separated by a plane shutter *7*. If the shutter is closed the atoms of hydrogen, according to experiment, do not penetrate through it. Note, that the quality of the shutter should be very high. The width of the flange should be no smaller than 10 mm, the contacting surfaces of the flange and disk should be perfectly polished. The small particles of glass, dust, deposited on these surfaces make the shutter penetrable for particles. Chamber *2* contains tantalum tray with antimony which underwent the vacuum melting. It is used to deposit antimony atoms on the walls of the vial and on target *5* which is the quartz substrate with platinum contacts. The measuring of resistance of such film during and after deposition enables one to monitor the kinetics of ordering of the antimony film. The same chamber contains tantalum or platinum filament necessary for pyrolysis of hydrogen molecules. The temperature of filament is controlled by a platinum-platinum rhodium thermocouple 9. In several experiments hydrogen atoms were obtained by a electrodeless discharge in hydrogen atmosphere. A thin film of zinc oxide 6 was used as a sensor. This sensor could be moved from chamber 1 into chamber 2 using an iron weight 10 covered by glass and magnet.

The experiment showed that after turning on the source of hydrogen atoms and evacuation of the experimental set-up down to pressure of 10^{-6} Torr the sensor shows no signals when the shatter is shut-down (there is no change in its resistivity) (Fig. 6.2). If the shutter gets open (the sensor being placed in chamber 1) one observes a weak change in its resistivity which is indicative of adsorption of particles – donors of electrons (donor signal). When the sensor is put into chamber 2 and it is brought closer to the surface of the antimony film there is a sharp increase in the donor signal. Transposition of the sensor into the chamber 1 results in the fact that again a weak donor signal is detected, etc. Such a phenomenon can be observed at room temperature up to several days from the moment of deposition of antimony film and adsorption of hydrogen atoms.

Obviously to understand the effect detected it is necessary to establish what particles generate the signals of the sensor. This problem was addressed by several specific experimental techniques characteristic of the method of semiconductor sensors.

From general considerations it should be mentioned that only atoms of hydrogen can be present in gaseous phase in addition to atoms of antimony, molecules of antimony hydride and antimony hydride radicals in this system. Therefore, it is necessary to analyze plausible effects of each of above particles on conductivity of semiconductor sensor under experimental conditions.

It is known [16] that at room temperature antimony evaporates as molecules. The molecules of antimony according to [17] do not affect conductivity of the sensor made of zinc oxide. Similar conclusion can be obtained from experiments with freshly reduced antimony films. It occurs that without initial adsorption of hydrogen atoms one fails to detect any signals from the sensor in contrast to experimental data (see Fig. 6.2). The resistivity of the sensor remains constant for any distance from the surface of the antimony film. Consequently, the signals of the sensor detected in experiment are not linked with effects of the antimony particles on the sensor.

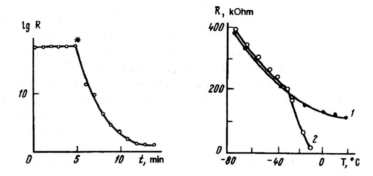

Fig. 6.3. The effect of antimony hydride on conductivity of the sensor. The temperature of the sensor is 23°C, at moment of time t = 5 min hydride was introduced.

Fig. 6.4. The temperature dependencies of resistivity of the sensor. 1 – in vacuum; 2 – in the vial with antimony.

Let us consider the effect of antimony hydride on the sensor. It was already shown by Panneth [18] that SbH_3 gets synthesized between antimony electrodes during glowing discharge in hydrogen. The experiments with SbH_3 [14, 15] using the method similar to that of [18] followed by mass-spectroscopic analysis of hydride showed that the effect of antimony hydride on the sensor dramatically increases its electroconductivity if the sensor is kept at a room temperature. We should note that in the experiment (see Fig. 6.2) the sensor was also kept at a room temperature. Figure 6.3 shows the change in conductivity of the sensor made of zinc oxide caused by interaction with antimony hydride: introduction of SbH_3 was made at t = 5 min. However, if the sensor is kept at a temperature lower than –25°C there is no donor signal observed during introduction of antimony hydride.

Figure 6.4 shows the change in the sensor conductivity as a function of temperature. Curve 1 shows the dependence of sensor resistivity with temperature when the sensor is positioned in evacuated installation. The introduction of antimony hydride was made at temperature – 75°C bringing about no change in resistivity. When the temperature of the sensor was increased up to – 20°C there were no effects detected on its resistivity caused by antimony hydride. Only at higher temperatures one can observe deviation of dependence $R(T)$ from curve 1 which is caused by decomposition of SbH_3 on ZnO. These results led to experiments on emission of H-atoms in a special vial when Sb-film treated by H-atoms was kept at a room temperature and sensors were kept at the temperature of – 80°C. Under these conditions, as is shown by above reasoning, the sensor is insensitive to antimony hydride but is still highly sensitive to hydrogen atoms. It occurs that in this case one can detect signals similar to those observed in experiment shown in Fig. 6.2. One can conclude, therefore, that in such experiments sensor readings are not connected to effects of antimony hydride.

Another justification of conclusions made is provided by the data on kinetics of the donor signal. It turns out that donor signals of the sensor can be observed even during four days from the moment of deposition of antimony film and adsorption of H-atoms on it. The time during which the signal of the sensor drops by 2 takes up to 33 hours whereas the half life-time of SbH_3 on antimony film is about 6 hours [19]. In papers [14, 15] antimony hydride was completely decomposed in a vial with clean walls (without initial deposition of a Sb-film) during 30 hours. Since decomposition of SbH_3 is a autocatalytical reaction which gets accelerated by precipitates formed during decomposition [19] it is obvious that in a vial with already deposited Sb-film, as it was in experiment (see Fig. 6.2) the decomposition would take less than 30 hours and the signals of the sensor if they were linked with antimony hydride would not have been detectable during several days.

Therefore, as a result of a series of experiments and analyzing the literature sources we proved that in experiment shown in Fig. 6.2 the signals of the sensor cannot be related to effects of antimony hydride.

Similarly to an experiment shown in Fig. 6.2 the donor signal of the sensor was observed in a vial with a Sb-film with adsorbed hydrogen atoms when the sensor was kept away from the surface of the Sb-film at a distance of 3 cm. This signal vanished completely when nitrogen or another gas passive to the film was led into the vial at a pressure of 100 Torr. The signals were observed again after evacuation of the gas (Fig. 6.5). Such a phenomenon is only characteristic for active particles which cannot penetrate the distance between the sensor and antimony film at experimental conditions because of heterogeneous and homogeneous recombination. (The molecular products would reliably reach the surface of the sensor under these conditions). Note, that according to the experiment, the molecular products freely penetrate through the polished shutter 7 (see Fig. 6.1). Therefore, if the signals of the sensor were controlled by effects of antimony hydride or any other molecular product the resistivity of the sensor would have changed at the shut shutter as well, which contradicts experimental data. Consequently, the change in resistivity of the sensor made of zinc oxide observed in experiment (see Fig. 6.2) can be related only to effects of hydrogen atoms or antimony-hydrogen radicals.

If we assume that donor signals of the sensor are controlled by adsorption of antimony-hydrogen radicals the radicals of SbH_3 adsorbed on the sensor similar to radicals of NH, $N\dot{H}_2$ [20] must operate as acceptors of electrons on a ZnO film (this parallel is justified because antimony hydride is an ammonium homologue). Consequently, this should result in increase of the sensor resistivity. In contrast, in experiment with above particles we observed the drop of resistivity of the sensor. Further, if one assumes that donor signals of the sensor are somehow related to effects of antimony-hydrogen radicals then the surface of the sensor detecting the donor signal in experiment (see Fig. 6.2) should feature traces of antimony which are detectable by any other experimental technique, e.g. by atom-adsorption analysis.

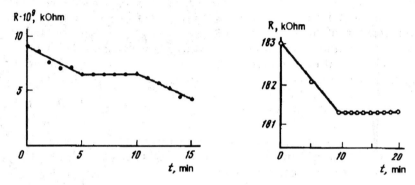

Fig. 6.5. The change in resistivity of the sensor in the vial with antimony film treated by hydrogen atoms: nitrogen was led inside the vial at $t = 5$ min; evacuation was conducted at the moment of time $t = 10$ min.

Fig. 6.6. The change in resistivity of the sensor: ethylene was led into the vial at the moment of time $t = 10$ min (0.1 Torr).

In order to increase the sensitivity of the atom-adsorption analysis we made a special sensor with the surface capable of accommodating the total amount of antimony during adsorption of antimony-hydrogen radicals, which might be assessed through the change in electric conductivity of the sensor. This area was accessible for atom-adsorption analysis. The analysis of the sensor indicated the absence of transposition of antimony with the particles analyzed. This means that in experiment (see Fig. 6.2) we observed the emission of initially adsorbed atoms of hydrogen from the surface of amorphous antimony during its ordering.

The above conclusion is confirmed by experiments were ethylene was led into the vial with an antimony film treated by hydrogen atoms. This experiment provides an example of a qualitative chemical analysis of atom hydrogen. In is known [21] that capturing of hydrogen atoms by molecules of ethylene develops with high rate resulting in creation of ethyl radicals. The latter get adsorbed on the deposited antimony layer during the very first collision [22] making the stationary concentration of radicals low to such an extent that they cannot influence the resistivity of the sensor. Figure 6.6 shows the change in resistivity of the sensor with time in the vial with Sb-film and adsorbed hydrogen atoms. At initial moment of time one observes a decrease in resistivity of the sensor caused by emission of hydrogen atoms. When ethylene gets introduced into the vial (pressure 10^{-1} Torr) the signals of sensors immediately disappear. We should note that introduction of such a small amount of nitrogen or other inert gas does not practically influence sensor readings.

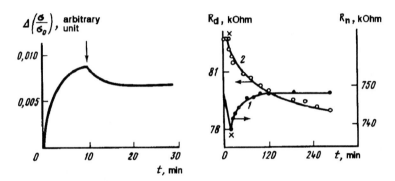

Fig. 6.7. The change in electroconductivity of a zinc oxide film due to hydrogen atoms emitted from the surface of the formed layers of platinum on the surface of fused silica during introduction of molecular hydrogen

The arrow head indicates the beginning of evacuation of hydrogen $P_{H_2} =$ 6.7 Pa, $T_{Pt} = 533$ K, $T_{ZnO} = 293$ K

Fig. 6.8. Emission of silver atoms: *1* – the change in resistivity of a sensor-substrate; *2* – the change in resistivity of a sensor-detector

Therefore, only using heterogeneous and homogeneous reaction known from the literature we managed to resolve the analytical problem put forward and identify particles emitted into the gaseous phase from the surface of an antimony film. The concentration of hydrogen atoms in gaseous phase in this case is low (the evaluation through the change in electric conductivity of the sensor yields the value of the order of 10^6 cm^{-3}). Therefore, this non-trivial phenomenon was observed only due to unique sensitivity of zinc oxide sensors.

Making use of these sensors it is possible to establish the mechanisms underlying emission of hydrogen atoms from the surface of amorphous antimony. It appears that the phenomenon is specific only for amorphous antimony during its crystallization. The gray crystalline antimony is passive from this stand-point. The adsorption of oxygen stops crystallization and, as a consequence, stops the emission of oxygen atoms. Heating amorphous antimony with adsorbed hydrogen atoms accelerates its ordering and, consequently, during heating the intensity of emission of hydrogen atoms increases. These experiments showed that there are exothermal stages in the process of ordering of amorphous antimony that are responsible for emission of hydrogen atoms. The energy released during such processes can be substantially high. For instance, during recombination of two antimony atoms the energy amounting to 71.8 kcal/mol gets released, the formation of three atom molecule from a two atom one results in release of 129 kcal/mol [23], the energy of brake down of the Sb-H bond amounts to 60 kcal/mol [24].

It is interesting to point out that emission of hydrogen atoms was observed in case of relaxation of platinum and palladium layers deposited in presence of hydrogen at low pressures as well [25]. In this case the authors draw the parallel with the change of electric conductivity of sensor caused by adsorption of hydrogen atoms emitted from metastable surface of a metal film (Fig. 6.7) with readings of the sensor caused by adsorption of H-atoms which in calibration experiment were obtained through pyrolysis on hot filament. Initially, this method made it possible to reliably identify the type of electron donor particles, and, second, assess the value of the flux of non-stationary emission of H-atoms, which made it possible to establish the correlation between the intensity of emission of H-atoms and the energy of aggregation of deposited metal.

6.1.2. Emission of silver atoms during their aggregation on solids

As it has been already mentioned in Section 6.1 the ordering of freshly deposited solid films is accompanied by numerous peculiar processes resulting in energy release during ordering (condensation-stimulated effects). The emission of adsorbed silver atoms from the surface of substrate into gaseous phase following completion of deposition provides an example of such a process which, presumably, can be detected and studied only with the use of semiconductor sensor. In this case the use of the sensor as a substrate-carrier of atoms makes it feasible to obtain experimental data leading to assessment regarding the mechanism underlying this process.

The experiments were conducted on similar experimental set-up as shown in Fig. 6.1. Substrate 5 had an internal heater – platinum film deposited on the back side of the substrate. The substrate was heated due to Joule heat while letting the current through the platinum film. The temperature of the film was controlled using Pt – PtRh thermocouple. Such arrangement made it possible to heat the substrate above 700°C. During this the temperature in chamber *1* was not substantially changed which means that there were no alterations in resistivity of the sensor (this is very important to interpret its readings). In several experiments the substrate could be rotated with respect to tray *3* using a glass-covered iron weight and a magnet. If the surface of various sides of the substrate was subject to different mechanical treatment, such installation made it feasible to evaluate the role played by the structure of the surface on emission of silver atoms.

The procedures of experiments were the following [15, 26]. After deposition of a specific quantity of silver on substrate the heating of a tray with silver was turned off, the shutter *7* was opened and the sensor was positioned opposite to the substrate in such a manner that the surface of the sensor was parallel to the surface of substrate. In these experiments we detected an irreversible donor signal of the sensor which can be related to adsorption silver atoms on the sensor made of a zinc oxide film. It is known [27] that silver atoms are donors of electrons. Note that the signals of the sensor were observed only when the sensor was positioned in front of a substrate. There were no signals detected in any other arrangement between sensor and substrate.

As it has been already mentioned the first issue to be resolved in such studies deals with identification of particles affecting sensor resistivity. In our case this task was simplified because the sensor was put into the adsorption chamber *2* only after completion of deposition; the silver atoms did not penetrate into chamber *1* through the shutter, moreover, deposition was conducted with the pressure of residual gases of 10^{-7} Torr, the temperature of silver evaporator not exceeding 850°C. This implies that at such conditions there is no generation of long-living active particles inside the vial which might affect the resistivity of the sensor and which can be present inside the vial when evaporator is cooled. Therefore, assumption that donor signals of the sensor are caused by emission of silver atoms can be considered substantially well ground.

This conclusion is proved by the element analysis of sensors which have registered more than 30 portions of silver emitted from the surface of substrate after completion of deposition of silver on its surface. The experiments on local analysis of the sensor indicated that such sensor has areas containing up to 0.3 wt.% of silver in 1 μm^3. The total amount of silver atoms incident on sensor was $7 \cdot 10^{12}$. This means that silver atoms get deposited on the surface of the sensor inhomogeneously, being localized close to defect allocation area which made it possible to detect silver on the sensor surface using X-ray microanalysis. It should be mentioned that control sensor made of the same batch did not contained silver at all.

Therefore these experiments showed a very interesting phenomenon, namely the emission of adsorbed silver atoms from the surface of a substrate after accomplishing the deposition process. In these experiments the semiconductor sensors were used in two ways: sensor-substrate onto which the silver was deposited from the tray, which made it possible to monitor the behaviour of silver atoms on the surface of adsorbent; and sensor-detector of emitted silver atoms.

Figure 6.8 shows the change in resistivity of a sensor-substrate (curve *1*) and the change in resistivity of sensor-detector of silver atoms (curve *2*) with time. As is shown in the figure the decrease in conductivity occurs during adsorption of silver atoms on semiconductor of n-type (zinc oxide is always an electron type semiconductor). This phenomenon was observed in [27]. During deposition of silver atoms the sensor-detector was placed in chamber *1* behind the shut down shutter therefore its resistivity was not changed. At a moment of time 5 min (asterisk in the figure) the source of silver atoms was turned off and sensor-detector was installed in chamber *2* opposite to the target. After switching off the source of silver atoms we observed relaxation of resistivity of the sensor-substrate (in compliance with results of [27]), which indicates that aggregation of adsorbed atoms develops on its surface. Silver aggregates are always electrically neutral on the surface of zinc oxide [28]. This term is applicable to all particles formed of several atoms not possessing the properties of a massive crystal. Simultaneously an irreversible change in resistivity of sensor-detector was registered, which, as it has been shown above, is determined by adsorption of silver atoms on its surface. The consecutive deposition of similar amount of silver atoms (usually the portion of deposited silver was about $3 \cdot 10^{15}$ cm^{-2}) there was a further decrease in resistivity of sensor-substrate which is indicative that there were silver atoms present on substrate in this experiment. Similar to the preceding experiment we observed relaxation of resistivity of sensor-substrate after turning off the source of silver atoms. Consequently, the aggregation of adsorbed silver atoms develops on the surface of substrate in this case as well. However, in this experiment the resistivity of the sensor-detector positioned opposite to the sensor-substrate was not changed. This means that there is no emission of silver atoms during subsequent deposition of silver on substrate.

The results of these experiments enable one to draw an unambiguous conclusion that there is no trivial desorption of weakly bound silver atoms because both in the first and the second experiments there were silver atoms adsorbed on the surface, yet, the emission was observed only in the first experiment [15, 26]. It should be pointed out that the centres with the highest adsorption heat get filled first during adsorption. Consequently, if conventional desorption of weakly bound atoms (whose number would have been much higher after second deposition) took place in these experiments, one could have expected more intensive signals of sensor-detector of silver atoms after the second deposition. Yet, in this case, as it has been shown by experiment, the emission was not observed at all.

These results can be understood if one assumes that during formation of silver aggregates composed of small amount of atoms a fraction of the energy released is spent on vibrational excitation of the bond of adatom with the surface which results in its release into gaseous phase. It is more probable during subsequent deposition that the adatom would get attached to available surface aggregates. The attachment of adatom to the aggregate is also accompanied by a release of substantial energy close to that of cohesion 68 kcal/mol [29], whereas the adsorption heat of silver atoms is 12 kcal/mol [30]. Therefore, the probability of vibrational excitation of the bond of adatom with the surface is substantially high in case of formation of small aggregate which enables one to observe the emission of silver atoms. At subsequent deposition we presumably deal with increasing size of available aggregates. The energy released during this process can be easily distributed over the bonds of a large particle and a fraction of the energy reduced to a single atom is insufficient to release it.

This supposition yields that for a given deposition rate the emission of silver atoms can be more intense only within a specific range of deposition of the surface by adatoms which is characteristic for a specific material and for a given deposition rate. Aggregation of adatoms does not occur even at heating of the substrate made of zinc oxide up to 200°C if the coverage of the surface by adatoms is small ($\sim 10^{-4}$ to 10^{-2}) [31]. Therefore, when the amount of silver atoms at the surface of detecting sensor registering the emission of silver atoms is small one does not observe the relaxation of resistance of the detecting sensor. At higher deposition rates the aggregation takes place but the probability of formation of aggregates composed of more than two atoms is low, therefore at such deposition the emission of silver atoms is small. Finally, if the deposition degree by adatoms is high the aggregation and the emission linked to the latter occurs in the case of deposition of adatoms (the well known re-evaporation of atoms [32, 33]) and amount of atoms released prior to the completion of deposition is small. This explanation of the emission mechanism makes it possible to understand the complex dependence of the re-evaporation rate on the value of the flux of incident atoms observed in [32, 33].

In [14, 26] it was shown that emission of silver atoms develops only in case when the surface concentration of adatoms lies within the range $10^{15} \leq [Ag] \leq 10^{16}$ cm^{-2} for deposition rates $10^{13} - 10^{14}$ cm$^{-2} \cdot$s^{-1}. Strictly speaking the above range of surface concentration should be dependent on the deposition rate of silver atoms.

Other experiments [14, 26] confirm the assumptions made on mechanism of emission of silver atoms. If one heats the substrate with adsorbed adatoms and silver aggregates the formation of microcrystals occurs as a result of aggregation of adatoms and growth of size of aggregates [34]. This process might result in a "purification" of the surface from adatoms and silver aggregates. Then, the adsorption of new portion of atoms ($3 \cdot 10^{15}$ cm^{-2}) would lead both to aggregation of adatoms and aggregation induced emission, as well as to the attachment of adatoms to the already available crystals. As it stems from the proposed emission mechanism the latter process would not result in emission of silver atoms. Consequently, during deposition of the same dose ($3 \cdot 10^{15}$ cm^{-2}) of silver on substrate containing microcrystals one should expect less intensive emission than in experiments with pure surface.

We heated the substrate of zinc oxide containing 10^{16} cm^{-2} of silver atoms (in this case there was already no emission after completion of deposition) at 300°C. Such thermal treatment results in formation of microcrystals, rather than evaporation adatoms on the surface of the substrate made of zinc oxide. In paper [34] it was shown that microcrystals with diameter 100 Å deposited on the zinc oxide surface are acceptors of electrons, therefore the formation of microcrystals results in increase of resistivity of a sensor substrate above the initial value (prior to silver deposition). In this case the initial value of the resistance of sensor-substrate was 2.1 MOhm, after adsorption of silver atoms it became 700 kOhm, and as a result of heating at 300°C and formation of microcrystals – acceptors of electrons it in increased up to 12 MOhm. If such a substrate is subject to deposition of $3 \cdot 10^{15}$ cm^{-2} silver again, then emission of silver atoms gets detected. From the change of resistivity of sensor-detector due to deposition of silver atoms one can conclude that in this case the emission of atoms is 4 times as low than in experiment with pure substrate made of zinc oxide, which confirms the supposition made on the mechanism of emission of adatoms.

Finally, if we heat the sensor-substrate with deposited silver atoms using internal heater (platinum film attached to the back side of the sensor-substrate) up to 700°C then the surface of zinc oxide gets completely cleaned of silver. This can be confirmed by the value of resistivity of sensor-substrate which comes back to the initial value of 2.1 MOhm (the silver during such treatment partially evaporates, partially migrates to the contacts). The experiment showed that as a result of such treatment and cleaning of the surface of sensor-substrate it completely restores its initial activity in respect to emission of silver atoms.

One can expect from proposed emission mechanism that this effect would be highly sensitive to the material and the state of the surface of substrate. Indeed, depending on the type of material of substrate and the manner to treat its surface conditions for aggregation of adatoms and transfer of energy of aggregation to a specific adatoms would vary. While verifying the supposition we carried experiments on emission from the quartz surface. The surface was subject to various mechanical treatment. It was fused, polished or grinded. In each experiment the same amount of silver atoms ($3 \cdot 10^{15}$ cm^{-2}) was deposited on the surface. It occurred that the highest activity with respect to emission of adatoms was shown by grinded quartz. Its activity, however, is one order of magnitude lower than that of zinc oxide. On the other hand, fused quartz is absolutely non-active in this process. Such a result can be linked to the fact that the aggregation of silver on the surface of fused quartz, presumably, develops already during the deposition process due to high mobility of adatoms. Therefore, there is no emission of silver atoms observed after turning off the source of silver atoms. Obviously, the surface of polished and grinded quartz provide different conditions for the rate of lateral diffusion and the energy transfer by different atoms which results in notable emission from the surface of substrate.

Therefore, the use of a combination of sensors (sensor-substrate and sensor-detector of emitted atoms) enabled us to obtain unique information concerning the mechanism of emission of adsorbed atoms of silver in the course of their surface aggregation. The data obtained makes it possible to provide an insight of mechanisms and energetics of condensation-stimulated phenomena [11].

The same method of analysis of emitted particles was applied to elucidate the origin of non-stability of piezoquartz resonators with silver electrodes [35].

6.1.3. Cold emission of silver atoms from the surface of silver films on quartz caused by ultrasound oscillations

These studies were carried out on industrially manufactured piezoquartz resonators with an AT-cut featuring silver thin-film electrodes. Oscillations with a frequency of 10 MHz (resonant frequency) were generated by generator of the TKG-3 type. The sensor of silver atoms (films of zinc oxide) were positioned in the same vial with resonator, the sensor was positioned parallel to the resonator plane; the distance between them was about 5 mm. Prior to the experiment the vial containing resonator and sensor was heated up to 473 K and kept at above temperature during 30 min under conditions of oil-free vacuum (10^{-7} Torr). Then, at a temperature of 296 K oscillations were excited in resonator resulting in continuous increase in electric conductivity of the sensor until the oscillations of resonator were stopped. The subsequent excitation of oscillations inside resonator resulted again in similar signals of the sensor. The Auger-analysis of the sensor after accomplishing the experiment confirmed the transfer of silver from the surface of resonator to that of sensor.

The evaluation of amount of silver atoms desorbed from resonator during 80 h operation was made applying the known relations [36] for silver atoms adsorbed on the sensor surface in charged form accounting for the fraction of atoms transferred from the source (resonator) to the target (sensor). The estimates indicate that intensity of the flux of silver atoms from the surface of resonator in this experiment was about $1.5 \cdot 10^{13}$ $m^{-2} \cdot s^{-1}$, i.e. during 75 h operations approximately 10% of the surface silver atoms left resonator (under condition that the amount of surface silver atoms is about $4 \cdot 10^{19}$ m^{-2}) explaining the shift of resonant frequency from its nominal value 10 MHz by 11 Hz.

Therefore, application of sensors enables one to elucidate the course of fluctuations of resonant frequency in piezoquartz resonators which is the emission of surface silver atoms from the surface of silver film due to ultrasonic oscillations. The non-triviality of this effect deals with the fact that additional energy supplied to adsorbed layer is small: 10^{-4} eV, i.e. it is much lower than the energy of binding of the surface atoms of silver with the surface. It is known [37] that the local mechanical stresses may attain 10^{8} $N \cdot m^{-2}$ in cavities on the surface acoustic waves operating in the 300 – 500 Hz frequency range. In the order of magnitude this coincides with the surface tangents localized along the perimeter of the thin film silver electrode AT-cut quartz resonator. Presumably, such internal stresses and elastic deformations caused by them would be accompanied by generation and annihilation of defects – vacancies and interstitial atoms. The energy released during annihilation of such defects is sufficient to stimulate the desorption of separate atoms.

6.1.4. Emission of alkyl radicals from disordered selenium surface

To conduct experiments of this kind it is very convenient to make use of disorder adsorbent provided by a film of amorphous selenium. During deposition under vacuum conditions, the pressure being no higher than 10^{-2} Torr, the amorphous modification of selenium is being formed [38]. There are two forms of amorphous selenium which differ in coordination numbers and radii of coordination spheres. The first form is formed during deposition of selenium atoms on substrate at 20°C, the other one – at 60°C. The contemporary studies consider amorphous selenium as a polymeric material which may be composed of helixes, rings of Se_8-type and chains, positioned in different planes and exhibiting different orientation in the volume of material.

With increase in the temperature the Se_8-rings of amorphous selenium get decomposed into chains with free radical-like ends. This short chains can be polymerized into long chains of hexagonal selenium. The crystalline lattice is disordered at the onset of the ordering process so partially ordered areas – micels – start forming. The chains in micels can have different radical form, different orientation or be wrinkled. The micels are immersed into amorphous substance. During crystallization micels grow due to amorphous selenium. Moreover, spherulites which are a special type of textured polycrystals growing from one centre get formed from selenium. The temperature range of spherulite crystallization lies within the interval 70 – 150°C, at the same time the optimum temperature of crystallization due to micel growth is about 210°C. Therefore, in amorphous selenium the ordering occurs over the wide range of temperatures (60 – 210°C). However, even after long heating of a Se-film the selenium exhibits a superposition of various crystalline structures.

Interaction of atomic hydrogen and selenium is accompanied by formation of volatile hydride, therefore the use of H-atoms as adsorbate is not convenient. From this stand-point it is better to use aliphatic radicals which are ready to interact with selenium [22].

The experiments on evaluation of plausible emission of alkyl radicals from the surface of amorphous selenium were carried out inside a vial shown in Fig. 6.1 [15, 39]. The experiments were carried out as follows. The amorphous selenium was deposited on target 5 from tray 3 inside a vacuum vial. Initially the selenium was purified and melted. Following that the vapours of preliminary purified diethylselenide were let inside the vial at a pressure of 10^{-1} Torr. The continuous value of the resistance of the ZnO semiconductor sensor after letting the vapours in was used as a proof of vapour purity with respect to oxygen content. Then with shutter 7 being shut down the pyrolysis of diethylselenide on hot filament 4 was conducted (in several experiments this filament was made of either Pt or Ta). Following this exposure the pyrolysis filament was cooled down and the vial was heated at a constant rate of 10 deg/min with monitoring the rate of change of electric conductivity of the sensor. The latter value as it has been already shown is proportional to the concentration of active particles in gaseous phase.

The experiment with films of amorphous selenium with adsorbed ethyl radicals [39] showed that during heating of these films in vacuum there is irreversible increase in resistivity of the sensor caused by adsorption of particles – acceptors of electrons. As it has been already shown in Chapter 2 alkyl radicals belong to such particles. These signals from the sensor were not observed if the selenium film with adsorbed diethylselenide was heated without prior pyrolysis on the hot filament. It should be mentioned that results were the same if heating of selenium film occurs both after pumping out diethylselenide vapours or in the presence of vapours. Consequently, the particles affecting the resistivity of the sensor are being formed only as a result of pyrolysis of diethylselenide. Further, the pyrolysis of $(C_2H_5)_2Se$ and adsorption of ethyl radicals were conducted when the temperature of pyrolysis filament was in the range of 500 – 650°C when the linear dependence of acceptor signals of the sensor $\ln(d\sigma / dt)_{t_0} \approx T^{-1}$ caused by effects of ethyl radicals on the sensor was observed. At such temperatures the probability of braking up the bond C–H is low. Therefore, in this experiment C_2H_5 provide the only active particles formed during pyrolysis.

If heating of amorphous selenium with absorbed ethyl radicals is conducted in presence of nitrogen or inert gas with pressure 100 Torr no signal is picked up from the sensor. This implies that in this case the signals of the sensor are controlled by adsorption of ethyl radicals on its surface. Consequently, the heating of amorphous selenium with adsorbed radicals and resulted crystallization lead to emission of radicals.

Figure 6.9 shows the initial rate of the change of electric conductivity of the sensor (this value is strictly proportional to the concentration of radicals in gaseous phase) as a function of the temperature of selenium film with adsorbed ethyl radicals. As the figure shows this dependence has a bell-shaped form. This profile, presumably, reflects two competing processes. The up coming branch of the curve reflects the increase in the processes of ordering with the rise in the temperature and, as a consequence, increase in intensity of emission. The down going branch is indicative of depletion of chemisorbed layer due to emission or recombination. The evaluation of intensity of emission of radicals using the sensor indicate that there is 0.01 part of ethyl radicals of the total of adsorbed from the surface of selenium film. This estimate was made as follows.

The reaction vial (see Fig. 6.1) was changed in order to make the distance between sensor and tantalum filament (generator of ethyl radicals) equal to the distance between filament (radical source) and selenium film as well as to the distance between the sensor and selenium filament. Dimensions of pipes linking them were also the same. Then, measuring the initial rate of the change in electric conductivity of the sensor during generation of radicals one can assess in arbitrary units the concentration of radicals incident on the surface of the sensor. Due to the same size of geometrical surface of sensor and the selenium film there would be the same amount of radicals on the surface of selenium as on the surface of sensor. (Here we assume that sticking coefficients of ethyl radicals for selenium are close to those of zinc oxide). Having measured the initial rate of the change in electric conductivity of the sensor during heating of selenium film with adsorbed ethyl radicals one can obtain the amount of radicals living the surface of selenium film in the same arbitrary units. Doing that one should account for a fraction of radicals delivered from selenium film to detector (which can be done if the heating takes place after evacuation of parent molecules). During calculations we neglected recombination of radicals in the course of treatment of selenium.

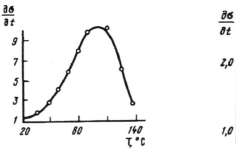 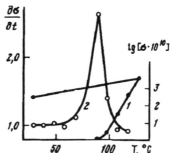

Fig. 6.9. The change of readings of the sensor as a function of temperature of selenium film with adsorbed alkyl radicals during its heating with rate 10 deg/min.

Fig. 6.10. The dependence of electric conductivity of the selenium film with adsorbed radicals as a function of temperature (1) at the rate of change of electric conductivity of sensor (2).

Emission of free radicals was observed in case of relaxation of disordered surface of selenium with adsorbed methyl radicals [18, 39] which were obtained during pyrolysis of acetone [40] over the temperature range 500 – 650°C, as well. As during pyrolysis of diethylselenide the above temperature range was chosen to ensure low probability of the brake up of C–H bond.

During heating of a Se-film with initially adsorbed methyl radicals we observed the readings of the sensor similar to those which had been registered during experiments with Se-film with adsorbed ethyl radicals. The control experiments dealing with introduction of inert gas, heating of clean film of amorphous selenium, heating of selenium film with adsorbed molecules of acetone and those carried in acetone vapours without pyrolysis enabled us to conclude that sensor signals observed in this experiment were solely caused by emission of initially adsorbed methyl radicals.

Figure 6.10 shows the change in electric conductivity of selenium film with temperature indicating the development of ordering process (curve *1*). Curve *2* reflects the dependence of initial rate of the change in electric conductivity of the sensor on the temperature of selenium film. Similar to experiments on the study of emission of ethyl radicals in this case the ascending branch of the curve reflects the increase in intensity of emission of radicals linked with growing ordering processes. The descending branch again is controlled by depletion of chemisorbed layer.

We should pay attention to the fact that the "flash up" in the change of electric conductivity of the sensor was marked at much lower temperatures than the change in resistivity of selenium film. Consequently, the maximum intensity of emission of radicals from the surface of selenium occurs at initial moment of crystallization of selenium, i.e. when cluster particles composed of several atoms get formed with higher probability than in case of phase transitions (which is caused by a higher release of energy per atom). The energy of braking up of bond Se−C is about 65 kcal/mol [24]; in case when a dimer is formed from different atoms the energy of 75.7 kcal/mol gets released, for trimer the energy amounts to 124 kcal/mol [41]. This energy is obviously sufficient not only to brake up bonds in the surface organometallic compounds but to emit their fragments into the gaseous phase as well.

Therefore, the use of semiconductor sensors in these processes made it possible to detect and identify alkyl radicals emitted into the gaseous phase. The rapid monitoring of concentration of these particles in gaseous phase made it feasible, in its turn, to establish several peculiarities of the mechanism and understand the energy features of these processes.

6.2. Emission of oxygen atoms during interaction of reduced silver with molecular oxygen

Interaction of pure metals with various chemically active gases is accompanied by numerous emission phenomena. Polyakov [42] observed interaction of palladium with molecular oxygen accompanied by glowing of free oxygen atoms emitted from the surface of palladium. Chemisorption of halogens (for instance chlorine and bromine) on several metals (yttrium, zirconium, titanium, hafnium) originates emission of ions and electrons [43]. In [5] the emission of charged particles during oxidation of Al, Ni, W, Zn was considered. In several cases the emission phenomena are linked both to adsorption and to initial stage of oxidation.

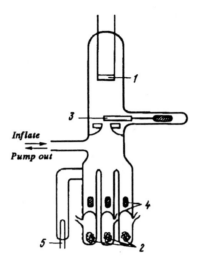

Fig. 6.11. The schematics of experimental set-up to study emission of atomic oxygen. *1* – sensor of oxygen atoms; *2* – samples of reduced silver; *3* – shutter; *4* – weights to brake membranes; *5* – platinum filament to calibrate sensor against the concentration of oxygen atoms.

In [44] the semiconductor sensors were used to show that interaction of freshly reduced silver with molecular oxygen results in substantial emission of oxygen atoms.

The experiments were conducted in a vial (Fig. 6.11) featuring a system of gas input and evacuation based on oil-free electric discharge and sorption pumps. The whole-made vial was separated into two chambers using a plain polished shutter *3*. The sensor of oxygen atoms *1*, a thin ZnO film was positioned in the upper chamber. The lower chamber contained ampoules with samples of silver *2*, separated by thin membranes from the rest of the vial which made it possible to treat the detector without oxidizing the silver. In these experiments [44] the samples of silver were different: a plate with an area of 40 cm^2, a grid with area 800 cm^2, fine particles deposited on SiO_2; the area of the surface of the metal phase in this case amounted to 7 m^2. In separate experiments freshly deposited or aged films of silver were used. The preparation to the experiment involved a long (ca. 30 h) reduction of silver samples in hydrogen at atmospheric pressure, the temperature being 350 – 400°C followed by evacuation to the pressure better than 10^{-6} Torr.

The experiments were conducted as follows. Oxygen was put into the vial (usually at pressures 10^{-1} Torr) and the system was kept during 12 – 15 hours. Over this time the resistivity of the sensor caused by adsorption of molecular oxygen was stabilized and the sensor itself became insensitive to the change in pressure of molecular oxygen. At the same time it retained high sensitivity to active oxygen particles, i.e. atoms, electron- and vibration-excited molecules. Following such exposition the silver samples were heated up to required temperature (usually 200 − 400°C). This accomplished, the glass membranes were broken and immediately after or at a specific moment of time following opening of the shutter the rate of change in electric conductivity of the sensor was monitored. This rate was proportional to concentration of active particles in gaseous phase.

Experiments showed [44] that the interaction of oxygen with clean silver surface always results in emission of acceptor particles. This conclusion is justified by decrease in electric conductivity of the sensor. Straightforward experiments indicate that signals of sensor in this case are caused by effects of active particles. When argon (400 Torr) was let inside the vial the sensor signals were immediately eliminated. In this case, as was the case with hydrogen atoms and alkyl radicals, the active particles cannot travel over the distance between the source and sensor due to heterogeneous and homogeneous recombination. If the sensor signals were generated by stable molecular products these particles would have freely reached the sensor. Moreover, as it has been already mentioned, the molecular fragments freely penetrate through closed polished shutter.

It is well known [45] that hydrogen is not being absorbed on silver, and consequently in this case only interaction of oxygen with clean silver surface takes place, which means that signals from sensor can be caused by effects of neutral oxygen particles (oxygen atoms, molecules of singlet oxygen and vibrationally excited oxygen molecules). Charged particles having different sign as well as photons cannot reach the sensor due to a complex geometry of the installation and substantially high pressure of molecular oxygen.

Using several physical and chemical techniques typical for the method of semiconductor sensors one can identify emitted particles.

It is known that vibration-excited oxygen molecules get desactivated through collisions with molecules of carbon dioxide with probability $1/5000$ [46]. Therefore, letting CO_2 at pressure 0.1 Torr at a chosen geometry of the vial should completely eliminate the signals from sensor. The experiments showed [44] that introduction of above amount of CO_2 does not result in notable change of signals from the sensor. This means that the signals observed in experiment are not related to effects of vibration-excited molecules of oxygen on a sensor.

It is known from several papers that deactivation of electron-excited molecules of oxygen on the surface of depleted in oxygen Co_3O_4 can develop effectively enough so that probability of this process can approach 1 [46, 48]. Therefore, if the inner side of the tube ($d_T = 2$ cm,

l_T = 10 cm) is covered with Co_3O_4 the whole distance from the source of singlet oxygen to the sensor, electron-excited molecules of oxygen will be completely desactivated and will not reach sensor at a pressure of 0.1 Torr. Separate experiments with the known source of singlet oxygen and Co_3O_4 covering used for analytical purposes confirm the efficiency of quenching of $^1O_2$ on its surface. Moreover, under these conditions the intensity of the flux of oxygen atoms from the source, for instance from hot platinum filament 5 (Fig. 6.12), drops by 2 times. In case described in [44] it occurred that application of Co_3O_4 to the walls led to decrease in the flux of particles emitted from the surface of silver no higher than by 2 times. These results indicate that under these conditions only oxygen atoms can be emitted from silver surface.

The above conclusion was confirmed by control experiment. It is well known [47] that freshly deposited silver films are very efficient in adsorption of atomic oxygen at room temperature. Therefore, the attachment covered by Ag-film similar to that with a Co_3O_4-covering should completely stop the flux of oxygen atoms. (The experiment with the known source of oxygen atoms – hot Pt-filament – confirms the efficiency of above attachment). It occurred [44] that in case of particles leaving the Ag surface such an attachment completely shuts down the flux of particles from the surface of the sample under investigation.

Consequently, during interaction of oxygen with freshly reduced silver surface the emission of atom oxygen takes place.

Application of semiconductor sensor combined with several other physical and chemical methods makes it feasible to obtain data concerning the mechanism of this process. Figure 6.12 shows the kinetics of emission of oxygen atoms during interaction of molecular oxygen at P = 0.1 Torr and 290, 320 and 365°C with the surface of silver samples reduced at identical conditions. The evaluation of intensity of the flux of oxygen atoms from the silver surface using the known source of oxygen atoms (hot platinum filament) show that maximum density of the flux of atoms reaches the value of $(4 \div 5) \cdot 10^{14}$ at/cm^2·s. The energy of activation of the process evaluated by the function of the maximum rates on the temperature is about 125 kJ/mol. Figure 6.13 shows the adsorption by a sample of silver reduced to a unit area (curve 4), at $P_0 = 10^{-1}$ Torr and 365°C. It is clearly visible that the maximum intensity of emission is observed for cover $5 \cdot 10^{15}$ at/cm^2, i.e. it is substantially higher than that provided by monolayer which amounts to $7 \cdot 10^{14}$ at/cm^2 if we assume that Ag_2O is being formed on the surface.

According to these data it is clear that first, the emission of O-atoms into the gas phase is provided not due to heat of oxygen adsorption (which drops with increase in deposition) because the maximum intensity of emission is observed for layers thicker than a monolayer one.

Second, the energy of emission activation is close to the energy of activation of adsorption of oxygen by silver – 146 kJ/mol [49]. This implies that emission of O-atoms occurs due to energy of defect annihilation in the surface-adjacent layers of catalyst due to adsorption of oxygen. From the stand-point of such assumption it is obvious that
"depletion" of emission is linked with sloroing down of the oxygen absorption rate (as shown, for instance, in Fig. 6.12).

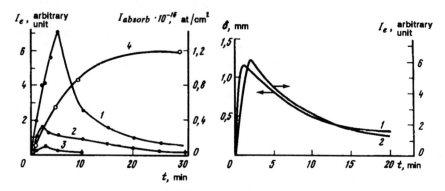

Fig. 6.12. The kinetics of emission of O-atoms and adsorption of molecular
oxygen during interaction of O_2 with silver at 365 (*1, 4*), 320 (*2*) and 290°C.

Fig. 6.13. The kinetics of bending of polyimide film with deposited silver (*1*)
and emission of O-atoms (*2*) when oxygen is let inside the vial.

In order to confirm such supposition we carried out a series of special experiments focusing our attention on adsorption of oxygen [50] by
samples of silver reduced through the method described in [44]. Using
the method of a free rod [52] we measured the kinetics of mechanic
stresses in thin silver films deposited on polyimide substrate. It occurred
that in this case the films became stressed (the substrate bending was
observed in experiment). Applying the well known relationship one can
evaluate the mechanic stresses using the value of the bending [51]

$$\varsigma = \frac{E_{Ag}(d_S + C_H d)^3}{6T_k d_S d(1 - \nu_P)} ,$$ (6.1)

where d_S is the thickness of substrate; d is the thickness of film; ν_P is
the Poisson coefficient of substrates; T_k is the curvature radius of the
sample; C_H is the ratio of Hook's modulus for silver and polyimide substrate. These results indicate that the maximum value of stress may attain 10^5 kg/cm^2 which is consistent with data given in [52]. As a result

of heating at 250 – 300°C the healing of mechanical stresses takes place (the rebound of substrate was observed at experimental conditions).

While letting oxygen inside ($P = 0.1$ Torr, $T = 300$°C) we again observed deformation of the rod. The kinetics of deformation was similar to the dependence of intensity of emission on time. Figure 6.13 shows the kinetics of the change in degree of bending and emission of oxygen atoms. One should note that the maximum bending occurs when emission of O-atoms has the maximum value (see Fig. 6.12). The value of deformation can be used to obtain the relative elongation of the sample, whose kinetics would be similar to deformation kinetics. A relationship linking the value of the relative transverse elongation of a thin Ag film having concentration of vacancies N_V was derived in [10]:

$$\Delta l = -\frac{1}{3}\left[\frac{4}{5(1-v)} - 1\right]N_V .$$ (6.2)

Therefore, the kinetics of generation of defects in surface-adjacent layers is similar to kinetics of emission of O-atoms. (The estimates indicate that the maximum concentration of vacancies in this case may attain the value of 10^{-5} for a sample with area 1 cm^2). If one assumes that the emission of oxygen atoms is caused by processes of annihilation of vacancies in the sample, then the coincidence in time dependence of stationary concentration of defects can be indicative that these processes are limited by generation of defects, which, in its turn, is controlled by processes of formation of oxide phase in surface-adjacent silver layers. Oxidation, especially at initial stage, is characterized by intensive formation of defects [54].

These data agrees with information on kinetics of dissolving of oxygen in silver obtained using the method of piezoquartz microweighing whose resolution with respect to oxygen may attain 10^{-3} of monolayer deposition [53]. Figure 6.14 shows the kinetics of absorption of oxygen by silver at $T = 200$°C, $P = 10^{-1}$ Torr, plotted in coordinates $\Delta m(\ln t)$, where Δm is the amount of oxygen absorbed. It is seen from this figure that starting with the moment of time $t = 180$ s this curve tends to be a straight one, i.e. the absorption of oxygen can be described by the Cabrer-Mott oxidation law in case of ultrathin oxide layers (of the order of several Å) when oxidation is limited by tunnelling of electrons to the surface through oxide layer [54]. In this case the process would occur in high electric field (~ 10^7 V/cm) and would be accompanied by intensive transfer of ions of metal and oxygen through oxide layer and, consequently, recombination of defects at respective surface centres at the phase interface, capturing of electrons which tunnel from metal through oxide film up to the surface. Emission of oxygen atoms can be caused by energy released in above processes. The energy of a vacancy in silver is 1.1 eV [8], the energy of interstitial atom being 3.5 eV [8]. Moreover, one can expect emission of oxygen atoms to be caused by charge recombination in the surface layers similar to situation observed in [55] for vibrational excitation of particles observed on the surface due to recombination of electron-hole pairs.

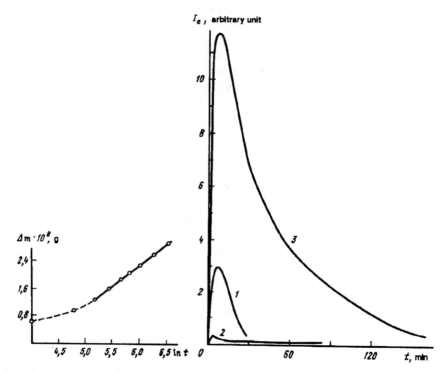

Fig. 6.14. Kinetics of absorption of oxygen by silver.

Fig. 6.15. Kinetics of emission of O-atoms during interaction of O_2 with silver: 1 – pure Ag; 2 – Ag + 0.2 at%Se; 3 – Ag + 0.2 at%K.

Then, from the stand-point of such consideration the period of time $t = 180$ s during which the logarithmic kinetics of adsorption of oxygen does not hold corresponding to the time of "flash up" of emission of oxygen atoms and "flash up" of deformation of the sample can be considered as a time of formation of oxide layer with thickness of several Angstroms. This conclusion agrees with results of volumetric studies (see Fig. 6.12) as well as sequential layer Auger-analysis according to which the thickness of layer containing oxygen amounts to 10 Å by the moment of oxidation process $t = 180$ s.

It is obvious that during deformation of the sample due to mechanical loading the creation and annihilation defects will also take place. Similar to preceding experiments in this case the value of deformation would determine the concentration of defects. However, in case of mechanical loading the defects will be evenly spread over the whole volume of samples, whereas in case of silver oxidation they remain localized only in the surface-adjacent layers. Therefore, emission of oxygen atoms under conditions of mechanical deformation of samples in oxygen atmosphere has low probability due to intensive annihilation of defects in surface-adjacent layers. Special experiments confirmed this conclusion.

Several other experiments justified the idea that emission of oxygen atoms is controlled by processes related to dissolving of oxygen in silver. It is well known that doping silver by electrically negative elements (sulphur, selenium, tellurium etc.) abruptly decreases the dissolving rate of oxygen in silver [49]. Doping by alkaline metals, on the contrary, increases the rate of dissolution due to internal oxidation [53]. As the proposed mechanism of emission of oxygen atoms suggests the doping with selenium should suppress emission of oxygen atoms whereas doping by alkaline metal would intensify this process. Figure 6.15 depicts the intensity of emission of oxygen atoms from pure silver (curve 1) of the sample containing 0.2 at%Se (curve 2) and sample containing 0.2 at%K (curve 3). Apparently, the doping by Se drops the intensity of emission of O-atoms by one order of magnitude in contrast to doping by potassium which intensifies this emission at least by 50 times.

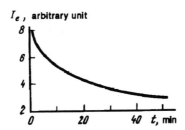

Fig. 6.16. Kinetics of emission of O-atoms during interaction of oxygen with nickel films

It should be mentioned that emission of oxygen atoms at initial stage of oxidation can be observed in case of oxidation of other metals (for instance nickel) as well. In this case due to high rate of oxidation the emission can be observed only as a result of oxidation of freshly deposited films of nickel at a room temperature. The Cabrer-Mott logarithmic law holds for nickel only after forming the oxide film with thickness 3 Å [58], which is pretty close to the values of thickness of oxide obtained for silver. Figure 6.16 shows the kinetics of emission of atom oxygen during oxidation of freshly deposited films of nickel by oxygen ($T = 20°C$, $P = 10^{-1}$ Torr). It is evident that duration of emission in this experiment which corresponds to most intensive structural changes in the surface-adjacent layers of nickel is equal to 40 min. Note that according to magnetometric measurements [56] oxygen remains in mobile state approximately the same time that oxygen interacts with nickel.

Therefore, the use of several specific techniques while implementing the method of semiconductor sensors makes it feasible to detect and analyze emission of oxygen atoms at initial stage of metal oxidation although in case of silver it should be noted that there are no phase of silver oxide formed due to its instability at such conditions [57]. Rather, the absorption of oxygen by silver would be related to dissolution and internal oxidation.

6.3. Emission of singlet oxygen from disordered surface of solids

One of most important problems of contemporary physics of the surface deals with the processes of electron excitation both on the surface of solids and in absorbed layers. A wide scope of phenomena is especially important from this stand-point. These phenomena are related to behaviour of electron-excited molecules of oxygen (singlet oxygen) in heterogeneous systems. It is well known [59] that molecules of singlet oxygen are widely spread in nature. They participate in development of several heterogeneous and photoheterogenous processes. Photodestruction of polymers, fading of photographic films, photooxidation of oil film on water belong to such processes. There are some grounds to assume that high selectivity of several catalytic reactions is explained by involvement of singlet oxygen. It is quite understandable that studies of peculiarities of formation and emission of singlet oxygen from surfaces of solids are of a substantial interest because they can explain mechanism underlying numerous heterogeneous processes. In Sections 6.3 and 6.4 we describe experiments on the studies of emission of I_{O_2} using SCS technique.

6.3.1. Emission of singlet oxygen from surface of mechanically disordered quartz

Mechanically activated quartz featuring active centres of various origin on its surface is a very interesting object to conduct the studies of this kind [60]. Some of these centres irreversibly annihilate during heating.

Experiments on the studies of possible emission of active particles were conducted in a vial similar to that shown in Fig. 6.1. Chamber *2* was linked to the vacuum grinder. The grinder involved thick wall quartz test tube and a quartz piston (glass-covered iron weight). When pulse signals were fed to the coil of an electromagnet the piston rose by 2 – 3 cm and fell down. Such an instrument made it possible to grind quartz with substantial efficiency in a thick wall test tube made of fused quartz [61].

It occurred that during heating of quartz ground in oxygen ($P_{O_2} = 10^{-1}$ Torr) the emission of particles – acceptors of electrons was observed with confirmation provided by the change in electric conductivity of ZnO sensor-film [61]. Using the method of identification of active particles shown above it became feasible to confirm that in this case one observes the emission of singlet oxygen. For instance, the distribution of oxygen atoms along the length of tube was measured, oxygen atoms being provided by platinum pyrolysis. The same was applicable to particles emitted from the quartz surface. It occurred that the ratio of coefficients of heterogeneous decay of these particles is close to that taking place for a singlet oxygen and oxygen atoms obtained in [62]. Moreover, if the surface of the tube linking the mortar with the rest of device is covered by a Co_3O_4 layer heated at 700°C, the particles emitted from the surface of quartz will not reach the sensor. Finally, pumping through the solution of diphenilbenzofurane in hexadecane with initial concentration of 10^{-4} mol/l during 2 hours resulted in a 30% decrease in concentration of the chemical. These results give the grounds to maintain that in this case we observe the emission of singlet oxygen [61].

These phenomena are of high interest to understand processes of the transfer of energy of electron excitation to oxygen in the surface-adjacent layers of solids. We should note that such generation is observed very rarely at the surface of solids and there are only few papers referring to it [63, 64].

It is known [65, 66] that oxygen is observed at the surface of freshly ground quartz in several forms which are characteristic of various conditions of formation, thermal stability, concentration, etc. Jointly with other physical and chemical methods application of semiconductor sensors enabled us to establish the origin of absorbed forms of oxygen responsible for $^1O_2$ emission [67].

Figure 6.17 shows results of measurements of dependence of intensity of emission of $^1O_2$ as a function of temperature and state of the surface of freshly ground quartz. The mechanical treatment of quartz in the inert atmosphere with subsequent heating does not result in formation of singlet oxygen. Thermoemission of $^1O_2$ occurs only in case when the molecular oxygen is absorbed on the surface of a sample. Chemisorption of oxygen can be conducted by two means: grinding quartz in atmosphere of this gas or treating the surface of a sample by oxygen (the sample being previously prepared by grinding in inert atmosphere). In both cases the intensity and shape of $^1O_2$ thermoemission curve are approximately the same. Emission is observed over the range 200 – 500°C, the maximum of intensity being situated at 300 – 320°C (see Fig. 6.17, curve 1). The subsequent heating of the same sample in inert atmosphere or in presence of oxygen is not accompanied by creation of singlet oxygen. The emission activity of the sample of quartz gets restored only in case of a new grinding session.

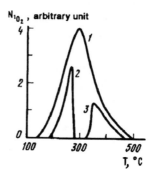

Fig. 6.17. The intensity of emission of singlet oxygen as a function of temperature in freshly ground quartz: 1 – grinding in helium atmosphere, chemisorption of O_2 at 20°C; 2 – same as above + treatment at 280°C; 3 – sample obtained as stage 2 followed by oxidation at 20°C.

Curve 2 in Fig. 6.17 corresponds to experiment which we started as usual but reduced the surface of quartz through treatment by carbon monoxide at 280°C. At this moment emission disappeared and further increase in temperature would not bring it back. The sample was cooled after reaching the temperature of 320°C, oxidized and heated anew. Such subsequent oxidation of quartz resulted in formation and emission of singlet oxygen (see Fig. 6.17, curve 2) but only at temperatures around 320°C, i.e. above the temperature of initial heating.

Therefore, the formation of molecules of $^1O_2$ is linked only to oxygen chemisorbed on the surface of freshly ground quartz and occurs during irreversible process of thermal annihilation of oxygen adsorption centres. The analysis of the literature data regarding the reaction capacity and thermal stability of forms of absorbed oxygen [65, 66] indicates that over the temperature range 200 − 500°C the decay of O_2^- anion-radicals occurs. The decay of this form of adsorbed oxygen is accompanied with annihilation of the centres on which it is generated. The control experiments on measurements of the temperature stability of O_2^- anion-radicals, whose concentration was determined by EPR technique assessing the amplitude of signal in point with $g = 2.032$ as paper [68] suggests, indicate that the temperature intervals of decay of anion-radicals and emission of singlet oxygen are fairly close. Thus, we can conclude that formation and emission of singlet oxygen on the surface of freshly ground quartz occurs during the process of irreversible annihilation of oxygen adsorption centres. The oxygen gets adsorbed on these centres in O_2^- form. Presumably, the energy necessary to excite this process is stored in the chemisorption centres.

It is interesting to note that quartz acquires the capacity to emit singlet oxygen not only as a result of mechanical grinding but due to exposition to conditions of burning the $H_2 + O_2$, as well [69]. In this case the particles can be identified using the same methods as molecules of singlet oxygen [67]. The comparison of distribution of oxygen atoms along the height of the vial and particles emitted from the surface of quartz walls showed that the ratio of decay coefficients is the same for these particles on the surface of glass as for atoms of oxygen and molecules of singlet oxygen [62]. Covering the vial walls by Co_3O_4 heated at 700°C along the way from reactor to sensor results in elimination of sensor signals. Finally, pumping oxygen containing particles emitted from the surface of quartz through solution of 1,3-diphenylbenzofurane in hexadecane of results in 40% bleaching of solution which is characteristic for singlet oxygen. Consequently, as a result of explosion of gas mixture the quartz wall acquires the capacity to emit singlet oxygen which can participate in the overall process, for instance in reaction with atom oxygen through reaction: $^1O_2 + OH \rightarrow O + OH$ (1). The rate constant of this reaction [70] is

$$K = 10^{-9.74\pm0.29} \exp\left(-\frac{6280 + 930}{RT}\right) \text{ cm}^3/\text{mol·s} .$$

For the sake of reference let us quote here the constant of reaction rate for triplet oxygen with atom hydrogen [71] $O_2 + H \rightarrow O + OH$ (2)

$$K = 0.94 \cdot 10^{-10} \exp(-15100 \, / \, RT) \; \mathrm{cm^3/mol \cdot s} \; .$$

The value of the flux of singlet oxygen emitted from the surface of quartz walls can be evaluated through bleaching rate of 1,3-diphenyl-benzofurane solution in hexadecane. This evaluation indicates that at low pressures (0.01 – 0.1 Torr) inputs of above reactions are comparable.

This result enables us to establish a reason leading to the change in activity of the walls of a quartz reactor in which burning of oxygen takes place at low pressures. From the stand-point of this conclusion it is easy to understand the change in oxygen burning rate at low pressures in a vial with walls covered by magnesium oxide [72] because the latter is a good quenching agent of singlet oxygen [73]. We should note that one can observe the emission of singlet oxygen from the surface of mechanically activated magnesium oxide only at high temperatures [73]. Consequently, the input of reaction (1) developing in a vial whose walls are covered with magnesium oxide will be negligibly small.

6.3.2. Emission of singlet oxygen from the surface of vanadium pentoxide

Applying the technique previously used to identify molecules of singlet oxygen emitted from the surface of quartz it was shown that heating partially reduced vanadium pentoxide in vacuum or in presence of oxygen one observes emission of the molecules of singlet oxygen [74]. Identification of $^1O_2$ was carried out using the reaction with rubrene solution in hexadecane. In this case the peculiarity of emission, i.e. its complete depletion over several hours of heating of catalyst at pressure of oxygen 0.1 Torr was noted. This results in the fact that emission of $^1O_2$ is controlled by non-equilibrium processes in surface-adjacent layers of catalyst. The validity of this conclusion is confirmed by results of paper [74] which showed that only partially reduced vanadium pentoxide is capable of emitting singlet oxygen during heating in vacuum or in oxygen atmosphere. On the other hand both pure V_2O_5 (without admixture of four valance vanadium) and lower vanadium oxides V_2O_4 and V_2O_3 are passive with respect to emission of singlet oxygen. Moreover, in vacuum this emission gets depleted due to reduction of oxide and resumes after a partial oxidation. The best way to prepare sample of vanadium pentoxide showing emission of $^1O_2$ is provided by its treatment in atmosphere of oxygen with naphthalene vapours under conditions of partial oxidation of naphthalene (under atmospheric pressure, the temperature being 300°C).

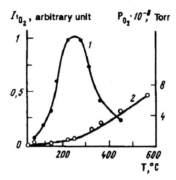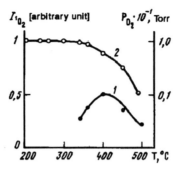

Fig. 6.18. Temperature dependence of $^1O_2$ emission (*1*) and triplet oxygen release (*2*) during linear heating of V_6O_{11} sample in vacuum.

Fig. 6.19. Temperature dependence of triplet oxygen absorption (*1*) and singlet oxygen emission (*2*) during the heating of V_6O_{11} sample in 0.1 Torr O_2.

It should be noted that vanadium oxides exhibit exceptionally high mobility of oxygen in lattice and capacity of oxides to form large number of phase oxides, for instance the Magneli phases with a common formula V_nO_{2n-1} [75]. During release of oxygen from the lattice of such oxide there is no oxygen vacancy formed and the deformation of the lattice is accompanied by formation of a new phase. At constant conditions (constant pressure of oxygen and temperature) homogeneous oxide is being formed, the time of formation of homogenized oxide attaining several days at temperature ~ 900°C. Therefore, heating of homogeneous mixture of oxides of vanadium or heating homogeneous oxide far from domain were homogeneity exists results in a new content and new structure of oxide which, in its turn, is accompanied by generation and annihilation of point defects and mechanical stresses in surface-adjacent layers of solids, i.e. processes in which generation and emission of active particles and molecules of singlet oxygen in particular, can take place.

It was shown in papers [76, 77] that creation and emission of singlet oxygen is controlled by a processes of mutual transformations of the Magneli phases. Figure 6.18 shows the temperature profiles of emission of singlet oxygen and emission of excited oxygen during linear heating in vacuum of V_6O_{11} samples; synthesis of oxide was carried out using the method described in [75] in a quartz ampoule with a thin membrane. Homogeneity of oxide was confirmed by an X-ray analysis. The ampoule with synthesized V_6O_{11} was fused with a vial containing ZnO sensor and shutter. After evacuation of the vial and making sure that the sensor is ready to operate the membrane was broken and the experiment commenced. Such a method made it possible to rule out plau-

386

sible change in oxide parameters during its loading into the experimental set-up prior to evacuation. Figure 6.18 indicates that emission of $^1O_2$ occurs over the temperature range featuring a substantial emission of non-excited oxygen as well, i.e. when the substantial change in the structure of surface-adjacent layers develops. Quenching of emission of $^1O_2$ with rising temperature can be explained by the drop in intensity of restructuring in the surface-adjacent oxide layers (for a particular material), whereas the singlet oxygen formed in deeper layers of the sample gets deactivated during its diffusion towards the surface. This conclusion is confirmed by a substantially lower emission of $^1O_2$ after cooling and subsequent heating of the sample.

Figure 6.19 shows the temperature profiles of emission of singlet oxygen and adsorption of non-excited oxygen during heating (the initial pressure of O_2 equals to 0.1 Torr) of V_6O_{11} sample. We should note that in this case the decay in emission is, presumably, related to slowing down of restructuring processes in the surface-adjacent layers of the sample, too. Similar result can be obtained for other Magneli phases, as well, e.g. for V_4O_7, V_5O_9, the difference being in the temperature range and intensity of emission of singlet oxygen. This result is dependent on different stability of oxides as well as on different probability of annihilation of $^1O_2$ on oxide surface.

6.4. Photoemission of singlet oxygen from the surface of solids

6.4.1. Photosensibilized formation and emission into gaseous phase of singlet oxygen in solid-dye-oxygen systems

An idea concerning the formation of singlet oxygen in heterogeneous systems during interaction of triplet dye molecules with oxygen through triplet-triplet energy transfer was initially formulated by Kautsky [78]. Kautsky observed oxygen induced quenching of fluorescence of tripaflavine adsorbed on silicagel accompanied by oxidation of leuco base of brilliant green dye on other grains of silicagel. Oxidation developed over a narrow range of oxygen pressures ($1.2 - 2.4 \cdot 10^{-3}$ Torr). Kautsky supposed that electron-excited oxygen molecules provide the oxidizing agent through reaction $T + {}^3O_2 \rightarrow S_0 + {}^1O_2$, where T, S_0 are the triplet-excited and ground state of the molecule of dye, respectively; $^3O_2$, $^1O_2$ are the ground and singlet-excited oxygen states.

Due to the fact that sensibilizer and oxidized substrate were separated Kautsky supposed that singlet oxygen is emitted into the gaseous phase. It should be noted, however, that there is no oxidation when sensibilizing dye and oxidized substrate are separated by a distance of several millimeters. Therefore, it rose a discussion while trying to explain the nature of oxidizing particles. Several other experiments were conducted later on [79, 80, 81] in which photosensibilized formation of singlet oxygen was monitored using spectral techniques (as well as through detection of specific oxidation products). In [82, 83] the photosensibilized formation of $^1O_2$ was studied by monitoring the concentration of the singlet oxygen emitted into the gaseous phase by the method of semiconductor sensors. This method is free from drawbacks characteristic of other techniques. It possesses high sensitivity with respect to singlet oxygen and enables one to study the processes of interest in absence of sensibilizer. The use of this method made it possible [82, 83] to obtain several absolutely new facts and make a new step towards understanding the process of photosensibilized formation and emission of singlet oxygen into the gaseous phase in heterogeneous systems.

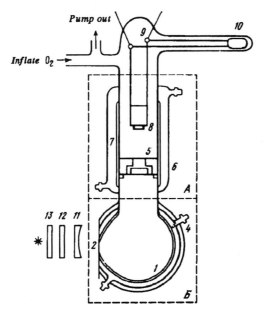

Fig. 6.20. The schematic diagram of experimental set-up to study photoemission of $^1O_2$. 1 − quartz vial; 2 − quartz window; 3 − thermostating jacket; 4 − aluminum shield; 5 − filter; 6 − thermostating jacket; 7 − blue glass; 8 − sensor; 9 − platinum rings; 10 − glass covered weight; 11 − 13 − lenses; A, B − jackets providing optical isolation of chambers.

Installation in which these studies were conducted is shown in Fig. 6.20. It was composed of two optically isolated chambers. The chamber to generate singlet oxygen contained a sensibilizer (tripaflavin, rose bengale) applied to quartz or silicagel. Another chamber contained a mobile sensor of $^1O_2$ molecules. Optical isolation of these two chambers was provided by the fact that a section of tube linking these two chambers was made of a special blue glass (C-50, doped with cobalt). The same glass was used to make a special labyrinth providing an assess for $^1O_2$ molecules but shutting down the light from mercury lamp DRSh-500. Because the flux of $^1O_2$ molecules can be easily stopped through shutting down the light beam incident on the surface of photosensibilizer, the use of the shutter was superfluous in several experiments. Identification of the molecules of singlet oxygen was provided by methods similar to those considered in [62]. Moreover, to evaluate concentration of $^1O_2$ in gaseous phase we used the differential cobalt calorimeter [83]. One of shoulders of the bridge contained the platinum thermoresistor — a wire with diameter of 30 μm and length 407 cm, another shoulder contained a similar wire covered by cobalt. The resistivity of shoulders of the calorimeter was measured using the bridge schematics. Initially we ran the calibration of calorimeter. The sensitivity of calorimeter with respect to singlet oxygen was about 10^{10} cm^{-3} if one assumes a priori that the heat effect of desactivation of singlet oxygen on cobalt is equal to the energy of electron excitation. As experiments with differential calorimeter showed the concentration of singlet oxygen which is formed during quenching of triplet excitation of sensibilizers is of the order of 10^{11} cm^{-3}.

Using sensor it became possible to verify several peculiarities of photosensibilized creation of singlet oxygen [82, 83]. It occurred that concentration of $^1O_2$ molecules in gaseous phase shows a linear dependence on intensity of incident light. This implies that emission of singlet oxygen into the gaseous phase develops as a first order process. Moreover, the results also indicate that the rate of photooxidation of sensibilizer is negligibly small if compared to the photogeneration rate. Further, the dependence of intensity of emission of $^1O_2$ plotted as a function of pressure of molecular oxygen was used to conclude that $^1O_2$ was formed in adsorbed layer. This conclusion is confirmed by additional adsorption of water which does not affect the phosphorescence of tripaflavine with simultaneous inhibition of $^1O_2$ photogeneration owing to removal of adsorbed oxygen from the dye surface.

The kinetic schematics proposed in [82, 83] describes the peculiarities observed fairly well especially the dependence of intensity of the flux of $^1O_2$ on the pressure of oxygen, which is controlled both by the rise in amount of triplets interacting with oxygen while increasing the pressure of oxygen and by the increase of the input of decay of excited molecules inside the volume and on glass walls during pressure growth. This schematics is also suitable to describe the temperature dependence of intensity of $^1O_2$ emission which can be explained by decrease in the life-time of dye-adsorbed singlet oxygen. The activation energy of the rate of photosensibilization induced formation of singlet oxygen can be reduced to the adsorption heat of singlet oxygen on dye because the phosphorescence quenching rate of sensibilizer is not temperature-dependent [84]. The processes of radiation − free degradation of adsorbed sensibilizer [84] and the transfer of sensibilizer into the substrate are negligible, and one can discard them from consideration. The increase in concentration of non-equilibrium chemisorbed oxygen with temperature does not take place because here we are dealing with equilibrium oxygen chemisorption on dye.

Therefore, the dependence of concentration of singlet oxygen with temperature can be explained fairly well with decrease of life-time of singlet oxygen on surface. This stand-point can be confirmed by observed absence of quenching of sensibilizer photophosphorescence in presence of oxygen at liquid nitrogen temperature [84]. This phenomenon can be understood from the view-point of the temperature dependence of the life-time of singlet oxygen in adsorbed state. The adsorption heat on tripaflavine assessed in [84] equals to 6 kcal/mol. This value agrees with the value of 10 kcal/mol obtained in [83].

It is important to have a suitable substrate on which sensibilizer is applied to provide photosensibilization-induced formation of $^1O_2$. Obviously, the geometric structure of substrate (the pore size, the specific surface value) would affect the amount of collisions of $^1O_2$ molecules with substrate during transport of these particles from pores into the volume of the vial. Therefore, several features (for instance the dependence on pressure) of the emission of singlet oxygen into gaseous phase for dyes applied to silicagel differ from those of dyes applied to smooth quartz.

The electron properties of the surface are also very important. It was already shown in [85] that apart from the excitation energy transfer from the dye to oxygen adsorbed there are processes of spectral sensibilization of non-organic semiconductors, i.e. the transfer of energy of excitation to solids. This implies that during photosensibilized generation of $^1O_2$ there are two competing processes: the transfer of energy from dye to substrate and to adsorbed oxygen. The use of zinc oxide as a substrate made it possible to elucidate the role of substrate electron subsystem in the mechanism of energy transfer (excitation) to solids, because the electron properties of this oxide can be easily controlled through oxidation or reduction. It occurred that in case of highly oxidized zinc oxide an efficient generation of singlet oxygen is observed, whereas there was no singlet oxygen created during the use of highly oxidized zinc oxide reduced in hydrogen.

These results can be understood if one takes into account the effect of spectral dye-induced sensibilization of photosensibilization conductivity of zinc oxide [86]. Oxygen adsorption on the surface of ZnO is accompanied by thermally activated capturing of electrons in the surface layers of adsorbed oxygen [87]. According to [88] the excitation energy of adsorbed dye is effectively transferred to these levels. In [89] it was shown that reduced zinc oxide exhibits a large amount of surface ionized super-stoicheometric zinc atoms and, correspondingly, oxygen adsorption creates a high density of surface levels. In such a case the effective transition of dye excitation energy (tripaflavine) into the substrate (reduced zinc oxide) occurs, therefore, the probability of the transfer of energy to oxygen is low which results in low efficiency of photosensibilization induced generation of $^1O_2$. After treatment of the surface of zinc oxide by molecular oxygen it gets post-oxidized. The amount of super-stoicheometric atoms of zinc sharply decreases and, consequently, the adsorption of oxygen brings about the decrease in density of surface levels [89]. As a result, the probability of transfer of excitation energy to oxygen increases which results in photosensibilization induced emission of singlet oxygen. During subsequent reduction of zinc oxide by hydrogen the process of energy transfer into the substrate begins to dominate again which phases out of singlet oxygen during photoemission.

This conclusion is in agreement with experiments in which a smooth quartz and cellulose were used as substrates. For above materials the transfer of excitation energy of the dye into the substrate is low which is confirmed by intensive luminescence of adsorbed tripaflavine. Note, that the activation energy of emission of singlet oxygen is close for zinc oxide oxidized by oxygen atoms, quartz and cellulose and amounts to 5 − 10 kcal/mol [83].

A very interesting result can be obtained while using amorphous films of a dye instead of adsorbed tripaflavine molecules as sensibilizers. In this case we also observe an emission of singlet oxygen into gaseous phase during illumination of films with adsorbed oxygen. It should be noted, though, that amorphous tripaflavine films do not feature phosphorescence [90]. However, the creation of $^1O_2$ can be explained by a triplet-triplet transfer of excitation energy of dye, as well. The molecules of dye in solid phase retain their individual properties to the greater extent because they are bound by weak dispersion forces which, for instance, is confirmed by similar spectral features of tripaflavine solutions and tripaflavine films [91]. Such a film can be considered as a highly adsorbing surface layer deposited on a certain "substrate". Then, the peculiarity of behaviour of tripaflavine molecules adsorbed on the surface of amorphous films determines the specificity of formation and emission of singlet oxygen. The observed experimental peculiarities [83] in this case can be easily explained if one assumes that the rate of energy transfer from an excited dye molecules to oxygen is close for molecules of dye adsorbed on quartz and for that adsorbed on a tripaflavine film. The degradation rate in the latter case would be two orders of magnitude higher if contrasted to the case of molecular dispersed tripaflavine. This can be attributed to vibrational degradation of excitation energy of the dye as a result of interaction with substrate. As a result, the activation energy of creation and emission of singlet oxygen-sensibilized by tripaflavine film is about 15 kcal/mol which is substantially lower than for molecularly dispersed tripaflavine deposited on quartz or cellulose.

Therefore, the application of sensors made it possible not only to detect and identify the molecules of singlet oxygen emitted during photosensibilization by dye molecules but to understand several peculiarities of redistribution of the energy of electron excitation and transfer to solids, as well.

6.4.2. Photosensibilization - induced formation of singlet oxygen on the surface of deposited oxides of transition metals

The formation of singlet oxygen through the Kautsky mechanism [78] was observed not only when sensibilization was induced by adsorbed dye molecules but also in case when sensibilization was provided by deposited oxides of transition metals [92, 93]. This is very important for understanding numerous heterogeneous and photocatalytic processes.

The studies dealing with opportunity of photosensibilized formation of $^1O_2$ from the surface of deposited oxides were studied in a vial similar to that described in [24]. The identification of $^1O_2$ was provided by similar techniques. The evaluation of concentration of $^1O_2$ molecules in gaseous phase involved the assessment of oxidation rate of 1,3-diphenil-benzofurane in hexadecane. It occurred that in this case the concentration of $^1O_2$ molecules amounts to 10^{10} cm^{-3}.

The analysis of experimental data [92, 93] makes it possible to conclude that photosensibilized formation of singlet oxygen in heterogeneous system V_2O_5/SiO_2 − gaseous phase involves the Kautsky mechanism. It should be mentioned that these are the surface triplet states of oxocomplexes with the charge transfer between V^{5+} ions and oxygen of oxide of lattice that generate $^1O_2$.

This conclusion is confirmed by numerous experimental data. It occurs that the photoluminescence of samples of aerosil promoted by vanadium ions (1.5% by weight) is almost completely quenched by oxygen (0.5 Torr). According to experimental results [94, 95] the photoluminescence observed is a phosphorescence with radiation centres provided by triplets of charge transferring oxocomplexes. Simultaneously, the same sample exhibited an emission of singlet oxygen. We should note that in contrast to the surface of quartz vial covered by quartz the studies of photoemission of $^1O_2$ from the surface of motivated aerosil are less convenient to due to geometrical factors mentioned in preceding section [92, 93].

It was established in [96] that during light excitation of vanadium ions deposited on SiO_2 electrons and "holes" ($O^-_{lattice}$) do not undergo a spatial separation in the charge transfer band. Therefore, the respective electron-excited state exhibit a small life-time and can be either triplet or singlet. We encountered another situation while exciting a massive V_2O_5 sample. As is known [96] this oxide is a semiconductor of n-type with the width of forbidden band of 209 kJ/mol. A transfer of an electron takes place from a twice negative charged ions of the lattice oxygen to the metal cation during excitation of the oxide inside the band of active light. However, the electron is not localized on cation but reaches the conductivity band of oxide. Therefore, during adsorption of a light quantum a spatial separation of positive charges (holes O^-) and negative ones (conductivity electrons) takes place. In this case there is no point to refer to formation of singlet and triplet electron-excited states. Owing to this reason [93] no emission of oxygen was detected from the surface of a massive vanadium oxide during irradiation by light within the band $\lambda > 400$ nm.

The adsorption spectrum of aerosil containing admixture vanadium ions exhibits a maximum within the band $290 - 380$ nm which was attributed by authors of [97] to the charge transfer transitions in oxygen-containing complexes of five valance vanadium: $V^{5+} = O^{2-} \rightarrow V^{4+}O^-$. On the other hand, photosensibilization induced formation of singlet oxygen on the surface of deposited vanadium oxide catalysts was observed within the band $\lambda < 400$ nm.

On the ground of the data obtained in experiments one can draft the schematics of photosensibilized formation of oxygen as follows

$$[V^{5+} = O^{2-}]_{S_0} \underset{\leftarrow}{\overset{h\nu}{\rightarrow}} [V^{4+} - O^-]_{S_1} \rightarrow [V^{4+} - O^-]_{T_1} + {}^3O_2 \rightarrow [V^{5+} = O^{2-}]_{S_0} + {}^1O_2 \, ,$$

where S_0, S_1 are the ground and the first excited singlet states in vanadium-oxygen complex with charge transfer, respectively.

During interaction of hydrogen or carbon monoxide photoexcitation centres of SiO_2 surface containing ions of five valance vanadium one observes a surface-adjacent photodissociation of hydrogen and photooxidation of carbon dioxide through the following scheme [93, 96]

$$[V^{4+} - O^-]_{T_1} + H_2 \longrightarrow V^{4-} + OH^- + H_{ads} \, ,$$

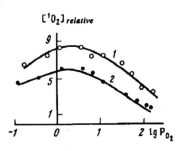 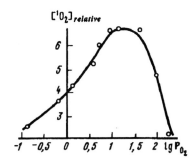

Fig. 6.21. The dependence of relative concentration of I_{O_2} on pressure of O_2:
1 – oxidized sample; *2* – reduced sample.

Fig. 6.22. The dependence of concentration of singlet oxygen on pressure of non-excited oxygen

$$[V^{4+} - O^-]_{T_1} + 2CO \longrightarrow V^{3+} - CO + CO_2 .$$

As a result of these reactions the drop in surface-adjacent concentration of ions of five valance vanadium occurs meaning that there is a decrease in concentration of active centres that generate singlet oxygen.

Figure 6.21 shows a dependence of relative concentration of singlet oxygen in gaseous phase as a function of oxygen pressure in a reaction vial. Curve *1* corresponds to the data plotted for oxidized sample, curve *2* was obtained for the same sample with 5 min treatment by hydrogen. It can be easily seen that over the whole interval of measurements the emission capacity of reduced sample was substantially lower than in initial one which can be explained by decrease in surface concentration of ions of the 5 valance vanadium. Additional experiment showed that 15-minute photoreduction of the sample at above conditions reduces its activity with respect to emission of singlet oxygen by 3 – 5 times, 60-minute exposure almost completely suppresses the activity. The subsequent phototreatment in hydrogen does not affect the residual emission capacity of the sample. This can be explained by the fact that a portion of vanadium ions can exist on the surface as clusters. As a result of piling up of reduced forms of vanadium in clusters their photoreduction rate sharply decreases because in such surroundings the life-time of excited charge transfer complexes also decreases due to quenching by paramagnet ions. Similar results were obtained during photoreduction in carbon monoxide.

Thereby, photogeneration of singlet oxygen molecules from the quartz surface containing molecular dispersed molecules of pentoxide occurs due to triplet-triplet transfer of electron excitation from triplet-excited complexes with resultant transferring of charge to oxygen.

Using semiconductor sensors we examined several peculiarities of photoemission of $^1O_2$. It occurred that concentration of $^1O_2$ in gaseous phase exhibits a linear dependence on intensity of exciting light which is completely consistent with the Kautsky mechanism (see the experimental schematics). Moreover, these results indicate that in this case the readsorption processes (as well as local heating of sensibilizer) are not important.

Figure 6.22 shows the dependence of relative concentration of singlet oxygen in gaseous phase on oxygen pressure inside the vial. The rise in emission of singlet oxygen during increase in oxygen pressure from 10^{-3} to $6 \cdot 10^{-2}$ Torr is caused by increase in formation rate of $^1O_2$ on the surface of illuminated adsorbent. It should be mentioned that in [94, 95] the opposite trend in phosphorescence of V_2O_5/SiO_2 – catalyst was observed at the same pressure of oxygen. The further rise in pressure brings about a decrease in concentration of $^1O_2$ molecules in gaseous phase which is controlled by the growing rate of molecular desactivation in gaseous phase and on the walls of the vial.

Using the schematics proposed the formation of singlet oxygen can involve both collision and adsorption mechanisms. However, as the experimental results indicate (see Fig. 6.22) in this case the chemisorption mechanism is more probable. This can be confirmed by experiments in which photoemission of singlet oxygen was observed after evacuation of oxygen from the vial. When the pressure of oxygen dropped from 10^{-3} to 10^{-5} Torr the intensity of emission decreased only by $3 - 5$ times which cannot be explained by collision mechanism of photogeneration of $^1O_2$. The amount of collisions with walls is known to be strictly proportional to pressure, so, the two orders of magnitude drop in pressure of oxygen would have resulted in the similar drop in concentration of $^1O_2$ in gaseous phase.

The studies of temperature profile of photogeneration of singlet oxygen from the quartz surface doped by five valance vanadium ions showed that emission of $^1O_2$ is the activated process with activation energy equalling to 6 kJ/mol. Considering this fact within the frame work of chemisorption mechanism in the same manner as when addressing photosensibilization induced generation of singlet oxygen by tripaflavine molecules, one can assume that above temperature dependence is controlled by decrease in life-time of singlet oxygen on the surface with growing temperature. When the temperature of photosensibilizer drops below 180 K no photoemission of $^1O_2$ is detected, i.e. $^1O_2$ molecules get stabilized on oxide surface.

The capability to photogenerate singlet oxygen is inherent not only to V_2O_5/SiO_2 systems but to several other non-organic heterogeneous systems as well. Long living local excitations of the surface with the energy exceeding 94.3 kJ/mol occur on the surface of such systems during illumination by UV radiation. Quartz doped by six valance molybdenum ions provides an example of such a system. It is known [96] that in this system a long living luminescence is caused by existence of triplet-excited states with a six valance molybdenum ion charge transfer oxygen surface complex. The light-induced activation of this process in this oxide can be written as follows:

$$[Mo^{6+} = O^{2-}]_{S_0} \xrightarrow{h\nu} [Mo^{5+} - O^-]_{S_1} \longrightarrow [Mo^{5+} - O^-]_{T_1} \,,$$

where S_0, S_1 are the ground and a first excited singlet state of the complex, respectively, T_1 is metastable triplet-excited state. The height of triplet level equals 272.4 kJ/mol in this case.

Using analytical technique similar to that applied during analysis of $^1O_2$ photogeneration from the surface of V_2O_5/SiO_2 system it was shown that the photogeneration of singlet oxygen was observed in case of Mo^{6+}/SiO_2 system as well. The mechanism of formation of these particles is similar to the former one. The peculiarities of photogeneration (dependent on the intensity of light, pressure of oxygen, quenching of phosphorescence by oxygen, etc.) confirm the applicability of the Kautsky mechanism to this process, i.e.

$$[Mo^{5+} - O^-]_T + {}^3O_2 \longrightarrow [Mo^{6+} = O^{2-}]_S + {}^1O_2 \,.$$

Photoreduction of SiO_2 doped by six valance molybdenum results in decrease of the surface concentration of Mo^{6+} ions and, as a consequence, reduces the capability of the sample to emit $^1O_2$. However, such a treatment does not lead to a complete suppression of emission capability because, according to [96], illumination reduces only 30% of Mo^{6+} ions in molybdenum oxide catalysts deposited on SiO_2.

Similar to molybdenum oxide catalyst the capability to emit singlet oxygen is inherent to SiO_2 doped by Cr^{6+} ions as well. Similar to the case of vanadium oxide catalysts in this system the photogeneration occurs due to the triplet-triplet electron excitation transfer from a charge transfer complex to adsorbed oxygen.

Thereby, the application of highly sensitive sensors jointly with several specific techniques can be recommended to study complex heterogeneous and homogeneous processes where an experimentalists are faced with an option to analyze small concentrations of active particles in gaseous phase.

References

1   Ya.B. Gorokhovatski, T.P. Kornienko, V.V. Shalya, Geterogene-ous-Homogeneous Reactions, Technika Publ., Kiev, 1972 (in Russian)
2   R.A. Marlbow, B. Lambert, Surface Sci., 74 (1978) 107 - 115
3   I.A. Myasnikov, E.V. Bolshun, E.E. Gutman, Kinetika i Kataliz, 4 (1963) 867 - 871 (in Russian)
4   A.A. Lisachenko, F.I. Vilesov, Zhurn. Fiz. Khimii, 39 (1969) 592 - 597 (in Russian)
5   I.V. Krylova, Uspekhi Khimii, 45 (1976) 2138 - 2157 (in Russian)
6   L.N. Grigorov, V.Ya. Munblit, V.B. Kazanski, Kinetika i Kataliz, 19 (1978) 412 - 416 (in Russian)
7   A.V. Sklyarov, V.V. Rozanov, M.U. Kislyuk, Kinetika i Kataliz, 19 (1978) 416 - 420 (in Russian)
8   A.N. Orlov, Yu.V. Trushin, Energy of Point Defects in Metals, Metallurgiya Publ., Moscow, 1983 (in Russian)
9   J.G. Fripiat, K.T. Chow, M. Boudart et al., J. Mol. Catal., 1 (1973) 59 - 67
10 L.S. Palatnik, M.Ya. Fuks, V.M. Kosevich, Mechanism of Formation and Substructure of Thin Films, Nauka Publ., Moscow, 1972 (in Russian)
11 A.A. Sokol, V.M. Kosevich, L.V. Zozulya, Poverkhnost, 2 (1987) 116 - 119 (in Russian)
12 S.A. Vekshinski, Zhurn. Tekhn. Fiziki, 10 (1940) 1 - 7 (in Russian)
13 Gmelins Handbuch. Stibium. Zellerfeld, 1943
14 S.A. Zavyalov, I.A. Myasnikov, L.A. Evstigneeva, Zhurn. Fiz. Khimii, 46 (1972) 2599 - 2601 (in Russian)
15 S.A. Zavyalov, Emission of Active Particles from the Surface of Solids, PhD Thesis, Moscow, 1980 (in Russian)
16 A.I. Nesmeyanov, Vapour Pressure of Chemical Elements, Nauka Publ., Moscow, 1961 (in Russian)
17 D.G. Tabadze, I.A. Myasnikov, L.A. Evstigneeva, Zhurn. Fiz. Khimii, 46 (1972) 2599 - 2601 (in Russian)
18 F. Panneth, Ber. der Deut. Chem. Ges., 59 (1922) 775 - 782
19 K. Tamaru, J. Phys. Chem., 59 (1959) 1084 - 1092
20 V.I. Tsivenko, I.A. Myasnikov, Zhurn. Fiz. Khimii, 47 (1943) 871-874 (in Russian)
21 H. Ocabe, J. Melesby, J. Phys. Chem, 61 (1962) 306 - 310
22 F.O. Rais, K.K. Rais, Free Alyphatic Radicals, USSR Academy of Sci. Publ., Moscow, 1937 (in Russian)
23 J. Kordis, A. Ginderich, J. Chem. Phys., 54 (1973) 5141 - 5177

24 Break Up Energy of Chemical Bonds. Ionisation Potential and Affinity to Electron, USSR Academy of Sci. Publ., Moscow, 1974 (in Russian)

25 N.N. Savvin, I.A. Myasnikov, N.E. Lobashina, Zhurn. Fiz. Khimii, 59 (1985) 2641 - 2643 (in Russian)

26 S.A. Zavyalov, I.A. Myasnikov, E.E. Gutman and N.V. Ryltsev, Zhurn. Fiz. Khimii, 54 (1980) 516 - 518 (in Russian)

27 I.A. Myasnikov, E.V. Bolshun and V.S. Raida, Zhurn. Fiz. Khimii, 53 (1973) 1619 - 1623 (in Russian)

28 I.A. Myasnikov, E.V. Bolshun and B.S. Aganyan, DAN SSSR, 220 (1975) 1122 - 1124 (in Russian)

29 A.P. Zeif, in: Elementary Physical and Chemical Processes on Semiconductor Serface, Siberian Division of USSR Acad. of Sci. Publ., Novosibirsk, 1979 (in Russian)

30 H. Poppa, J. Appl. Phys., 38 (1967) 3833 - 3839

31 Yu.A. Cherkasov and I.B. Zakharova, Uspekhi Khimii, 56 (1977) 2240 - 2261

32 V.A. Bely, A.M. Krasovsky and A.V. Rychagov, Doklady of Bielorussian Acad. of Sci., 21 (1977) 417 - 421 (in Russian)

33 V.A. Bely, FTT, 17 (1975) 3430 - 3432 (in Russian)

34 I.V. Miloserdov, Effects of the Surface Microcrystalls of Metals on Electrophysical and Adsorption Properties of Oxides, PhD Thesis, Moscow, 1977 (in Russian)

35 E.E. Gutman, A.V. Kashin, I.A. Myasnikov et al., Zhurn. Fiz. Khimii, 59 (1985) 1296 - 1298 (in Russian)

36 K.A. Petrzhak and M.A. Bak, Zhurn. Tekh. Fiziki, 25 (1955) 636 - 642 (in Russian)

37 J. Latham and Surevet, Thin Solid Films, 64 (1974) 9 - 15

38 G.B. Gabdullaev and D.M. Abdinov, Physics of Selenium, Nauka Publ., Moscow, 1975 (in Russian)

39 S.A. Zavyalov, I.A. Myasnikov and E.E. Gutman, Zhurn. Fiz. Khimii, 53 (1979) 1303 - 1305 (in Russian)

40 I.A. Myasnikov, Mendeleev ZhVKhO, 22 (1975) 19 - 38 (in Russian)

41 J. Drowart and B. Coldfinger, Angew. Chem., 6 (1967) 581 - 589

42 M.V. Polyakov, in: Proceedings of the Physico-Chemical Conference, USSR Acad. of Sci. Publ., Moscow, 1929 (in Russian)

43 M.P. Cox, J.S. Foord and R.M. Lambert, Surface Sci., 129 (1983) 399 - 407

44 S.A. Zavyalov, I.A. Myasnikov and E.S. Solovyova, Kinetika and Kataliz, 26 (1985) 1506 - 1509 (in Russian)

45 P.V. Zimakov, Uspekhi Khimii, 28 (1959) 1343 - 1350 (in Russian)

398

46   L. Libscomb, R.G.W. Norrish and B.A. Thrush, Proc. Roy. Soc. London. A, 233 (1956) 455 - 462

47   L. Elias and E.A. Ogryzlo, Canad. J. Chem., 37 (1959) 1680 - 1687

48   S.J. Arnold, M. Cubo and E.A. Ogryzlo, Adv. Chem. Ser., 77 (1968) 133 - 137

49   R.P. Kasomov, N.V. Kulkova and M.I. Temkin, Kinetika and Kataliz, 15 (1974) 157 - 162 (in Russian)

50   S.A. Zavyalov, I.A. Myasnikov and E.S. Solovyova, Zhurn. Fiz. Khimii, 64 (1990) 2749 - 2754 (in Russian)

51   G.A. Kurov, V.F. Markelov and I.O. Svitnev, Mikroelektronika, 15 (1986) 277 - 279 (in Russian)

52   M.H. Francombe and H. Sato, Single Crystal Films, Pergamon Press, N. Y., 1964

53   L.M. Zavyalov, I.A. Myasnikov and S.A. Zavyalova, Khim. Fizika, 6 (1987) 473 - 474 (in Russian)

54   K. Kauffe, Reactions in Solids and on the Surface, Foreign Literature Publ., Moscow, 1963 (translation into Russian)

55   V.F. Kiselev, G.S. Plotnikov, V.A. Bespalov et al., Kinetika and Kataliz, 28 (1987) 20 - 27 (in Russian)

56   A.A. Slinkin, A.V. Kucherov and A.M. Rubinshtein, Izv. AN SSSR, Ser. Khimiya, 28 (1) 17 - 21 (in Russian)

57   J. Benar, Oxidation of Metals, Metallurgiya Publ., Moscow, 1968 (translation into Russian)

58   V.D. Borman, E.P. Gusev, Yu.Yu. Gusev at al., Poverkhnost, 11 (1988) 138 - 144 (in Russian)

59   S.D. Razumovski, Oxygen. Elementary Forms and Properties, Khimiya Publ., Moscow, 1979 (in Russian)

60   V.A. Radtsig, Kinetika and Kataliz, 20 (1979) 448 - 452 (in Russian)

61   S.A. Zavyalov and I.A. Myasnikov, Gas Analysis in Environment Protection, Proceedings of the Conference, Leningrad, 1981 (in Russian)

62   J.P. Cuillory and C.M. Shiblom, J. Catal., 54 (1978) 24 - 27

63   A.H. Khan, Chem. Phys. Lett, 4 (1970) 567 - 572

64   M. Anpj, I. Tanahashi and Y. Kubokava, J. Phys. Chem, 84 (1980) 3440 - 3449

65   V.A. Radtsig and A.V. Bystrikov, Kinetika and Kataliz, 19 (1977) 713 - 719 (in Russian)

66   I.V. Berestetskaya, A.V. Bystrikov, A.N. Streletskii et al., Kinetika and Kataliz, 21 (1980) 1019 - 1025 (in Russian)

67   S.A. Zavyalov, A.N. Streletskii and E.V. Karmanova, Kinetika and Kataliz, 26 (1985) 1005 - 1009 (in Russian)

68 V.A. Radtsig, Kinetics and Mechanism of Chemical Reactions in Solids, Chernogolovka, 1981 (in Russian)

69 S.A. Zavyalov, I.A. Myasnikov and E.V. Karmanova, Khim. Fizika, 2 (1984) 297 - 300 (in Russian)

70 V.Ya. Basevich and V.I. Vedeneev, Khim. Fizika, 4 (1985) 1102 - 1106 (in Russian)

71 L.V. Karmilova, A.B. Nalbandyan and N.N. Semyonov, Zhurn. Fiz. Khimii, 32 (1988) 1193 - 1199 (in Russian)

72 V.V. Azatyan, Khim. Fizika, 1 (1982) 491 - 503 (in Russian)

73 S.A. Zavyalov, I.A. Myasnikov and E.V. Karmanova, Khim. Fizika, 4 (1985) 984 - 985 (in Russian)

74 S.A. Zavyalov, I.A. Myasnikov and L.M. Zavyalova, Zhurn. Fiz. Khimii, 58 (1984) 1532 - 1536 (in Russian)

75 A. Anderson, Acta. Chem. Scand., 8 (1954) 1959 - 1968

76 S.A. Zavyalov, I.A. Myasnikov and L.M. Zavyalova, Zhurn. Fiz. Khimii, 58 (1984) 2117 - 2120 (in Russian)

77 S.A. Zavyalov, I.A. Myasnikov and L.M. Zavyalova, DAN SSSR, 284 (1985) 378 - 391 (in Russian)

78 H. Kautsky, Trans. Faraday Soc., 35 (1939) 217 - 223

79 G.S. Egerton, J. Soc. Dyers. Col, 65 (1949) 764 - 769

80 G.S. Egerton, Nature, 204 (1964) 1153 - 1159

81 E.C. Blossey, D.C. Wecker, A.L. Thayler et al., J. Amer. Chem. Soc., 95 (1981) 5820 - 5826

82 E.I. Grigoriev, I.A. Myasnikov and V.I. Tsivenko, Zhurn. Fiz. Khimii, 55 (1981) 2907 - 2910 (in Russian)

83 E.I. Grigoriev, Photosensibilized Formation and Annihilation of Singlet Oxygen in Solid - Gas Heterogenous Systems, PhD Thesis, Moscow, 1982 (in Russian)

84 J.L. Rosenberd and D. Shombert, J. Amer. Chem. Soc., 82 (1960) 3257 - 3263

85 A.N. Zakharov and V.E. Aleskovski, Izv. Vuzov. Khimiya i Khim. Tekhnologiya, 7 (1964) 517 - 524 (in Russian)

86 A.N. Terenin, Photonics of Dye Molecules, Nauka Publ., Leningrad, 1967 (in Russian)

87 I.A. Akimov, in: Spectroscopy of Molecular Photoconversions, Nauka Publ., Leningrad, 1977 (in Russian)

88 I.A. Akimov, Yu.A. Cherkasov and M.I. Cherkashin, Sensibilized Photoeffect, Nauka Publ., Moscow, 1980 (in Russian)

89 G. V. Malinova and I.A. Myasnikov, Kinetika i Kataliz, 11 (1970) 715 - 718 (in Russian)

90 A.P. Terenin, Mendeleev ZhVKhO, 5 (1960) 498 - 510 (in Russian)

91 A.T. Vartanyan, Zhurn. Fiz. Khimii, 24 (1950) 1361 - 1366 (in Russian)

92 M.V. Yakunichev, I.A. Myasnikov and V.I. Tsivenko, DAN SSSR, 267 (1982) 873 - 876 (in Russian)

93 M.V. Yakunichev, Photostimulated Emission of Singlet Oxygen from the Surface of Deposited Oxides of Transit Metals (Photocatalysts), PhD Thesis, Moscow, 1986 (in Russian)

94 M. Anpj, I. Tanahashi and Y. Kubokava, J. Phys. Chem., 84 (1980) 3440 - 3449

95 M. Anpj, I. Tanahashi and Y. Kubokava, J. Phys. Chem., 86 (1982) 7 - 17

96 S.A. Surin, A.D. Shuklov, B.N. Shelimov et al., Photosorption and Photocatalytic Phenomena in Heterogenous Systems, Siberian Brunch of USSR Acad. of Sci. Publ., Novosibirsk, 1974 (in Russian)

97 A.M. Gritskov, V.A. Shvets and V.B. Kazanski, Kinetika i Kataliz, 14 (1973) 1062 - 1069 (in Russian)

98 A.A. Konnov, A.N. Pershin, B.N. Kazanski et al., Kinetika i Kataliz, 24 (1983) 155 - 167 (in Russian)